Fish and Shellfish Health Management

NIPA® GENX ELECTRONIC RESOURCES & SOLUTIONS P. LTD.
New Delhi-110 034

Fish and Shellfish Health Management

K.M. Shankar, **MFSc, PhD (Microbiology, Canada)**
Former Professor
Aquaculture-Fish Health Management & Biotechnology
Former Dean (Fisheries)
Karnataka Veterinary, Animal and Fisheries Sciences University (KVAFSU)
Mangalore, Karnataka, India
and
Former Emeritus Scientist ICAR, New Delhi
Fellow National Academy of Agricultural Sciences, New Delhi

NIPA® GENX ELECTRONIC RESOURCES & SOLUTIONS P. LTD.
New Delhi-110 034

NIPA® GENX ELECTRONIC RESOURCES & SOLUTIONS P. LTD.

101,103, Vikas Surya Plaza, CU Block
L.S.C.Market, Pitam Pura, New Delhi-110 034
Ph. +91 11 27341616, 27341717, 27341718
E-mail: newindiapublishingagency@gmail.com
www: www.nipabooks.com

For customer assistance, please contact
Phone: + 91-11-27 34 17 17
Fax: + 91-11-27 34 16 16
E-Mail: feedbacks@nipabooks.com

Print ISBN: 978-93-58871-61-6

ebook ISBN: 978-93-58875-09-6

Composed and Designed by NIPA®.

S. Ayyappan
Former Director General
ICAR, New Delhi

Foreword

Aquaculture has grown steadily at 6% annually worldwide including India during the last four decades, contributing today close to 60% of total fish production. It is growing with increase in number of species cultured, area of cultivation in conventional and non conventional waters and innovative culture technologies. However, diseases are the major stumbling block in the production and sustenance of aquaculture causing losses in billions of dollars. As in other animal husbandry, diseases existing and new emerging ones due to microbes and parasites are a serious concern in aqua health management. Additionally, non infectious diseases due to poor water quality, industrial and agricultural pollutants/toxins and nutritional deficiency/excess have affected both fish and the consumer. Losses due to parasites causing death and morbidity in aquatic animals is huge, and not sufficiently documented. And furthermore effective control of parasites by large scale application of chemicals/pesticides considered now risky to fish and its consumers, requiring urgent preventive alternatives including vaccines using modern technologies.

With increase in number of indigenous and exotic culture species and their products and their movement within the country as also across national and international borders, spread of diseases is rapid and complex. This has necessitated urgent measures for biosecurity of stocks, with quarantine certification, and contingency plans to contain epidemics. Zoonotic potential of microbial pathogens and parasites has amplified with aquaculture activities and hence need for creating awareness and measures for the safety of humans. Furthermore, a new trend in aquaculture ethics, an increased global awareness on welfare of the fish is gaining importance which should be appreciated. R&D in new areas of aquaculture particularly biotechnology for diagnosis, vaccine development and more importantly, the aquatic medicine exclusively for cold blooded animals need emphasis for the growing aquaculture. An important feature of aquaculture health management different from that of land animals is that it involves delicate cold blooded animals with lesser developed immune systems, that are highly influenced by ambient temperature and environment and hence easily susceptible to diseases. Currently, good amount of research is going on with regard to immune systems of Asian fish and shellfish species which contribute 80% to global aquaculture production. Against this large background, seven emerging areas can be identified in aquaculture health management, *viz.*, Non infectious diseases, Demand for biosecurity measures, Awareness on zoonotic diseases, Welfare of aquatic animals in aquaculture, Immunology and prophylaxis, Biotechnological interventions through development of diagnostics and vaccines, and more importantly, aquatic medicine for the cold blooded animals.

Fisheries education in health management today has undergone a sea change, with increasing information through R&D, emerging new areas, technologies and trends in aquaculture. In this backdrop, the present book "Fish and shellfish health management", with 20 chapters including seven separate chapters on the emerging areas, *viz.*, 1. Introduction to aquaculture health management; 2. Pathological process and Disease development; 3. Epidemiology and Biosecurity; 4.Viral diseases of fishes; 5. Viral disease of shellfishes; 6. Bacterial diseases of fishes; 7. Bacterial diseases of shellfishes; 8. Fungal diseases of fishes; 9. Parasitic diseases of fish and shellfishes; 10. Non infectious diseases; 11. Toxicology in aquaculture; 12. Zoonotic diseases in aquaculture; 13. Conventional diseases diagnosis; 14. Nucleic acid based diagnosis; 15. Antibody based diagnosis; 16. Biotechnology in aqua health management; 17. Welfare of fish and shellfish in aquaculture; 18. Defence mechanisms in fish and shellfish; 19. Prophylaxis in aquaculture; and 20. Pharmacology and Aquatic medicine, assumes great significance.

A dedicated team of teachers involved in teaching, research and extension in health management for the last 30 years, with good number of peer reviewed international publications and technologies has contributed to this book. I compliment Prof. K.M. Shankar, Former Dean, College of Fisheries, Mangaluru, a Fish Health Expert, for his efforts in bringing out this book that would be a valuable reference material for students of Fisheries Science, as also aqua managers and policy makers.

(S. Ayyappan)
Former Director General
ICAR, New Delhi

Preface

I have edited and also contributed 16 chapters to this book on "Fish and shellfish health management". My colleagues in universities in India and abroad and the Indian Council of Agricultural Research(ICAR), New Delhi have contributed rest of the chapters. This book has my experience of 40 years long career in research, teaching and extension in Aquaculture and Fish and shellfish health management at the Karnataka Veterinary, Animal and Fisheries Sciences University (KVAFSU), India. My six year long exposure and experience in North America during doctoral programme in microbiology / fish virology has further helped in writing the book. Experience in Teaching UG and PG programmes in aquatic health management, contributed immensely to carve chapters to this book. I had an opportunity to serve as Chairman (Fisheries), 5th Deans Committee ICAR, New Delhi 2014 for revising syllabus for the four years fisheries programme (BFSc) in the ICAR and the State Agricultural Universities(SAU). Experience gained through interaction and discussion with the colleagues for introduction of Pharmacology, Aquatic medicine, Immunology and Biotechnology for the first time in India has been helpful in writing this book. My research experience in important key areas of fish health management- Production and application of monoclonal antibodies for pathology, development of diagnostics and vaccine, Development of a novel biofilm oral vaccine model for fishes, Artificial substrates for boosting fish health and production through microbial biofilm, besides work on general fish and shellfish diseases has been useful for completion of this book. National and international grants with research collaboration at national /international level helped me to a great extent in quality research with publications and technology developments. Experiences of extensive field visits to carp, freshwater prawn, shrimp and aquarium fish breeding farms in India over the four decades, interacting with farmers, demonstrating farmer level diagnostics are worth mentioning which also contributed to this book.

In the last four decades aquaculture is growing at 5-6% annually world over, contributing to total production exceeding 50% of total fish production in some countries. Aquaculture has expanded in area, number of species cultivated, technologies, movement of species and marketing which however, also has big challenge from disease outbreaks, which is a stumbling block for its sustenance. Conventionally, we have been teaching health management of these cold blooded vertebrates and invertebrates following those principles and practices for warm blooded animals and birds from the veterinary sciences which is not adequate and appropriate and needs to be improved and changed at the earliest. Teachers and scientists are aware of this lacunae and research has been initiated world over in the last 3 decades on fish and shellfish diseases, immunology, water qualities, bio-security, zoonosis and several biotechnological interventions. Against this backdrop, in this book an attempt has been made to develop chapters using up to date research findings in health management for the cold blooded vertebrates and invertebrates

Diseases are the major stumbling block in the production and sustenance of aquaculture in the world causing loss in billions of dollars. As in other animal husbandry, diseases existing and new emerging ones due to microbes and parasites are a serious concern in aqua health management. Additionally, non infectious diseases due to poor water quality, industrial and agricultural pollutants/toxins and nutritional deficiency/excess have affected both fish and the consumer. Loss due to parasites causing death and morbidity in aquatic animals is huge, and not sufficiently estimated and documented. And furthermore, effective control of parasites by large scale application of chemicals/pesticides, is risky to fish and its consumers, requiring urgent preventive alternatives including vaccines using modern biotechnologies/ technologies. With increase in number of indigenous and exotic culture species and their products, their movements within the country and across international borders, spread of diseases is rapid and complex. This has necessitated urgent measures for biosecurity of stocks, with quarantine, certification, and contingency plans to contain epidemics. Zoonotic potential of microbial pathogens and parasites has amplified with aquaculture activities and need for creating awareness and measures for the safety of humans. Furthermore, a new trend in ethics in aquaculture, an increasing global awareness on welfare of the fish in aquaculture is gaining importance which should be appreciated. R&D in new areas of aquaculture particularly biotechnology for diagnosis, vaccine development and more importantly, the aquatic medicine exclusively for the cold blooded animals needs emphasis for the growing aquaculture. An important feature of aquaculture health management different from that of land animals is that it involves delicate cold blooded animals with less developed immune systems, that are highly influenced by ambient temperature and environment and hence easily susceptible to diseases. Currently, good amount of research is going on with regard to immune systems of Asian fish species which contribute 80% to global aquaculture production. Against this large background/status,seven emerging areas have been identified in aquaculture health management, viz., Non infectious diseases, Demand for biosecurity measures, Zoonotic diseases, Immunology/Defense mechanisms in fish and shellfish and prophylaxis, Biotechnological interventions through development of diagnostics and vaccines, and more importantly, Aquatic medicine for the cold blooded animals and Welfare of aquatic animals in aquaculture. These current developments /issues ,are reflected in different chapters of this book. The book has comprehensive updated information on fish /shellfish health management in 20 chapters with eight new chapters on the above emerging area in aquaculture health. Overall, fisheries education in health management today has undergone a sea change, with increasing information through R&D, emerging new areas, technologies and trends in aquaculture.

This book can easily serve as a textbook for undergraduate fisheries students in India and the Asia Pacific who in general have similar fisheries education programme with syllabus on health management. In India, the book is very useful to fisheries students of the 67 Agricultural universities. Furthermore, the book is useful for a large number of post graduates taking ICAR NET, SRF, JRF examinations, besides researchers, teachers, extension workers and policy makers. I hope the book is easily accessible for fisheries students in India and the Asia pacific.

I acknowledge with thanks computer assistance received from Dr Prakash Patil, Scientist at the Nitte University, Mangalore and Mrs. Soumya Dhananjay, Senior Assistant at the College of Fisheries, Mangalore for writing this book.

(K.M. Shankar)
Author and Editor

Contents

List of Contributors

Abhiman, MFSc, PhD, Assistant Professor, Department of Aquaculture, College of Fisheries Kishanganj-855107, Bihar Animal Sciences University, Patna, Bihar, India

K.M. Shankar, MFSc, PhD (Microbiology, Canada), Former Professor of Aquaculture-Fish Health Management & Biotechnology, Former Dean(Fisheries), Karnataka Veterinary Animal and Fisheries Sciences University (KVAFSU), Mangalore-575002, India, and Former Emeritus Scientist, ICAR, New Delhi, Fellow, National Academy of Agricultural Sciences, New Delhi, India

K.S. Ramesh, Former Professor of Fish Pathology, Karnataka Veterinary, Animal and Fisheries Sciences University(KVAFSU), Mangalore-575002, Karnataka, India

M.N. Venugopal, MFSc, PhD, Former Professor of Fisheries Microbiology and Former Dean(Fisheries), Karnataka Veterinary, Animal and Fisheries Sciences University (KVAFSU), Mangalore-575002, Karnataka, India.

Naveen Kumar, B.T., MFSc, PhD, Assistant Professor, College of Fisheries, GADVASU, PAU Campus, Ferozepur Road, Ludhiana, Punjab -141 004, India

Neeraj Sood, PhD, Principal Scientist, Exotics and Aquatic Animal Health Division ICAR-National Bureau of Fish Genetic Resources, Canal Ring Road, P.O. Dilkusha Lucknow – 226 002, Uttar Pradesh, India

Omkar V. Byadagi, MFSc, PhD(Taiwan), International Degree Programme of Fish Technology and Aquatic Animal Health, International College, National Pingtung University of Science and Technology, Pingtung, Taiwan

P.K. Pradhan, MFSc, PhD, Principal Scientist, Exotics and Aquatic Animal Health Division National Bureau of Fish Genetics, ICAR, Canal Ring Road, P.O. Dilkusha, Lucknow – 226 002, Uttar Pradesh, India

Prakash Patil, PhD (Medicine, Japan), Post-Doc (Pharmacology, Republic of Korea) Senior Scientist G-I, Central Research Laboratory and Member Secretary, Central Ethics Committee, K.S.Hegde Medical Academy (KSHEMA), NITTE (Deemed to be University) Mangaluru, Karnataka - 575 018, India

Raj Reddy, MSc PhD, General Manager, TPP India Private Limited. Bengaluru 560 098 Karnataka, India

Colour Plates

1

Introduction

K.M. Shankar

An Overview of Fish and Shellfish Health Management in Aquaculture in India

Aquaculture in India in particular freshwater fish culture comprised mainly of Indian major carp cultivation has a long history. Since 1980s commercial carp culture expanded, particularly in Andhra Pradesh and in few years became an Asia model. Then on, it is slowly and steadily expanding to different parts of the country particularly Punjab, Haryana, West Bengal, Orissa and Tamil Nadu and Karnataka. Of late, culture of *Pungasius* (Basa) has come up on large scale in AP. In addition, there is also growing small scale commercial cultivation of catfish, murrels, tilapia and freshwater prawn(*Macrobrachium rosenbergii*). Aquarium fish breeding is also expanding to establish itself as an industry in India.

Brackish water culture has been practiced traditionally in India in Pokkali fields of Kerala, Bheries of West Bengal and Ghaznis of Karnataka since hundreds of years. From 1990 onwards, commercial shrimp culture came into existence along the coasts of India. High export value, support from government and short culture period encouraged private entrepreneurs to pursue shrimp farming in a big way. There is vast potential for fish culture in brackish water and at the present efforts are on to popularize culture of Asian seabass, mullets, *Eutroplus* and milk fish. Mariculture is still at infancy but increasing at a slow and steady rate.

Aquaculture health management has to be viewed from the incidences and outbreak of diseases due to growth and expansion of the industry. Aquaculture has expanded tremendously in the last three decades, growing at 6-7 % annually. Contribution of fish production from aquaculture in the country is around 60 % compared to 10 % it was 40 years ago. In few years to come, aquaculture contribution may go up to 80 % of total fish production. Today

2.43 m ha of freshwater and 1.15 m ha of brackish water are under culture. In addition, 8.62 m. ha of irrigated inland area mainly in Haryana, Maharastra and Karnataka which have become saline and unsuitable for agriculture are slowly being reclaimed for aquaculture. Farmers are growing acclamatized saltwater fish and shrimp spp in inland saline waters and a shift like this may favour pathogens with different virulence to cause disease. Overall, it can be seen that in addition to increased conventional areas in fresh and brackish water, aquaculture is being practiced in unconventional areas like saline soils. Performance of fish and shellfish in these non conventional water bodies in the long run needs to be studied from their health point of view.

Until seventies, aquaculture practiced mostly in freshwater was extensive type with stocking density close to fish density in nature. With the beginning of brackish water shrimp culture high density intensive culture became common. Today aquaculture in farms and in large number of shrimp hatcheries (320 No) and recirculatory systems is high density culture. Culture practices with high density (semi and intensive stocking) involves heavy use of fertilisers and feeds leading to accumulation of waste creating high demand for oxygen. Availability of less space for movement and accumulation of waste leading to stressful life is common in aquaculture. Furthermore, these hatcheries could be breeding ground for parasites and pathogens and have become a source of diseases to aquaculture system and natural waters as well either due to accidental escape or during ranching of rivers, reservoirs and seas. In addition, ornamental fish breeding and marketing is growing steadily to an industry. High density poorly managed hatcheries of aquarium fishes is a major concern where more often stocks are lost to diseases during transport or when stocked in new place after transport by the customers. Fish food organisms including bacteria, rotifers, cladocerans, copepods, chironomid larvae and tubificid worms are cultured at high density in hatcheries can also serve as carriers of microbial pathogens and parasites.

Health management has also to be viewed from the number of species of fish and shellfish grown in aquaculture. In India today there are more than 40 species reared for food, sport and ornamental purposes compared to less than 10 species 40 years ago. Several species of carps, catfishes, murrels, perches, mullets, shrimps, prawns, mussels and aquarium fishes are cultured which have a wide range of pathogens and diseases. Furthermore, diverse nature of immune system of these fish and shellfish species in aquaculture has to be considered carefully for husbandry and health management. Fishes posses both specific and non specific immune system, the former with a memory component. They have well developed non-specific immune system comprising 50 % of their total immunity. However, their adaptive immune system although has specificity and memory component, can not be compared with that of well developed higher vertebrate immune system. Furthermore, fish have only IgM like immunoglobulin compared to well developed IgG, IgM, IgA, IgD in the higher vertebrates. On the other hand, we have shrimp and prawn in aquaculture having only well developed non specific immune system which comprises 75 % of their total immunity. Specific immunity with memory in the crustaceans and molluscs is not yet demonstrated. The cold blooded aquatic animals with unique low level of immune system are subjected to environment to a great extent compared to warm blooded vertebrates and hence are very susceptible to diseases. Disease in aquaculture, therefore is the outcome of interaction between environment, host and the pathogen. In general, aquatic animals are very delicate, need special care and management in husbandry different from that of husbandry of land animals.

Movement and marketing of fish and shellfish and their products on a large scale across national and international borders is a critical weak point in aquaculture for disease management as parasites and pathogens also move along with them to new areas. Spread of shrimp white spot virus(WSV) in the world through seed and frozen products is a classic example. Escape of fishes from culture system to wild is also a source of pathogens. White spot virus infected shrimp released from farms to open waters after epidemics have spread the virus to 50-60 % of wild shrimp in the ocean.

Loss due to diseases in aquaculture is increasing. Communicable diseases due to viruses, bacteria, fungi and parasites are taking heavy toll with significant economic loss. Diseases are the main impediment for expansion and sustainability of aquaculture and have great impact on the economy. In India, it is estimated that annual loss due to the white spot virus disease in shrimp alone is more than Rs. 400 crores. Parasites such as *Argulus* and *Lernaea* inhibit growth and sometime cause mortality resulting huge loss in carp culture. In addition, non-communicable diseases related to water chemistry and nutrition also cause loss in aquaculture. The outbreak of Epizootic Ulcerative Syndrome (EUS) in 1986 on a pandemic scale in fresh and brackish water fishes and the more recent outbreak of White spot disease (WSD) in cultured shrimp in 1994 have demonstrated the potential devastating impact of diseases on aquaculture production in India and Asia. The impact of these two diseases has given a new impetus to fish and shellfish health management and reflect the need for scientific health management in aquaculture. In all, it is estimated that we are losing annually Rs 1000 crores due to microbial and parasitic diseases in aquaculture system of India.

Overall Cause for Diseases in Aquaculture: Aquaculture is breeding, rearing and husbandry of fishes, crustaceans and molluscs for food, ornament and sports. It is a unique husbandry involving special animals- the cold blooded vertebrates and invertebrates, whose growth, metabolic activities and overall performance dependent heavily on ambient temperature. Furthermore, the water the very medium which supports the animal, influences the physiology, growth and other performances unlike the influence of environment on land based animals. The intensification of aquaculture operation results in water quality deterioration, oxygen depletion, and accumulation of solid waste and toxic metabolites such as hydrogen sulphide, methane, ammonia, nitrites, that cause stress to the cultured animals leading to diseases.

Disease outbreak occurs when this delicate balance between host, pathogen and environment is upset (Fig. 1). Intensive culture systems offer an ideal environment for disease outbreak, because such systems stress the host and favour virulent pathogens. Depending on the nature and severity, the disease may cause mass mortality of the affected population in a short time, produce protracted small scale mortality, reduce growth, or make cultured fish unsuitable for human consumption. Overall, health management in cold blooded aquatic animals considering their poor immunity, disease resistance and high influence of the environment on them is very delicate, tricky and difficult entirely different from that in higher vertebrates.

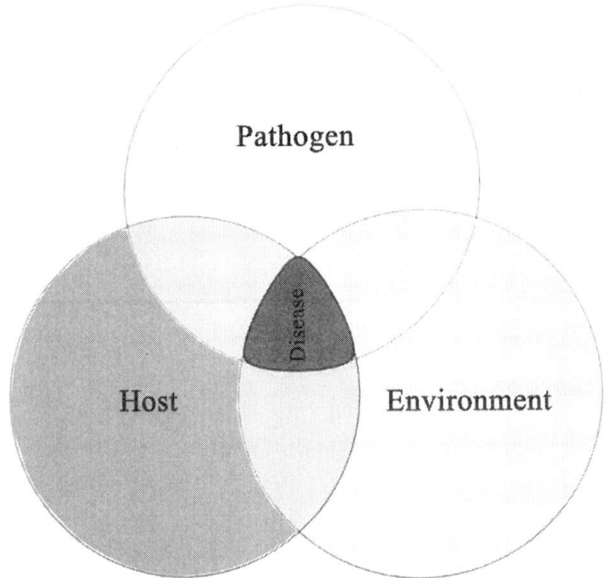

Fig. 1: Interaction of host, pathogen and the environment in disease development

Diseases caused by viruses, bacteria, fungi and parasites (protozoa to crustaceans) are common in aquaculture systems. Viral pathogens of fish in India fortunately are few such as Herpes virus of carp (HPC), Tilapia lake virus(TLV) and Viral nervalecrosis virus(VNV) of sea bass. However, in shrimp good number of viral diseases recorded in spite of short history of shrimp culture. Fifteen shrimp viruses have been reported worldwide including India of which important being white spot virus(WSV), infectious hematopoitic hypodermal necrosis virus(IHHNV), Taura syndrome virus(TSV) and yellowhead virus(YHV).

Common bacterial diseases in fish include aeromonosis by *A hydrophila*, vibriosis by *Vibrio* although these are considered secondary pathogens often causing diseases due to stress in fish. Vibriosis is common in shrimp in hatcheries and grow out ponds.

Fungal disease are due to *Saprolignia* mostly in hatcheries and young weak fish in nurseries. Recently, *Saprolignia* has been reported in mass infection and mortality of *Pungasius* (Basa) culture in Andra Pradesh. Epizootic ulcerative disease (EUS) by *Aphanomyces invadans* has caused havoc in natural fresh and brackish water bodies.

Parasitic infections constitute bulk of the disease, loss due to mortality and morbidity. Particularly in freshwater carp culture, of all the pathogens, two parasites namely *Argulus* and *Lernaea* cause heavy loss and damage to crop.

Non-infectious disease due to pollution, algal blooms and toxins, feed contamination, water quality alterations such as DO, pH also occasionally lead to crop loss.

Emerging new diseases: Aquatic animals are very delicate compared to higher land vertebrates with well developed immune system. Further no animal on land experiences changes in its environment as fishes. Temperature, DO and fish metabolites play odd in culture system. Exploiting the delicate nature of aquatic organisms, pathogens take upper hand, and new ones emerge with growth and expansion in aquaculture activities. Emerging diseases are more in crustaceans compared to that in fishes due to the low level of immune system of the former. In 40-50 years history of modern fish culture four or five viral pathogens have emerged. However, in a short span of two decades of shrimp culture nearly 15 viruses have been recorded to cause diseases.

Health Management

The term health management is very broad and encompasses wider areas like soil and water quality maintenance, providing proper nutrition, selection of good quality brood stock and seed, timely disease diagnosis, prophylaxis and chemotherapy. A proper understanding of the process of disease development role of abiotic and biotic factors in disease development, importance of case history and clinical signs will go a long way in evolving scientific health management package.

Understanding the process of disease due to infectious or non-infectious agents is vital for disease management. Developmental biology, pathogenecity and transmission mechanisms of pathogens, disease resistance mechanisms of the host, and the role of environment in influencing the disease process are the important aspects. By knowing these aspects, it is possible to devise strategies to manage the pathogen, host or environment either singly or in combination.

In disease treatment, the first task would be diagnosis and understanding the disease process. In diagnosis, both conventional and modern molecular based methods such DNA and antibody are in use. Biotechnological approaches are very efficient are becoming popular in diagnosis. In recent years, epidemiological approaches (population medicine) are also becoming popular for understanding the disease at the population level.

Prophylaxis

Preventive strategies include all measures to prevent the occurrence of disease. This could be directed at the environment (drying, liming, etc) and host (vaccination, immunostimulation). In addition, biosecurity measures preventing entry of the pathogen through seed, carriers and all other routes are important

Immune system of fish comparatively not well developed and there exists even difference in immune system among fishes. As there is memory and specificity components in fish, vaccines can be developed. However, memory in fish is short lived compared to that in higher vertebrates. Vaccines against viral and bacterial pathogens have been developed elsewhere in the world. In India several strategies for bacterial vaccine developed particularly against *A hydrophila* such as whole cell and sub units but no commercial vaccine is available yet. A novel whole cell biofilm oral vaccine has been developed and evaluated in terms of antibody titre and protection upon challenge is on field trial for commercialisation.

At the present large quantity of pesticides are used to control parasites which is not healthy. Against this background, few attempts to develop vaccine against parasites of fish- *Lernaea* and *Argulus*, on the lines of successful vaccine developed against animal tick parasite. Although this approach has potential but it is a very difficult proposition.

Immune system of shrimp/prawn is lacking specificity and memory. And hence it is difficult to develop vaccine to shrimp. However, immunostimulants which boost immunity non- specifically have been developed for managing microbial diseases in shrimp and prawn. A good number of probiotics have been developed and commercialised. But well designed indigenous probiotics are preferred compared to commonly available imported ones.

Biosecurity and Epidemiological Measures. For sustainable aquaculture development, health management should be broad based, followed at various steps viz, pond, farm, regional, national and also international level. In this direction, health certification and quarantine programme aimed at preventing entry and spread of exotic and indigenous pathogens to new areas is important.

Disease surveillance, certification for movement of stock, Quarantine, Contingency measures for pandemic are standard protocol required for any organised scientific animal husbandry, which has been used and largely helped in animal health maintenance. In aquaculture these biosecurity measures have just begun in India. Specific pathogen free (SPF) and specific pathogen resistant (SPR) stocks have been developed for shrimp farming as a new frontier area in aquatic animal health management

In therapy, strategies on use of chemicals to target pathogens at different developmental stages is important. Application of the drugs and chemicals in aquaculture at the present is largely based on the experience of medicine in animal husbandry comprising warm blooded animals. Fish and shellfish being cold blooded animals, drugs effect, mechanism, kinetics have to be studied systematically in detail. Furthermore, as each species differ in immunity, there is need to study chemotherapy for each cultivable tropical fish and shellfish. The effect of these drugs on soil and water and their chemistry in water needs to be understood and hence comprehensive chemotherapeutic study is required in aquaculture on a war foot basis.

Nevertheless, large quantity of drugs, chemicals, antibiotics, hormones and anesthetics are used in aquaculture. Some products are even available/ supplied to farmers without appropriate information on the content. Although the drugs, chemicals and pesticides give relief, continuous use in large quantities is bad both for fish and fish consumers. These chemicals with their residue are either directly harmful to consumers or indirectly favour development of resistant microbes and parasites. Application of most of these chemicals is through water indirectly controlling the disease. Therefore, understanding the effect of these chemicals on water and soil chemistry is of paramount importance.

Welfare of Fish and Shellfish in Aquaculture

"Fish also feel pain and crave for comfort from a doctor"

Fish and shellfishes are cold blooded animals which are subjected to environmental changes to a great extent than the warm blooded vertebrates. In the wild, fish enjoys living at low density with plenty of place to move around, adequate oxygen and food of choice. Fish in the natural waters lives with least stress surrounded with few pathogens. On the contrary, fish in aquaculture live at high density and although fed well, suffer from inadequate space for movement, low oxygen, accumulated waste and loads of pathogen likely to live a stressful life

There is strong increasing evidence that finfish, like other vertebrates are sentient animals. This means that fish are self-aware; they can feel pain and distress, they have long term and short-term memory and to some extent, they can experience emotions. Therefore, ethical treatment of fish in culture is very important. And therefore it is appropriate to say that fish also feel pain and crave for comfort from a doctor. In this direction, aquaculture husbandry should be handled by a well trained fish doctor with adequate training in fish biology, culture practices and aquatic medicine. Fortunately, there is increasing awareness in consumers about status of welfare of fish in aquaculture who would like to know how well their animals are reared. Till recently fish was not listed for consideration of ethical treatment, only largely mammals were under the list. It is a good trend where in the world has recognised the stress fish undergo during intensive culture and there are attempts to grow them more humane. Fish grown under such system are certified and there are markets recognising this value. Furthermore, welfare measures should also be viewed from health point of view as preventing diseases also reflects humane treatment of fish with increased production and profit.

References

FAO. 2020. The State of World Fisheries and Aquaculture 2020. Sustainability in Action. Rome.

K. M. Shankar, 2016 , Importance and Scope for Aquatic medicine in India, In Aquatic medicine-a training manual, Special publication No 01/2016, Editor Prakash Patil, Aquatic Animal Health Management Laboratory, College of Fisheries, Karnataka Veterinary, Animal and Fisheries Sciences University, Mangalore,575002

K. M. Shankar and C.V Mohan,2002, Fish and Shellfish Health Management , ISBN 81-7525-328-2, Fish Pathology and Biotechnology Laboratory, Dept of Aquaculture, UAS, College of Fisheries, Mangalore-575002

S Ayyapan, 2020, Drivers of Aquaculture Development in India, In Indian Aquaculture.

S Ayyapan and K.M. Shankar Agricultural Higher Education, Research and Extension in India-Fisheries, 131-148 In Agricultural Higher Education, Research and Extension in India Editor Debabrata Dasgupta. Published by AGROBIOS(India), Jodhpur-342003

V.V Sugunan, V.R. Suresh and C. K. Murthy, The Society for Indian Fisheries and Aquaculture, Hyderabad

2

Pathological Processes and Disease Development

K.M. Shankar and K.S. Ramesh

Diseases are among the greatest deterrents to the sustained production in aquaculture. White spot disease (WSD) in penaed shrimp and epizootic ulcerative syndrome (EUS) in fresh and brackish water fishes are the well known devastating diseases with serious impact in aquaculture. Aquaculture medicine which broadly encompasses prevention and management of diseases in cultured aquatic organisms becomes a vital requirement for a sustained industry.

The four K's essential for scientific aquaculture health management are knowledge about the disease process, knowledge about the pathogen, knowledge about the host and the knowledge about the environment. Disease development process is often complicated and involves host-pathogen-environment interactions. Knowledge about the pathogen, attaching and entering the host, deriving nourishment, reproduction, transmission, overcoming host defence barriers, etc are very essential. Susceptibility of a host to a pathogen is important which depends on host species, age, size, immuno competence and stress response. Besides knowledge about how the temporal and spatial aspects of environment stress the host and favour the pathogen to cause disease is vital

Disease Development Process

There exist a delicate balance between the host, pathogen and the environment. When this delicate balance gets upset, disease can result. Aquaculture environments can stress the host, favour the pathogen and result in disease development. To appreciate the process of disease development understanding the pathogenicity mechanisms of the pathogens, disease resistance mechanism of the host and the role of the environment is essential. Furthermore, role of various pathogens, their adaptive modifications, the functional importance of the target tissue, interaction between pathogen and host at the target tissue level, pathogenicity mechanisms of pathogens, etc are necessary to gain insight into process of disease development. Only when a pathogen can establish on or in the host, proliferate, overcome the non-specific and/or specific defense barriers of the host, produce the pathogenic factors, cause cellular and tissue damage, produce significant pathological changes, impair the function of the target tissue cause clinical disease often with mortality. This process is very complicated which often get accelerated by stress and environmental factors. However, the sequence of disease development will to a large extent depend on the nature and load of the pathogen (parasite, bacteria, fungi, virus), its intensity per unit area or unit weight of the host, size of the host and environmental factors.

Role of Stress in Disease Development

The importance of stress in predisposing the fish/shrimp to infections is widely recognised. It is now well known that many of the routine aquaculture practices and pollutants subject the fish to stress and predispose them to disease. Stress is a non-specific response and it involves series of changes in the animal in trying to adapt to the changed situation. Stress is produced by environmental or other factors which extends the adaptive responses of the animal beyond the normal range or which disturbs the normal functions to such an extent that the chances of survival are significantly reduced. The series of changes termed "stress response" tries to help the animal restore the normal homeostasis. This process has both advantages and disadvantages. During stress, hypothalamus-pituitary-inter renal axis (HPI axis) gets stimulated and increases the output of stress hormones called corticosteroids. These stress hormones help to mobilize additional energy during the response to regain the internal homeostasis. The concentration of corticosteroids comes down to basal level after the withdrawal of the stimuli. On the other hand, stress hormones are basically immuno-suppressive in nature. This immunosuppressive properties of the stress specific hormone can significantly lower the efficiency of both non-specific and specific immune system of fish and can render the animal more susceptible to disease.

It is well known that common husbandry practices like handling, netting, transportation and the normal features of an intensive culture system like suspended solids, low oxygen, high organic matter, overcrowding, high ammonia, etc. can elevate the level of corticosteroids in the blood. During these phases, the fish may easily get infected if there are sufficient number of pathogens in the water. Similarly, many of the pollutants at very low levels can stress the fish and elevate the level of corticosteroids in the blood and make the fish relatively more susceptible to infection. Many of the stressors encountered in intensive culture systems are of chronic nature and can keep the level of corticosteroids above basal levels for longer duration. Therefore, minimising stress factors is the key to successful health management.

Pathological Processes

Pathology is the basis of medicine. It tells us what is happening at the tissue level (Plate 1). Only through pathology, host-pathogen interactions can be best understood and appreciated at structural, functional, microscopical and ultrastructural levels. Pathology in simple is the outcome of three basic processes:

(a) cellular responses to pathogen induced injury,

(b) inflammatory response exhibited by the host

(c) and the pathogenicity mechanisms.

The interaction of these three processes in the host target tissue produce series of pathological changes. Failure of the target tissue in its function leads to clinical manifestation of the disease followed by morbidity and mortality. Pathology will thus provide an insight into many of these processes and help to evolve scientific health management package.

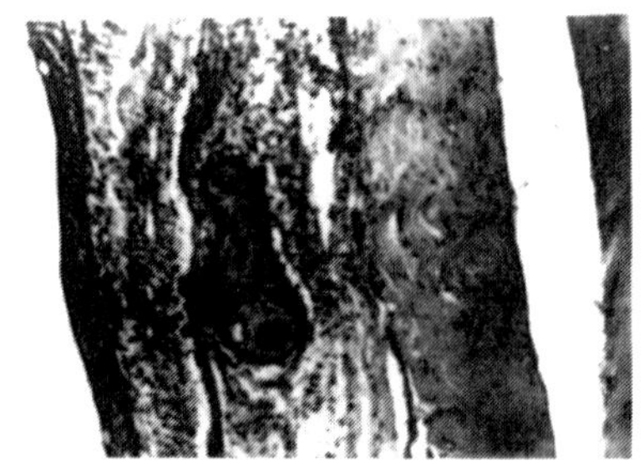

Inflammatory response in the dermis to germinatin fungal spore.

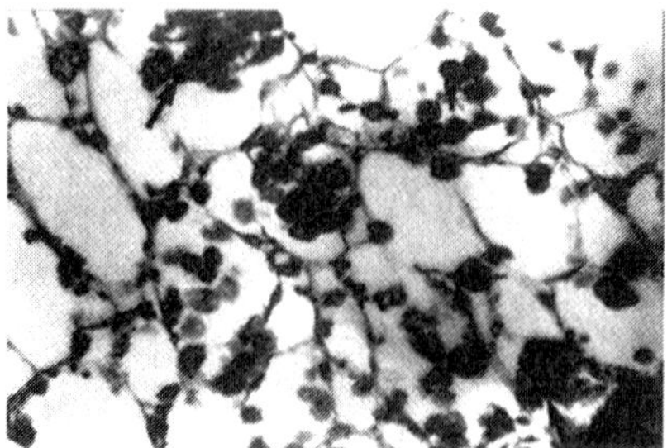

Sarcolysis accompanied with acitve myophagia and accumulation of macrophages

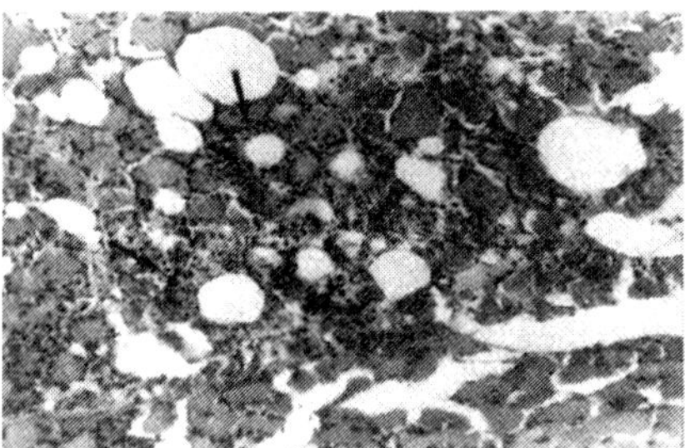

Accumulation phagocytic cells around adjuvant droplets

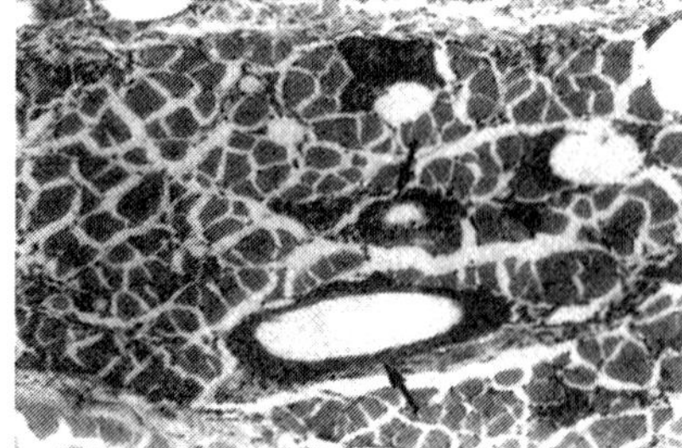

Macrophage induced encapsulatory response

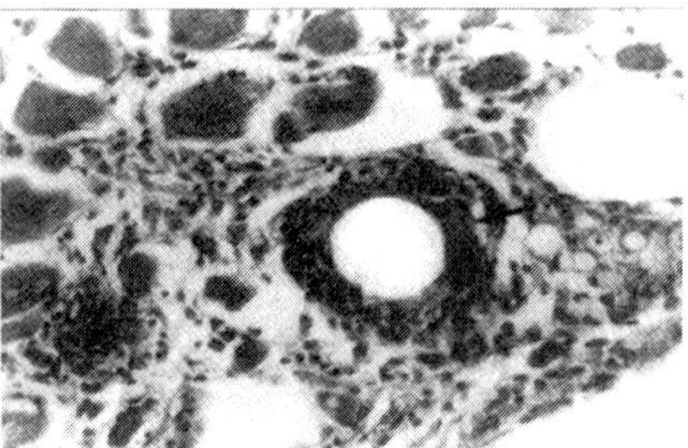

Mature granuloma around on adjuvant droplet

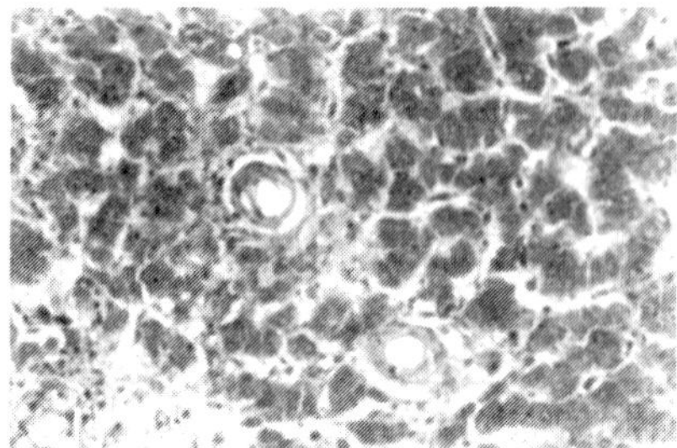

Epethelioid granuloma around fungal hypase

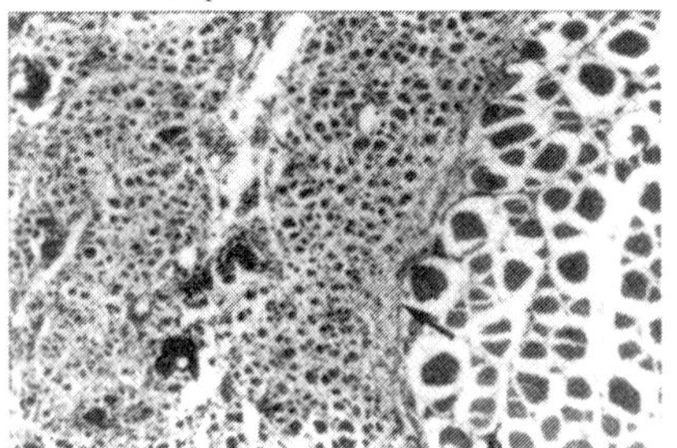

Demarcation of inflammatory lesion by the perimysium from the adjacent normal mscle

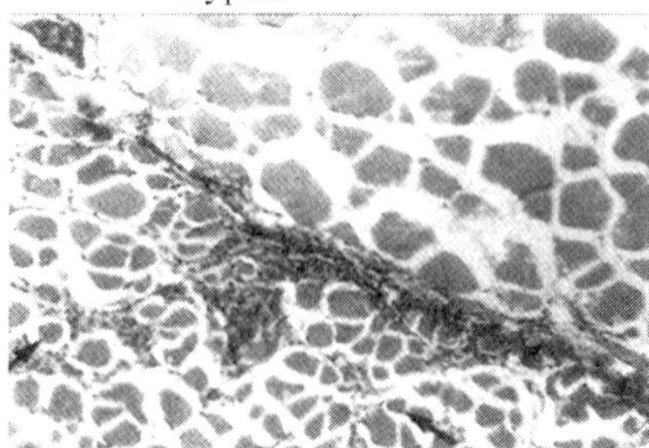

Healed lesions with regeneration muscle cells

Plate 1: Pathological Processes

Cellular Responses to Injury

The cellular environment is constantly changing and as a consequence, cells have to make continuous adjustments to accommodate these changes. The cells have a great capacity for adaptation to their environment and are able to respond to changes in the internal and external environment by alterations in both their structure and function. For example, adipose tissue cells respond to prolonged excessive food intake by increasing their synthesis of storable fat (increase in the size of adipocytes -hypertrophy). Skeletal muscle fibres increase in size in response to increased work load as seen in athletes. These are examples of cellular responses (physiological hypertrophy) to physiological stimuli.

Certain changes lie outside an acceptable physiological range. Such adverse changes may be termed as pathological stimuli. Cells may respond to pathological stimuli by extending their normal physiological adaptive processes. These adaptive processes include:

(a) an increase in cellular activity

- i) hypertrophy(increase in cell size)
- ii) hyperplasia(increase in cell number)
- iii) enzyme or metabolic induction

(b) a decreased activity of cell

- i) cell atrophy (reduced cell size)
- ii) tissue or organ atrophy (reduced number of cells)
- iii) decreased cellular metabolism

(c) Change in morphology and function of mature cell type

- i) modification in morphology and function to a cell type more suited to the changed environment (metaplasia)

Failure of Adaptation of a Cell

Cells which are intrinsically unable to adapt, or which have reached their limit of adaptability, begin to show structural changes which indicate their failure to withstand the changed environment. These cells show cytoplasmic and nuclear changes as a manifestation of organelle failure (cloudy swelling hydropic degeneration, fatty change and cell necrosis). Both hydropic and fatty degeneration are reversible if the deleterious stimulus is short lived. If adverse conditions persist however, or if the initial pathological stimulus was severe then these changes continue and progress into a sequence of events leading to cell death (cell necrosis). The onset of these irreversible changes is heralded by distinct morphological changes in the cell cytoplasm and the cell nucleus.

The initial responses of cells to injury are manifested at a sub cellular level by morphological changes in the various cytoplasmic organelles. Membrane enzyme systems such as those constantly engaged in maintaining ionic gradients and membrane transport are particularly vulnerable to pathological influences. One of the earliest consequences is the loss of efficiency of the sodium pump permitting ingress of sodium ions and water, resulting in swelling of the cell. This change is potentially reversible. If the cell sustains irreversible damage, mitochondria become more swollen and disruption of oxidative phosphorylation deprives the cell of aerobic metabolism. The resulting collapse of many intracellular homeostatic mechanisms leads to progressive disintegration of nuclear and cytoplasmic organelles and release of lysosomal enzymes, leading to autodigestion of the cells.

Acute Inflammation

Living tissues sustain injury from a wide variety of physical, chemical, microbial or immunological causes, and the response in dealing with resulting tissue damage or destruction is known as inflammation. Whatever the cause or type of tissue involved, the initial series of processes which classically ensues is described as acute inflammation and is directed towards neutralising the injurious agents and to restoration of tissue to useful function. Inflammation is a non-specific, protective, vascular response. Damage of mast cells during an injury releases vaso active amines which have an influence over the microcirculation of the area. The characteristic feature of acute inflammation is the formation of an inflammatory exudate which has three principal constituents: fibrin, serum and leucocytes, predominantly neutrophils. The formation of the inflammatory exudate involves three vascular processes and functional blood supply is an essential requirement for inflammation to proceed:

i) dilation of local blood vessels leading to engorgement with blood (hyperaemia)

ii) increased capillary permeability permitting plasma proteins to pass into the tissue

iii) migration of leucocytes from blood vessels into the area of injured tissue

The outcome of acute inflammation depends on three major factors: the degree of tissue injury, the nature of the injurious agent and the type of tissue involved.

When tissue damage is minimal, the exudate is reabsorbed into nearby vessels leaving no subsequent evidence of injury, this process is known as resolution.

More frequently, the exudate undergoes a process called organisation and repair in which dead tissue is removed by phagocytosis and the defect is filled by a highly vascular connective tissue called granulation tissue, this then progressively undergoes fibrous repair with the formation of a dense fibrous scar at the site of the original tissue destruction. Depending on the type of tissue damage, there may be also some degree of regeneration of original tissue which depends on the ability of cells of mature tissue to undergo division.

When tissue damage is caused by certain types of bacteria, then large numbers of dead and dying neutrophils accumulate with fibrin and fluid of the acute inflammatory exudate to form a localised collection of pus known as an acute abscess.

In some circumstances, an injurious agent persists over a prolonged period causing continuing tissue destruction whilst at the same time the body is attempting to deal with the previous tissue damage by the process of acute inflammation, organisation and repair, in such a case the damaged area may exhibit tissue necrosis, acute inflammatory exudate, granulation tissue and fibrous scar tissue concurrently. This process is known as chronic inflammation.

Chronic Inflammation

In most cases the injurious agent is destroyed or neutralised in the earlier stages of the acute inflammatory reaction, and the rest of the changes follow in sequence. Sometimes, however, the damaging stimulus persists despite the tissue responses directed at destroying or neutralising it, and further episodes of tissue destruction may result. In such circumstances the changes of tissue damage, acute inflammation, granulation tissue formation and attempts at fibrous repair may all proceed concurrently instead of sequentially as they do when the injurious agent is eliminated early in the sequence. This phenomenon is known as chronic inflammation and it most commonly follows previous acute inflammation where that process had failed to eradicate the damaging stimulus.

Chronic inflammation also follows many acute abscesses; bacteria frequently survive and proliferate in the pus at the centre of the abscess cavity and are not easily accessible to endogenous or therapeutic bacteriocidal substances. The wall of the chronic abscess is composed of fibrous granulation tissue, with an inner zone of acute inflammatory exudate bordering the pus filled cavity.

Less commonly, chronic inflammation occurs virtually de novo in response to specific damaging agents which are resistant to destruction by neutrophils during the acute inflammation or which fail to excite a strong acute reaction. When the damaging agent is not destroyed by the neutrophils, the initial neutrophil response is usually sparse and short lived, and is quickly followed by a macrophage response which persists and dominates the histological picture. This local macrophage accumulation produces a discrete lesion called a granuloma, and the chronic inflammation characterised by this type of response is known as chronic granulomatous diseases. The outcome of the chronic inflammation thus depends on whether local and systemic factors favour the injurious agent or alternatively, the attempts at healing and fibrous repair. Chronic inflammations are often of considerable duration, in which immunological mechanisms play an important role.In summary, chronic inflammation is marked by continuing tissue damage, acute inflammatory exudation, organisation and fibrous repair occuring concurrently rather than sequentially.

Pathogenicity Mechanisms

Diseases of aquatic animals are caused by parasites, fungi, bacteria, viruses and non-infectious agents. The mechanisms ectoparasites use to establish on the host and derive nourishment, induce necrotic changes at the cellular and tissue level. Adaptive and inflammatory response exhibited by the host adds to the pathological picture in the case of ectoparasitic diseases. Necrosis and proliferative response are common pathological features associated with ectoparasitic protozoans. Ectoparasitic metazoans (worms and crustacean parasites) cause attachment and feeding injuries resulting in skin and gill necrosis and tissue proliferation in the host. In the case of endoparasites, mechanisms of penetration and migration, mode of deriving nourishment, mode of reproduction and route of exit from the host, will all contribute to the pathology. Tissue necrosis, enteritis, fibrous encapsulation, nodule formation, xenomas, etc are some of the common endoparasite associated pathological changes. Some of the endoparasites like histozoic and celozoic sporozoans despite causing massive tissue destruction do not elicit any inflammatory response from the host.

In the case of bacterial diseases, enzymes, toxins, haemolysins and other histolytic factors produced by the proliferating bacteria induce severe pathological changes. Surface ulcerative bacterial diseases (vibriosis, columnaris) show necrotic and proliferative changes at the surface leading to development of ulcers. Acute systemic bacterial diseases (bacterial haemorrhagic septicaemia, furunculosis, enteric red mouth) produce areas of focal to general necrosis in affected organs. Chronic granulomatous type of bacterial diseases (bacterial kidney disease, fish tuberculosis) produce granulomas in various organs and tissues. In shrimp, bacterial infections like oral, cuticular, enteric and systemic vibriosis produce massive necrotic changes in the affected organs. Characteristic haemocyte led inflammatory response in the form of melanised haemocytic nodules are a common feature of shrimp vibriosis.

In the case of fungal diseases, the histolytic properties of the invading fungal hyphae induce necrotic changes in the concerned target tissue. In dermal mycosis, the invading fungal hyphae brings about severe necrotic changes in the integument. Some of the fungal pathogens (eg.*Saprolegnia*) are able to invade the host tissue and cause destruction without stimulating any inflammatory response. Many of the systemic fungal pathogens (*Ichthyophonus*, *Exophila*) produce characteristic mycotic granulomas. *Aphanomyces invadans*, the necessary cause of EUS, produces mycotic granulomas in all the affected target tissues. In shrimp, fungal infections like *Lagenidium*, produce necrotic tissue pathology without any significant inflammatory response, while *Fusarium* infection induces typical granulomatous response.

Viruses are intracellular obligatory pathogens entirely dependent on host cell machinery for their replication. Viral replication induces areas of focal to general necrosis of cells in the target tissue. Apart from complete destruction of cells, viral infection can lead to enlargement of cells (giant or syncytial cells) or partial to complete loss of function. Viral infections produce intracytoplasmic or intranuclear viral inclusion bodies, which are of diagnostic significance. For example, the WSSV produces basophilic intranuclear inclusion bodies in the cells of ectodermal and mesodermal origin. Of significance is the fact that viral infections do not normally produce any inflammatory response.

What is not Pathology?

While using pathology for diagnostic purpose or understanding disease process, it is very important to appreciate what is true pathology and what is not. Physiological changes in the cell structure and function to physiological stimuli are not pathological changes. It is very essential to differentiate post-mortem changes from true pathological changes. In this context, it is necessary to emphasise the need for using only live or moribund aquatic samples for all histopathological diagnostic purposes. Artifacts resulting from poor sample fixation, processing, sectioning and staining should be clearly differentiated from true pathology. It is very common to see artifacts being read and published as pathology. This is largely due to the lack of training in histopathology.

Applications of Pathology in Aquaculture Medicine

Pathology is in importance, at least first among equals, being the basis of medicine on which health management is built. An accurate appreciation of what is happening in the tissues will form the basis of health management and aquaculture medicine. Histopathology provides the simplest and easiest means of learning about interactions going on between the host and the pathogen at the cellular level. The advantage of histopathology is its being a non-specific diagnostic tool. It can be used to diagnose parasitic, bacterial, fungal and viral diseases of aquatic organisms. Histopathological inferences can be put to various uses such as:

- For understanding the disease process
- For diagnosis of parasitic, bacterial, fungal and viral diseases
- For screening, certification and quarantine purposes
- For epidemiological analysis
- For understanding multiplication and transmission issues of pathogens
- For assessing the accessibility of a pathogen for chemotherapy
- For assessing the inflammatory response of the host to the pathogen
- For assessing whether immuno-prophylactic strategies can be used for a pathogen
- For formulating scientific and comprehensive health management strategies.

Through rational interpretation of histopathological findings, it is possible to arrive at conclusions on the pathogenicity mechanisms of pathogens, functional status of target organs, severity of a disease, cause of morbidity and mortality. However, histopathology being a non-specific diagnostic tool suffers from certain limitations, but the advantages of using histopathology for aquatic animal health diagnostics and management overweigh its limitations.

The pathological picture one sees in a tissue is therefore the combined effect of pathogenicity mechanisms of pathogens, cellular response to injury and the host inflammatory response. Understanding the pathology helps to appreciate the severity of the disease, the likely clinical signs, the possible approaches to management, etc. In summary, pathology and the disease process form the foundation for scientific health management.

References

https://doi.org/10.1002/9781118222942

K. M. Shankar and C.V Mohan,2002, Fish and Shellfish Health Management, ISBN 81-7525-328-2, Fish Pathology and Biotechnology Laboratory, Dept of Aquaculture, UAS, College of Fisheries, Mangalore-575002

Murray A.G., Peeler E.J. 2005. A framework for understanding the potential *for emerging dis*eases in aquaculture. Prev. Vet. Med. 67:223–235. doi: 10.1016/J.Prevetmed.2004.10.012.

Roberts, R. J. (2012). Fish Pathology: Fourth Edition. Fish Pathology: Fourth Edition, 1–581.

Yun-Zi Liu, Yun-Xia Wang and Chun-Lei Jiang 2017 Inflammation: The Common Pathway of Stress-Related Diseases Front. Hum. Neurosci., Volume 11 https://doi.org/10.3389/fnhum.2017.00316

3

Parasitic Diseases of Fish and Shellfish

K.M. Shankar, Naveen B.T. and K.S. Ramesh

An Overview of Parasitic Diseases in Asian Aquaculture

Parasitic induced diseases bring about significant economic losses in fish culture system. Estimated loss due to mortality and morbidity of fish due to parasitic infection is much more than that from microbial diseases. Protozoan ciliates (*Ichthyopthirius*, *Trichodina*) and flagellates (*Ichthybodo*, *cryptobia*) are some of the common ectoparasites of fish. Certain ectocommensal ciliates (*Epistylis*, *Vorticella and Zoothamnium*) attach to the external surface of fish leading to fouling. These protozoan ectoparasites are often associated with mortalities of younger stages of cultured fish. The situation becomes worst in water with low oxygen and high organic matter. Most extoparasitic protozoans have simple and short life cycle and reproduce by binary fission. Most protozoan ectoparasites are readily detected in direct microscopic examination of skin and gill scrapings. Histopathological changes in the integument following ectoparasitic protozoan infection are outcome of two counteracting cellular processes – hyperplasia of the epithelial cells, including mucus cells and chloride cells, versus a progressive cellular destruction. Cellular necrosis primarily occurs due to the feeding and attachment activity of the parasites. Generally, immersion therapy using suitable chemicals can control ectoparasitic protozoan infections

Disease caused by endoparasitic sporozoans are a serious threat to fish farms. These sporozoan spores present in the pond soil are normally ingested by the fish. The infective element of the spore (sporoplasm) is released in the gut. The sporoplasm reaches the target tissue using the vascular route. Once inside the target tissue, the vegetative trophozoites cause massive destruction of the target tissue and produce large spore containing cysts. There are two types of sporozoan infecting fish and shell fish namely myxosporidian and microsporidian. Myxosporidian occur in fish either as histozoic (intercellular tissue parasites) or as coelozoic (inside the lumen of various ducts and tubules) parasites. Microsporidians are intracellular obligatory parasites and lead to massive hypertrophy of infected cells (xenomas). There is no effective treatment for sporozoans. However, treating pond soil with chemicals and lime is the best option.

Monogenetic trematodes like *Dactylogyrus* (gill fluke) and *Gyrodactylus* (skin fluke) with their well developed attachment haptor and feeding apparatus can cause extensive pathology and mortality in early developmental stages of fish. Majority of digenetic trematodes (flat worms) occur in fish either as free or encysted metacercaria in various organ systems. The infective cercaria are released from snail (first intermediate host). Cestodes (tape worms) occur as adults normally in the gut and as free or encysted larvae in various organ system. As larvae and adults, these helminths bring about lot of pathological changes in target tissues. Zooplanktonic copepods serve as the first intermediate host. As digenetic trematodes and cestodes are endoparasites, there is no cost effective chemotherapy. Controlling intermediate host and prevention of contamination of aquatic system with eggs of these parasites from the excretory product of final host (birds and mammals) are the possible preventive measures.

Larger crustacean ectoparasites like *Argulus* (fish lice), *Lernaea* (Anchor worm), *Ergasillus* (Gill maggot) and isopods with their well developed attachment organs and mouth parts inflict considerable damage at the site of attachment and feeding. These ectoparasites can also cause significant primary damage in the form of skin lesions predisposing the fish to secondary systemic bacterial and fungal pathogens. The macroscopic grossly visible parasites can be easily identified. Broad spectrum pesticides are increasingly used in immersion treatments to control crustacean parasites. However, as residues of these chemicals are harmful to fish and its consumers in the long run, serious attempts are going on for development of safe vaccines against the parasite, although it is a herculean task.

Commonly occurring ecto and endoparasites of fish, their site of infection, pathology and associated clinical signs are described briefly (Plates 2 and 3). A detailed account of important parasites in aquacultural systems such as, *Argulus*, *Lernaea*, Flukes and *Enterocytozoonhepatopenaei* (EHP) is provided as they are common and economically important.

Common Parasites of Fish

I. Ectoparasites

A. Protozoans

a) Ciliates

i) *Ichthyophthirius multifiliis:* It is a parasite of gills and skin which do not penetrate below the dermis. Major pathology includes necrosis, lifting and separation of respiratory and skin epithelium, sloughing, infiltration of EGCs (Eosinophilic granular cells) and leucocytes to primary and secondary lamellae and proliferative response (hyperplasia). Clinical signs include parasites visible as white grit-like spots on skin, gills, fins. erratic swimming movements, skin detachments, mucus secretion (excessive), gasping for air, rubbing the body against hard surface.

ii) *Trichodina* sp.: It is a parasite of gills and skin with major pathology of necrosis and proliferation of skin and gill epithelium.Clinical signs include greyish-blue veil like coating over the body surface, darkening of skin, disintegration of delicate skin between the fin rays, progressive destruction of epidermis and excessive mucus production.

iii) *Chilodinella* sp.: The parasite infects skin and gills with major pathology including necrosis and proliferation of skin and gill epithelium. Major clinical signs are Pox-like white opaque raised lesions on skin, detached scales and gasping for air.

b) Flagellates

i) *Ichthyobodo necator:* A parasite ofskin and gills with pathology of necrosis of epithelial cells, lifting of scales, sloughing and , hyperplasia in skin. In gills pathology includes necrosis, hyperplasia, thickening of gills and fusion of secondary lamellae. Clinical signs includes Acute anorexia, lesions on the caudal fin, blue slime on the surface, excessive mucus production, chronic mortalities.

B. **Helminths**

a) Monogenetic trematodes

i) *Dactylogyrus* spp (gill fluke): The parasite infects gills, with major pathology involving necrosis, lifting and proliferation of respiratory epithelial cells, infiltration of EGCs to primary and secondary lamellae and fusion of secondary lamellae. Common major clinical signs are fish becoming restless, gathering near the inflow of water and gasping for air, dark coloration, gills covered with thick mucus layer, appear anemic, opercula opened, secondary infection and tiny lesions.

ii) *Gyrodactylus* sp. (skin fluke) **:** It is a parasite of skin with pathology involving necrosis, proliferation and hyperplasia of skin epithelial cells.

Major clinical signs include skin becoming spotted, necrotic and dark, covered with bluish grey mucus layer and damaged tissue between fin rays. Heavily infected fish swims listlessly and gather near water inflow to gasp for air.

C. **Crustaceans**

a) Branchiurans

i) *Argulus* sp. (Fish lice); The parasite infects skin. Major pathology includes necrosis, loss of blood, secondary infection, hemorrhagic ulcerative lesions around bite wounds. Major clinical signs are irritation, dashing wildly through water, splashing, jumping, tiny red spots, open wounds, ulceration and bite lesions

b) Copepods

i) *Lernaea* sp. (Anchor worm).The Site of Infection **is** body surface, eyes. Major Pathology includes inflammation and necrosis at the site of attachment, loss of blood, secondary infection, large scale infiltration of macrophages, fibrosis/encapsulation around the hold-fast organ leading to nodule formation. Major Clinical signs are **a**berrant swimming, marked emaciation and loss of weight, in severe cases fishes float with belly side up.

ii) *Ergasilus* sp. (gill maggot). The parasite infects gills.Major Pathology includes necrosis of respiratory epithelial cells, proliferation and sloughing. Major clinical signs are **e**xcessive mucus secretion, fusion of gill lamellae and gasping for air

c) Isopods. The site of infection is body surface, oral and branchial cavity. Major pathology includes necrosis of epithelial cells and hemorrhagic bite lesions. Major Clinical signs include mechanical obstruction of oral and branchial cavities leading to gasping and erratic swimming behaviour

D. **Annelids (Leeches).** Leeches infect body surface. Major Pathology includes bite wounds, necrosis and hyperplasia around bite wounds. Clinical signs include tiny red lesions, open ulcers, irritation and hemorrhages.

II Endoparasites

A. Protozoans

a) Myxosporidian sporozoans

i) *Myxobolus* (Histozoic).The parasite infects cartilage, skull and spine, gills, muscle and, kidney integument. Major pathology includes acute lesions, necrosis and destruction of target tissue and replacement of target tissue by developing vegetative trophozoites and fully developed cysts. Major clinical signs include loss of balance, typical whirling movements, malformations of vertebral column, cranium, jaw, visible cysts on surface of gills and internal organs. severe infection leads to mortalities.

ii) *Sphaerospora* sp. (celozoic).Site of infection of the parasite include kidney tubules with pathology of tissue proliferation and mechanical inhibition of several kidney functions. Clinical signs include inflated belly in the region of pectoral fins with large scale mortality.

b) Microsporidian sporozoans (intracellular). The parasites infect musculature and, ovary. Major pathology includes liquifactive necrosis in the muscle tissue, necrosis of oocytes and fibrosis. Major clinical signs include appearance of white spots in the musculature of the body or head, scoliosis disturbances in equilibrium, jerky swimming movements, sterility.

c) Flagellates

i) *Trypanoplasma* sp. and *Trypanosoma* sp. These are blood parsites which are usually mild, causing anemia under severe conditions. Clinical signs are fish are slack, sluggish, emaciated and anemic, pale gills, sunken eyes, pale skin and gasping for air.

ii) *Hexamita* sp. The sites of Infection of the parasite are intestine and gall bladder. Major Pathology include purulent inflammation of the gall bladder, atrophy and dilation of bladder. Clinical signsarenon-specific with abnormal swimming movements, emaciation, abdominal distention.

B. Helminths

a) Digenetic Trematodes (flat worms)

i) *Clinostomum* sp. (Yellow grub.) The flat worm infects many organs mainly integument and muscle causing tissue destruction, necrosis around the cysts, displacement of tissue in vital organs, gill damage and fibrous nodules encysting metacercaria. Clinical signs are yellow pigmentation in the muscle and distended abdomen.

ii) *Diplostomum spathaceum* (Eye fluke). The parasite infects eye leading to damage of the eye, free metacercaria in the lens, retina and humor causing blindness. Clinical signs include exopthalmia, haemorrhage, opacity of lens, ulceration of outer surface and , blindness

iii) *Sanguinicola* sp. (blood fluke). The fluke infects circulatory system (major blood vessels). Eggs of the fluke block the capillaries in the gills and kidney. Escaping miracidium produce necrotic pathology in gills and kidney. Major clinical signs are fish gathering at the water inlet gasp for air, damaged and pale gills and decolourised spots on the gill.

iv) Centrocestus sp. The parasite infects gills and integument. Major Pathology includes metaceracaria encysted with the cartilagenous chondrocytes. Gasping and respiratory distress are the clinical signs.

v) *Ichthyocotylurus* sp. Infects heart leading to tissue destruction due to encysted metacercaria and hemorrhages. Major Clinical signs include: Refusing to take food and protruded eyes are the clinical signs.

b) Cestodes (Tapeworms)

i) *Ligula intestinalis:* Larval Pleurocercoids of the tapeworm infects body cavity causing inflammation of the peritoneum, contraction and inflammation of the internal organs, vascular ruptures, exudates in the visceral cavity, necrosis, gonad degeneration due to the free pleuocercoids. Fish appear at the surface of water, disturbances in the equilibrium, decreased feeding, distended abdomen and belly burst are the usual clinical signs.

ii) *Bothriocephalus* (Asian fish tape worm). The tapeworm infects gut leading to enteritis in adults, blockage of gut, sloughing of gut epithelium. Clinical signs include weak emaciated fish, gut blockage and chronic mortality

iii) *Triaenophorus* sp. Larval pleurocercoids of the parasite infects internal organs leading to hyperemia of liver, shrinkage of gall bladder, intestinal inflammation, enlarged liver due to encysted larvae. Major Clinical signs include sluggish movements, distended abdomen due to dropsy, white capsules (of cysts) on the liver.

iv) *Diphyllobothrium latum* (Broad fish tapeworm of man). Larval pleurocercoids of the tapeworm Infect musculature, visceral organs causing clumping of visceral organs, peritonitis, decrease in lymphocytes, loosening of muscle due to encysted larvae. Clinical signs include fish floating on the surface with head bent to one side due to muscular cramps.

C. Acanthocephalans (Thorny headed worms) infecting intestinal wall causing mechanical damage to the intestinal wall, deprival of nourishment when infection is heavy, intestinal inflammation and perforation of gut. Clinical signs include striking reddening of the fish intestine and emaciation.

D. Nematodes (Round worms). Round worms infect body cavity, gut and visceral organs. Major Pathology includes encapsulated larvae in the visceral organ and muscle. Free larvae and adults are found in the body cavity and gut. Pathology is not significant unless parasite burden is heavy. Major Clinical signs are weight decrease and presence of larvae in internal organs free or encysted.

Economically Important Common Parasites of Fish

Argulus sp

Argulosis is a disease caused by crustacean parasite *Argulus* sp. commonly known as "Fish louse". There are about 140 species of *Argulus* recorded in marine, brackish, and freshwater environments. Juveniles feed on mucous and skin cells and with age they become blood feeders.Infections by *Argulus* are rarely a serious threat to natural populations of fish, but clearly a major problem in farmed fish populations causing morbidity, poor growth and mortality in aquaculture. *Argulus* also known to be intermediate hosts of nematodes in cichlids, and poecilids. *Argulus* can cause secondary infections with virus, bacteria and fungi. *Argulus* is a mechanical vector for spring viraemia of carp virus (SVCV) which is highly contagious and rapidly spreading to economically important cyprinid fishes in the Americas and Asia. Infection intensity in farms is enhanced by algal blooms, reduced water level and low rates of stock turnover. Under poor water, the parasite multiply rapidly affecting fishes with high morbidity. Infection in fish are high in poor water quality, varies with season -more in rainy season. The parasite can reproduce rapidly at a temperature range of 20–28 °C.Indian major carps, particularly brood are susceptible hosts with high incidence in rohu. Transmission occurs by direct contact or via fomites. The life cycle is direct and eggs are laid on the substrate. Without a fish host adults can survive about two weeks while newly hatched larvae for one or two days.

Pathology

Argulus attaches to the body of the fish by means of suckers and hooks. The wound made during feeding is sealed by the labrum and labium and the proboscis is used to suck the blood into the oral cavity. It also injects anticoagulant or other lytic compounds, to consume the blood, mucus, and tissue at the puncture site. Heavy infection with 100-150 parasite /fish can kill a medium size fish easily. In Indian major carps with high infection the skin showed abnormally pigmented while gills, kidney, liver and spleen are pale in colour. The most significant histopathological lesions in the gills are the secondary lamellar hyperplasia, hemorrhages in the tips of the primary lamellae and fusion of both secondary and primary gill lamellae. Skin epidermis is lost but the epidermal cells in the vicinity of the of attachment are hyperplastic, muscular layer showing degenerative and necrotic tissue. Kidney showed multifocal enlargement of glomeruli, glomerular tufts in many places shrunken, fragmented and necrosed. *Argulus* infection can modulate the immune system by immunosuppression suppressing macroglobulin and other protein levels. has

Clinical signs: Prominent clinical signs include stunted growth, peeling off of the scales and the appearance of red spots at the sites of infection. The affected fish become very weak and emaciated. Fish show clear signs of irritation and behaviour changes, such as jumping, and scratching against different objects. They may shoal tightly together or congregate at the margins of tank or in areas of high water movement.

Management

Short formalin bath and long term baths and chitin inhibitor diflubenzuron (at 0.01 mg/L up to 6weeks) controlled the parasite effectively. Application of Lindane (benzene hexachloride) at 0.02 mg/L shown to effectively control the parasite. Application of organophosphate and chlorines were found to be very effective control initially. However, gradually the parasite develops resistance to these chemicals which are also considered are toxic to fish consumers. Against this background, efforts are on to determine immunogenic antigens of larvae and adult to develop recombinant vaccine but still long way to go for success., Providing vertical wooden bamboo poles inside the water so fish may rub their bodies against them to get rid of their ectoparasites. Ponds showing severe infection with Argulus should be drained, applied with lime (0.1 to 0.2 g/L) and dried for 24hrs to control the parasite.

Lernaea cyprinacea (anchorworm)

Lernaea species, commonly known as "anchorworms," infect and cause disease and mortality in many wild and cultured freshwater fishes. *Lernaea cyprinacea*, is one of the common species worldwide, and most common in cyprinids, including koi, common carp, goldfish and Indian major carps.

However, it can also infect other species of fish. The parasite has direct life cycle not involving an intermediate host and spread directly between fishes. Anchor worms mate during the last free-swimming copepodid stage of development. After mating, the female burrows into the flesh of a fish and transforms into an unsegmented, wormlike form, usually with a portion hanging from the fish's body. Eggs are released from the egg sac attached to posterior tails into the water, where they hatch within 24 to 36 hours. The naupli will go through three stages before molting into copepodids, which associate with fish gills. After a further five stages and mating, the male leaves the host and dies, while the female transitions into the anchored stage.The parasite needs 18 to 25 days to complete requiring a fish or an amphibian to develop from egg to mature adult. The optimal temperature range for *Lernaea* is 26°C–28°C while below 20°C,females will not reproduce and juveniles are unable to complete their development. However, adult females can o*verwinter on* the fish, producing eggs in warmer spring. Adult females hardier than younger stage can survive 30 days on a fish. *Lernaea* infestations are most prevalent in the summer particularly in stagnant or slow- moving water bodies.

Pathology

The skin, fins, gills, and oral cavity are common sites of *Lernaea* infections. The parasites, burrow deeply into the tissues and embedding anterior anchor into the fish's body. Deep red colour with ulceration at the attachment site is common sight due to intense local inflammation and hemorrhage. Although infestation with small number is not fatal but cause irritation. *Lernaea* infection, normally leads to secondary bacterial (e.g., *Aeromonas hydrophila*) and fungal infections, which sometimes worsen and kill the fish. Larger numbers of adult and copepodite stage on the gill can interfere with respiration, leading to death. Chronic conditions frequently result in poor growth, although fish can survive the infection.

Clinical Signs

The adult female, appears on fish as a small, thin "thread" or "hair" approximately 25 mm long. Under the microscope, the long, tubular body has an anchor on the anterior end and paired egg sacs on the posterior end. The anchor located head, is typically embeds into the host's tissue. The posterior end, with egg sacs, extends out into the water. When examined under microscope, juveniles and copepodite stages can also be seen on skin, fin, or gill.

Management

Depending on situations and species of fish several treatment options are available to control *Lernaea.* In aquaria Potassium permanganate is ideal for treatment either in tank or as "dip". Other treatments include dip in salt or in formalin solution. A 30-minute bath with 25 mg/L potassium permanganate will kill larval lernaeids, but adults may survive. Other treatments such as addition of salt in the aquarium at 1 to 2 tablespoons may help prevent secondary infections. Prolonged immersion with an organophosphate such as trichlorofon is an effective treatment for ornamental fish. Diflubenzuron (at.066 mg/l) a pesticide which interferes with growth of the parasite can kill molting adult and larval stages.

In ponds with food fishes *Lernaea* can be controlled with salt with variable results. Organophosphates and chlorines have been used successfully. However, with time the parasite develops resistance. Furthermore, the residues of the pesticides is harmful to the consumers. Against this background development of safe vaccine against the parasite is a better option. Studies have shown that immune responses induced by *L. cyprinacea* protein extracts depicts a pattern of specific responses, in which the local humoral responses dominate the systemic humoral/cellular response. Several studies searching for immunogenic and protective antigen s of larval and adult *Lernaea* have been explored for developing recombinant DNA vaccines. Although results are encouraging, it has a long way to go for a final vaccine development against the parasite.

Fish Flukes

Flukes infecting the skin (*Gyrodactylus* spp.) and gills (*Dactylogyrus* spp.) common in freshwater fish such as carps, catfish, goldfish, and cichlids. Flukes feed on skin cells and mucus, attaching themselves with their hooked mouths. Poor water quality is often conducive to the introduction of parasites. Dirty water makes fish feel stressed, weakening their immune system and making them more susceptible to infection. Improper diet suppress immune system favours infection. Crowding favours infection. The fluke's life cycle is much quicker in live-bearing species and are easily spread among a crowded tank of fish. Warmer temperatures and poor water quality (higher nitrogen levels organic loads) can increase the reproductive cycles of these parasites

Lifecycle

The life cycle of the monogenean fluke does not have an intermediate host; therefore, it can perpetuate in a closed system indefinitely. These flukes can be livebearers or egg producers; the livebearers can reproduce at a phenomenal rate in a closed system. As adults, gill flukes are hermaphroditic possessing both male and female reproductive structures. Adult gill flukes are capable of laying anywhere from four to ten eggs each day, which hatch into either live or nonliving lärvae. This free-swimming water movement is made possible by cilia, and the hatchlings have less than 8 hours to find a suitable host. The time it takes for a creature to develop from a larva to an adult is greatly affected by the temperature of the water it lives in.Warm water temperature accelerates the reproductive process

Clinical Signs

The parasite is small (0.1 to 0.3 mm), invisible to the naked eye but can cause severe irritation to a fish's skin. Clinical signs vary but often include red spots, excess mucus, and difficulty breathing. Symptoms of flukes in fish are generally non-specific. Missing scales, red spots, excess mucus, hazy look to the skin. flashing behaviour, lethargy, decreased appetite. fish experiencing redness or inflammation on its skin, accompanied by itchiness. Fish may scratch itself by rubbing against surfaces in its tank, resulting in scale loss. When the flukes feed on the blood on the gills, the oxygen supply is affected, and fish may breathe rapidly or spend more time at the water's surface. The lack of oxygen may also cause lethargy. Flukes may destroy a fish's fin if they go untreated. Fish affected with skin flukes typically have clamped fins and increased mucus covering their body, while those affected by gill flukes present for difficulty breathing

Gill flukes cause high mucous secretion, swollen gills with spreaded opercula. Fish become restless congregating near inflow gasping for air. Fish appear dark, cease to feed, lose weight often swimming with high speed, scrapping against objects and jumping out of water. Epithelial growths and swollen gills are commonly observed. Serious epizootics and mortality. Spots or redness close to the gills are the most obvious sign of this parasite illness. These areas of skin will be noticeably redder than the rest of your body, and they may also have a slight purplish color. As the condition worsens, secondary infections in the respiratory tract might cause an increase in mucus production. This is because the gills have been damaged by the flukes, causing an increase in mucus secretion. Depending on how much mucus has grown up around your fish's mouth and body, it will seem yellow or white. In extreme cases of infection, you may see your pet swimming with its head pointing downward due to the weight of the mucus.

Diagnosis

Flukes are diagnosed using a biopsy or skin scrape of the fish's mucus or gills and examined under a microscope. Microscopic examination is the only definitive way to diagnose flukes as clinical signs are non-specific. The best method to confirm the presence of these flukes is a skin scrape and/or gill biopsy. Histologic examination of a sample will reveal elongate flukes that have a row of hooks on their opisthaptor. It is important to note that gill flukes can infest the skin and skin flukes can infest the gills too. In severe infestations fish flukes may cross over or both be present in the same sample. This is why it is necessary to look at skin and gill samples during physical examination of fish.

Management

Several treatments have been successfully used in aquaria and grow out fish farms. In aquaria a cocktail of formaldehyde, malachite green and methylene blue (1 liter of Formaldehyde (37%), 3.7 gm of Malachite Green oxalate, 3.7 gm of methylene Blue) at 1.0-1.2 ml/100 l water give good result. Treatment with Salt (5-10mg/liter aquarium water; 10gm/l for1-2 hours or 5gm/l for 5-7 days) known to give good relief. Organophosphates such as praziquantel 300-500 mg/100L for 1-3 days or 150 mg/10L for 2-3 hours, mebendazole, and toltrazuril have been used to manage monogenean fluke. Masoten bath 25 to 30 g/l for 5 to 10 min, Quinine hydrochloride 30 ppm bath for several days and Trichlorfon (0.25 –3.0 mg/1 liter water for 3 days) have controlled the parasite effectively. Effect of Trichlorofon depends on hardness of water, hard water requiring higher dose. Flubendazole (Dose: 100-200 mg/100L for 1-2 days with 50% water change), Mebendazole (at: 100-200 mg/100L for 2-3 days with water change) and Levamisole (100mg/100L for 1-2 days.) have effectively controlled flukes. Recovery of fish after treatment is quick and successful within days as confirmed by their behaviour reaching normal. Gill flukes, however are more challenging to treat than skin flukes.

Economically Important Parasites of Shrimp Microsporidians

Enterocytozoon hepatopenaei (EHP)

Enterocytozoon hepatopenaei (EHP) is a microsporidian parasite of shrimp infecting hepatopancreas causing hepatopancreatic microsporidiosis (HPM). EHP belongs to the family *Enterocytozoonidae*, phylum Microsporidia which comprises highly reduced and specialized, spore-forming, unicellular parasites of animals. EHP an endemic to Australia is first reported from Australia in *P japonicus* which is now widespread in several Asian countries.

EHP although does not appear to cause mortality, but is associated with severe growth retardation. EHP infection disrupts cells of hepatopancreas tubules. EHP infections in *P. vannamei* with more than 1×10^3 EHP copies per ng total HP-DNA detected by quantitative PCR could be correlated with growth retardation, reduced body weight together with high size variation. It was also frequently been reported that EHP infections may be associated with a white feces syndrome in cultured *P. vannamei*. EHP known to cause disease of the hepatopancreas of four known species of farmed penaeid shrimp i.e., *P. monodon*, *P. vannamei* and *P. stylirostris P. japonicus*. EHP outbreaks are occurring widely in China, Indonesia, Malaysia, Vietnam and Thailand Philippines, Mexico, Brunei and India. Despite observations of low prevalence and low intensity infections of EHP in early descriptions HPM caused by EHP has now reached epidemic proportions in the Asian penaeid shrimp culture industry.

Co-infection of shrimp hepatopancreatic microsporidian with AHPND and SHPN has been reported. EHP also increases shrimp susceptibility to infection by opportunistic *Vibrio*. EHP favours *Vibrio*. already present to colonize the sloughed cells and the exposed basement membrane of hepatopancreas.

Pathology

Externally visible signs of EHP are often absent, apart from retarded growth over time. EHP is parasitic in the lumen of the hepatopancreatic epithelial cells, affecting the normal absorption and storage of nutrients by the hepatopancreas. EHP hurts the tubule epithelial cells of the hepatopancreatic tissues of a shrimp, thereby interfering with its normal nutrient absorption. Although it does not necessarily lead to death of the shrimp, but recognized as the main cause of slow growth. EHP drives a degenerative cyclic pattern in the hepatopancreas microbiome of the shrimp,affecting the beneficial microorganisms and their beneficial functions. The EHP modulates the microbiome (bacteria/fungi) and its predicted functions over the course of disease progression. According to the infection intensity, three disease-stages, early, developmental and late, have been identified. During the early-stage, EHP is not consistently detected, and a high diversity of potentially beneficial microorganisms related to nutrient assimilation are present. In the development-stage, in most of the shrimp a decrease in beneficial microorganisms with increase in opportunistic/pathogenic fungi. During late-stage, shrimp display different infection intensities, show a displacement of beneficial microorganisms by opportunistic/pathogenic bacteria and fungi

Diagnosis

Ideally EHP can be detected by light microscopy (x100) with either stained tissue sections or smears of HP. In H&E staining, inclusion bodies in the cytoplasm can be identified, with oval or elliptical spores of 1.1±0.2~0.6-0.7±0.1 μm. Because of the time limitation histology is not recommended for routine diagnostic purposes, especially in juvenile and later shrimp life stages. In electron microscopy, a single nucleus in the spore, a posterior vacuole, and anchoring disk attached to the polar filament can be seen. Five to six polar filaments are seen wrapping around the thick electron-dense wall. Microscopy/hitopathology usually have limitations as spores are sometimes produced only in small numbers, even in heavily infected specimens and thus, the PCR detection is preferred. A nested PCR and a LAMP method are available to check whole PL and feces of brood stock.PCR is preferred as it has high sensitivity and specificity, and needs less preparation time than required for histology.

Management

Treatment and control of EHP is very difficult. However, EHP can be prevented at two stages i) hatchery ii) grow out pond

i) **In hatchery:** Hatchery should be dried for seven days and the tank should be rinsed with acidified chlorine (200 ppm chlorine solution at pH <4. 5). All brood shrimp should be washed by cleaning briefly using 2.5 per cent sodium hydroxide solution (25 gms NaOH/L fresh water) followed by washing for three hours. This treatment should also include all equipment, filters, reservoirs and pipes. Avoiding live feed (e.g., live polychetes, clams, oysters, etc.) for maturing brood stock is strongly suggested.

ii) **In grow out pond:** Appropriate preparation of ponds between cultivation cycles is essential, especially when a pond has previously been affected by EHP. Earthen ponds can be disinfected of EHP spores, by applying CaO (quicklime, burnt lime, unslaked lime or hot lime) to soil at 6 ton/ha followed by soaking and ploughing the sediment (10-12 cm). Soaking the sediment with water to activate the lime for one week before drying or filling. With application of CaO, the soil pH rise to 12 or more for a couple of days and then fall back to the normal range as it absorbs carbon dioxide and becomes CaCO3. PCR screening of PL for EHP is very essential in prevention of the parasite in grow out system.

Other Microsporidians in Shrimp

Microsporidians are ubiquitous in wild penaeids. The disease caused by microsporidians in shrimp are known by various names like cotton shrimp disease, milk shrimp disease and nosema disease. The common microsporidians which cause this disease condition are *Agmasoma* spp, *Nosema* spp *and Plistophora* spp.

In the infected shrimp the developing vegetative trophozoites replace the striated muscle causing it to become opaque and white. The gonads enlarge, turn opaque and become white. Large numbers of whitish tumour like swellings may be seen in the muscle and gonads. Very interestingly microsporidians while replacing the host tissue will not induce any host inflammatory response. The spores are the infective stages and are normally found in pond soil. It has been documented that the spores require "conditioning intermediate host" a fish. The conditioned spores passes through the feces of fish. There is no therapy for this disease. Avoidance and better pond soil management can reduce the incidence of this disease.

Gregarine Infections in Shrimp

At least 3 genera of gregarines infect the penaeid shrimp. These are *Nematopis* spp., *Cephalobus* spp, *Lecudina* spp. Gregarine trophozoites or gametocysts are found in mid-gut contents. Severely affected shrimp may show reduced growth rates, elevated FCRs and grossly visible yellowish discoloration of the midgut. Massive infections may show yellow coloration of the midgut, severe lesions of the midgut and physical blockage of the lumen by masses of the parasite. Damage to the midgut mucosa may provide a route of entry for potentially lethal bacteraemia by opportunistic *Vibrio* spp.

Fungal Diseases in Shrimp

Several species of fungi are known to cause diseases in larvae, juveniles and adults of cultured penaeid shrimps. Larval mycosis is very common in shrimp hatcheries and nurseries and are caused by the phycomycetes fungus *Langenidium* spp. Affected larvae contain an extensive, non-septate, highly branched fungal mycelium throughout the body and appendages. The invasive hyphae virtually replace all the larva's or PL's tissues and produce a systemic progressive necrosis that is accompanied by little or no host inflammatory response (i.e., hemocytic encapsulation or melanisation of the fungal hyphae). Typical lethal infections may also be accompanied by vibrosis. Specialised hyphae or discharge tubes with or without terminal vesicles on their distal ends, may be seen protruding from the body of moribund or recently dead larvae. Important clinical signs which are of diagnostic significance are the sudden onset of mortalities in the larval and early PL stages.

Larval mycosis can be easily diagnosed by: 1.Wet-mount demonstration of hyphae within the bodies and appendages and sporangia with diagnostic discharge tubes and motile zoospores. 2.Histological demonstration of hyphae and discharge tubes projecting from the shrimps body surface. 3.Isolation and culture of the fungus from infected shrimp.

The fungal disease of juveniles and adult penaeid shrimp is caused by the imperfect fungus *Fusarium solani* and several other species of the *Fusarium*. This disease is known by different names; Fusarium disease, fusariosis, fungus and black gill disease. *Fusarium* spp. lesions are typically marked by intense host inflammatory response and intense melanization. Such melanised lesions often with red fully necrotic distal portions are more prevalent on the appendages such as the antenatal blades and flagella, the eye stalks, the periopods, the uropods and telson. Histologically these lesions appear as prominent granulomatous lesions with multicentric hemocytic nodules with melanised centres. Close examination may reveal hyphae in such nodules or hyphae in sinuses or on the surface of the lesions. Advanced lesions consists only of necrotic tissue, melanised foci and at the lesion surface, fibrosis and invasion of adjacent tissues by the advancing hyphae.

Wet mount and histological diagnostic procedures can easily demonstrate the diagnostic features of *Fusarium* infection. Chemotherapy developed against larval mycosis is being successfully implemented in shrimp hatcheries. Effective therapy for *Fusarium* infection of adult shrimps is yet to be developed.

Fouling in Shrimp

Surface and/or gill fouling disease is very common in shrimp reared in high density culture systems or in systems of poor water quality. Nearly all of the organisms involved in gill and surface fouling syndromes are free living and are not true pathogens. These fouling organisms are called epicommensals or epibionts because they use shrimp as a substrate for attachment. These organisms do not cause any damage to the shrimps directly, but instead cause problems indirectly by attaching to gill or cuticular surfaces. They may kill shrimp by interfering with water flow over the gills, gas exchange across the gill surfaces, molting, feeding and exotoxins that may cause some degree of tissue damage.

Fouling in shrimps is known by different names; gill disease, fouled gills black gills, brown gills, filamentous gill disease and bacterial gill disease.

Numerous species of bacteria, *Leucothrix*-like organisms, algae and protozoa (*Zoothamnium* sp., *Epistylis* sp., *Vorticella* sp.) may be involved in fouling.

Gill and surface fouling in shrimps can be easily recognised based on gross clinical signs. Detritus, silt or other extraneous material present in the water are often trapped by fouling organisms on the gills or appendages giving the infected area a black, brown or green discoloration. In the absence of such materials the appendages and/or cuticle of the carapace may appear fuzzy with a mat like growth of fouling organisms.

Confirmatory diagnosis of fouling disease is rather simple and straight forward. Whole animal wet-mounts (larvae and PL), wet-mounts prepared from gill, epipodites, mandibular palp (juvenile and adults) examined under microscope will show the epicommensal organism(s) present. Fouling organisms of the general body, the gills and appendages may be detected and often identified in routine H & E stained paraffin sections. Through water quality management and accepted immersion treatments it is possible to contain the damaging effects of fouling organisms.

Freshwater Prawn *M rosenbergii*

Microsporidians infection

Microsporidia infections have been observed in *M. rosenbergii.* A microsporidian parasite *Potasporamacrobrachium* has been recorded in *Macrobrachium nipponense* causing progressive whitening of musculature and reduced survivability during holding and transportation of the animal. The description of this novel disease-causing microsporidian is an important discovery in freshwater prawn culture and should be considered as a potential threat to the culture of *M. rosenbergii*. Furthermore, *M. rosenbergii*co cultured with *P. vannamei* were positive by PCR for *Enterocytozoon hepatopenaei* (EHP) which however could not be demonstrated by histology.

Fouling in *M rosenbergii*

M. rosenbergii could be infested by a range of epi- and endo-parasitic infections. All life stages of the prawn are affected. Broadly, fouling organisms include microbes and protozoans

Microbial epibionts are commonly observed etiological agents in *M. rosenebrgii*. Epibiont-fouling organisms include algae, filamentous bacteria, or protozoa common to the aquatic environment. Nevertheless, the presence of epibionts can be deleterious to the giant freshwater prawn. Fouling by various microorganisms can provoke natatory and feeding impairment during the larval and post-larval phases. Heavy infestation of filamentous algae, protozoans and miscellaneous microbes over the body surface of reared animals in clear ponds diminish the market value of the prawns. Heavy fouling by the chlorophyta *Oedogonium* & a cyanophyta *Lyngbya* can cause lethargic behaviour in *M. rosenbergii*. These organisms are non-invasive and their presence on the cuticle surface does not trigger the activation of the proPO system, unless a localized mechanical trauma is present.

The most common protozoans found in freshwater are the peritrich ciliates *Zoothanium*, *Epistylis*, *Vorticells*, *Opercularia*, *Vaginicola*, *Cothurnia*,and Lagenophyrs and the suctorians *Acineta*, *Podophyra*, *Tokophyra* and *Ephelota*. Larvae are more susceptible to protozoan infections than adults. Generally, *Zoothannniumsp*., *Epistylis* sp., *Vorticella* sp., and *Acineta* sp. cause slight opaqueness in the body color of the organism. Shedding of the exoskeleton during ecdysis temporarily frees the organisms from these fouling microorganisms. Berried females may be a source of introduction of epibionts into hatchery culture systems, where they cause greater issues with feeding and swimming impairment due to the small size of M. rosenbergii larvae and post larvae. Presence of ciliates, with *Zoothamnium* spp. present at high prevalence on appendages and gills as well as *Acineta* spp., *Epistylis* spp. and *Vorticella* spp. The latter two were found both externally and within the gut. Aapostome ciliate cysts were commonly seen attached to gill lamellae, with heavy infections frequently observed, causing a melanization response at points of attachment. Gut-dwelling gregarines were also present in *M. rosenbergii* populations. Other eukaryote parasites reported include bopyrid isopods and the larval digeneans Opecoelid metacercariae attached to pleopods, periapods and antennules and microphallid metacercariae in the musculature. Improvement of water quality during the outbreak of the diseases very necessary. Treatment with 20-30 ppm formalin is effective and safe in controlling *Zoothamium* infection of the larvae. Repeated treatment with 2 ppm acetic acid as a 1-min dip is recommended for *Epistylis* sp. Formalin, merthiolate and copper based algicides have been found to be effective against these parasites

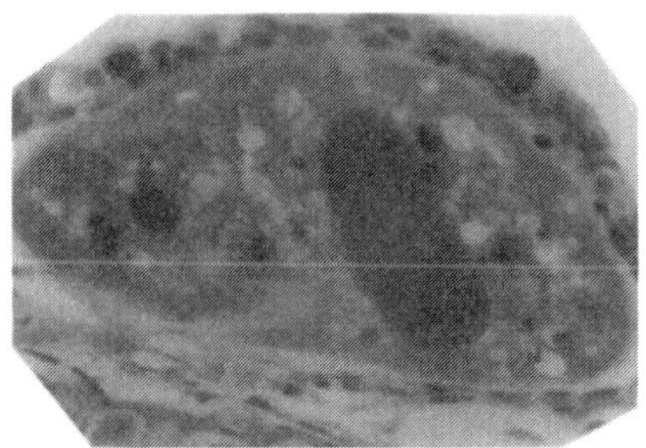

Adult *Ichthyophthirus multifiliis* with gill epithelium

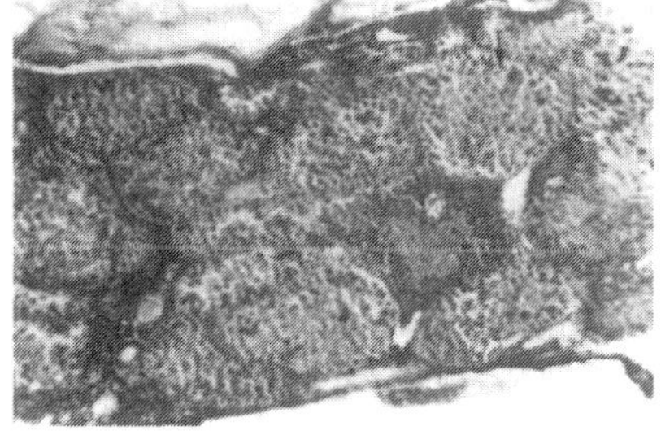

Cysts of *Myxobulus* sp. in the kidney of catla

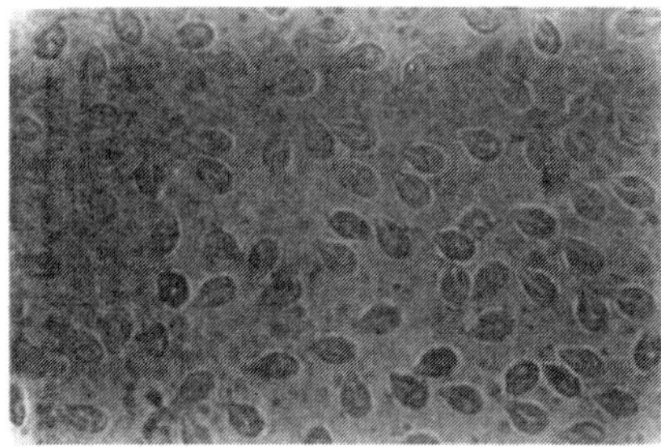

Spores of *Myxobulus* sp.

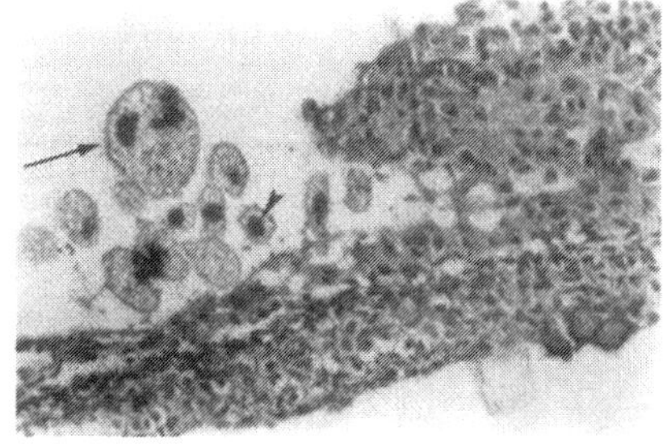

Piscinoodinium sp. in the epidermis of Tilapia

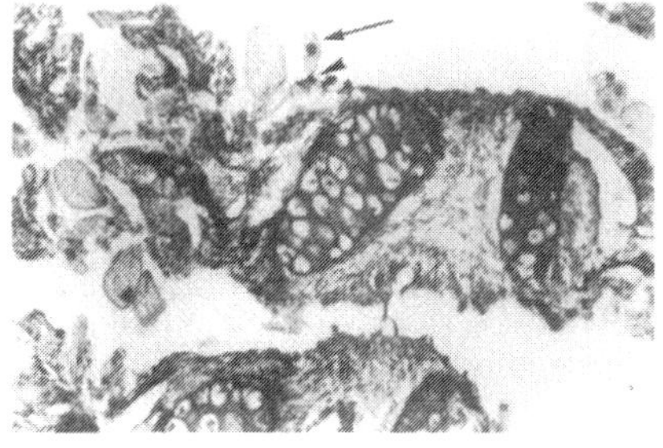

Flagellates in the gills of carp.

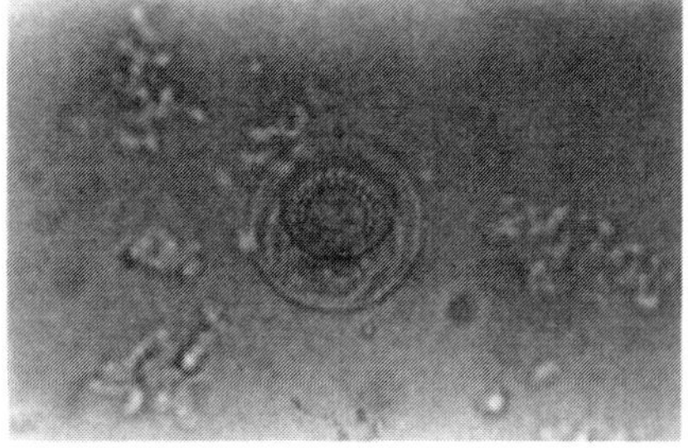

Trichodina in fresh samears of fish skin

Plate 2: Parasites of fish (See colour version on page 371)

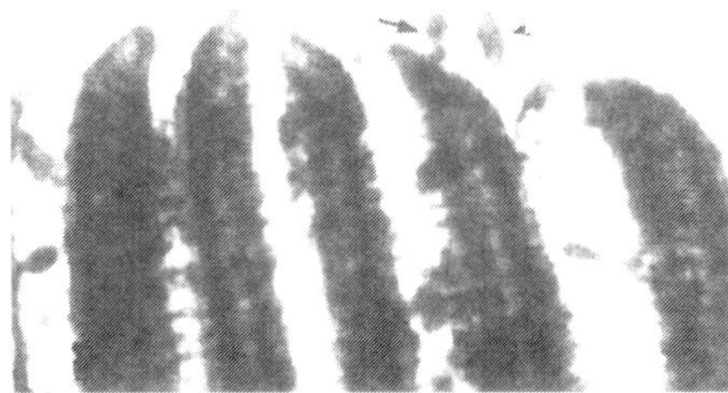

Ectoparasitic gill flukes

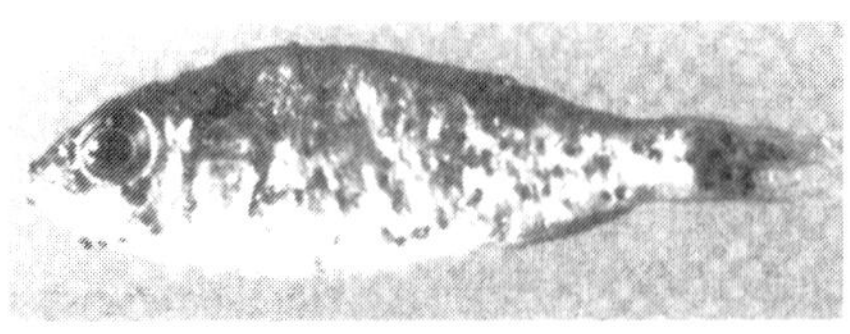

Mataceracruae if digenetic trematodes encysted in skin of carp

Pigmented cysts due to encysted metacarcarria

Encysted metacercarine within the cartilagenous tissue of primary lamellae

Pleurocercoid larvae of tapeworm in the body cavity

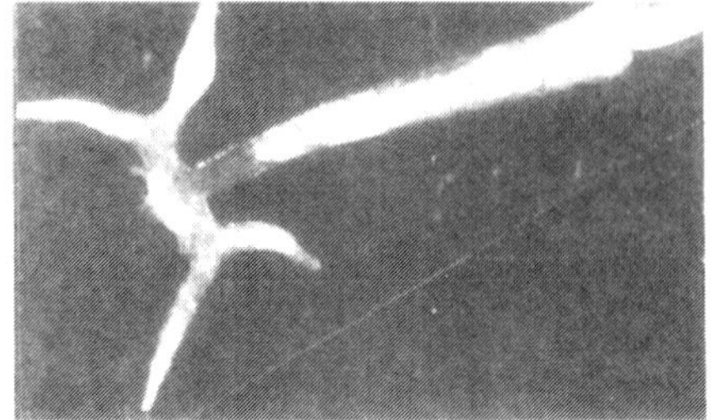

Adult *Lemaea* (anchor worm).

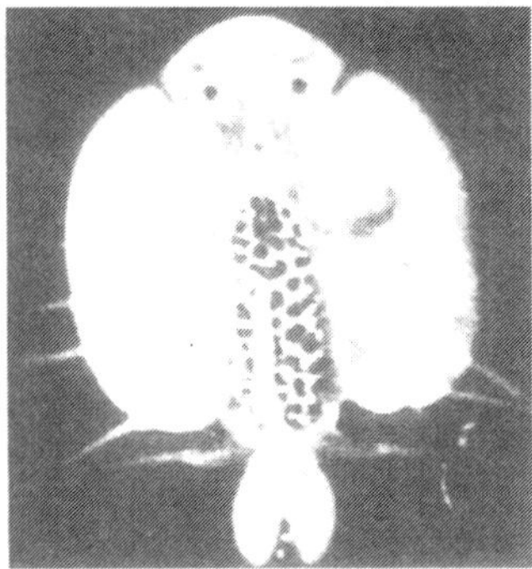

Argulus (Fish lice)

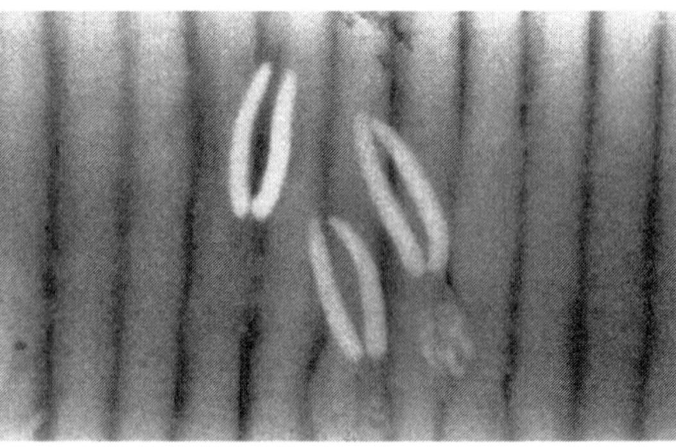

Ergasilus sp (gill maggot) on the gills of carp

Plate 3: Parasites of fish

References

K. M. Shankar and C.V Mohan,2002, Fish and Shellfish Health Management, ISBN 81-7525-328-2, Fish Pathology and Biotechnology Laboratory, Dept of Aquaculture, UAS, College of Fisheries, Mangalore-575002

K.P. Jithendran 2014 Parasite and Parasitic diseases in fish culture system, CIFE, Mumbai, publisher ICAR, New Delhi

Mohammed h Kotob et al 2017, The impact of co infection on fish ; a review, Veterinary Research, 47

Roberts, R. J. (2012). Fish Pathology: Fourth Edition. Fish Pathology: Fourth Edition, 1–581. https://doi.org/10.1002/9781118222942

4

Bacterial Diseases of Fishes

Naveen Kumar B.T. and K.M. Shankar

With increasing fish culture activities several bacterial diseases causing morbidity and mortality in fish have been reported in tropical and subtropical aquaculture systems. Common and major diseases are due to motile aeromonads, *Vibrios* and *Pseudomonas*. However, bacterial diseases of minor nature have also been reported from aquacultural systems. In the last two decades several new diseases due to *Streptococcus* sp, *Edwardsiella* and Ricketsia have also emerged.

Bacteria can cause disease either as primary pathogens or as secondary opportunistic invaders. Bacterial diseases in fish can be broadly classified as surface ulcerative, acute systemic and chronic granulomatous types (Table 1, Plates 4 and 5). Surface ulcerative types of diseases are characterized by hemorrhagic surface ulcers and are caused by species of *Aeromonas, Pseudomonas, Vibrios, Flexibacteria and Myxobacteria.* Surface ulcerative disease conditions at times develop to acute systemic disease. Acute systemic disease are characterized by the presence and proliferation of bacteria in internal organs like kidney, heart, spleen, blood and other visceral organs. These diseases produce significant necrotic changes in all affected organs and can cause mortality in a short time scale. Bacterial hemorrhagic septicemia caused by *Aeromonas hydrophila* is a major problem in carp farms. Chronic granulomatous type of disease conditions are characterized by the formation of granulomas or nodules with the initiating bacteria in the centre. These conditions will not produce mass mortality but can produce chronic low levels of mortality and can significantly reduce growth.

In acute systemic (septicemia) diseases, bacterial pathogen is found throughout blood stream and internal organs. Various predisposing factors for this conditions are i) bacterial flora is similar in fish and the environment, ii) poor water quality and organic pollution, iii) temperature fluctuations and extremes, iv) wild fish reservoir, v) husbandry factors such as overcrowding, poor nutritional status and trauma due to transport, weather, handling, cannibalism due to size variation and predators. Clinical and external signs of acute systemic diseases are a) sudden onset, b) depressed appetite, c) increased mortality and d) behavioural changes especially swimming abnormalities, lack of shoaling, collecting or crowding at net edges and fins clamping, e) colour changes especially darkening and f) hemorrhages. Post-mortem appearance (internal) of acute systemic disease include a) hemorrhages, b) congestion, c) ascites and d) swollen kidney and spleen. Histology of acute systemic conditions are often nonspecific with focal to generalised necrosis and occasional colonies of bacteria. Examples of causative agents are *Aeromonas salmonicida, A. hydrophila, Pseudomonas, Vibrio, Cytophaga, Pasteurella, Edwardsiella, Flavobacteria, Yersinea, Hemophilus, Streptococcus and Eubacterium.*

In surface ulcerative diseases, predisposing factors are similar to those as in acute systemic diseases. Here, differential diagnosis is important, since surface ulcers can also be related to trauma, parasites and fungi. Often mixed infections are common. Clinical signs are obvious with ulcer like loss of skin and epithelial covering. Ulcer may be superficial or deep, healing often prevented by accumulated necrotic material in the centre of ulcer. In histology there is extensive inflammation in dermis and muscle, loss of surface epithelium, necrosis at surface and hyperplastic epidermis at leison edges. Examples of causative agents are species of *Vibrio, Pseudomonas, Aeromonas, Cytophaga, Haemophilus and Myxobacteria.*

Chronic granulomatous conditions are common in old fish and seldom in young fish. Lack of hemorrhages, with nodules of varying size and colour throughout organs and often in skin and gills are the important features. Usually variable number of fish are affected and affected fish are often thin and emaciated. Occasionally, peritonities and ascites are seen. In histology, granuloma due to chronic inflammation is common. There will be a mixture of cell types in the granuloma, like epithelioid cells, macrophages, lymphocytes and granule cells. Granuloma will have an outer fibrous layer. Examples of causative agents of chronic granulomatous condition are *Mycobacteria, Nocardia, Flavobacteria, Renebacterium, Pasteurella and Streptococcus.*

The concept of pathogenecity and pathogenesis of bacterial pathogens in higher animals have parallel in fish. Mechanisms, such as, initial events of attachment and entry, site of entry, mode of spreading of infection in the host, mechanisms of cell and tissue damage involving exo and endotoxins and persistence in host are important. It is important to note that majority of bacterial fish pathogens in tropics are ubiquitous and cause disease as secondary opportunist under stressful conditions to host. Therefore, it is difficult to study their pathogenesis by experimental infection.

In general, fish bacterial diseases are diagnosed by both conventional and rapid molecular biology tools. As majority of bacterial fish pathogens are ubiquitous and secondary extreme care should be taken while sampling fish for bacterial disease diagnosis. Isolation of bacteria from surface lesions is not a good practice. Attempts should be made to isolate the bacteria from systemic organs such as kidney, heart, spleen and blood. Furthermore, it is common in bacterial disease management to identify the dominant isolate and check its antibiotic sensitivity before recommending chemotherapy. This is important from efficiency point of view of the drug, besides for preventing their misuse in aquaculture. Pathogenecity mechanism of some of the common bacterial pathogen of fish in tropical aquaculture are discussed below.

Accumlation of exudate in the body cavity of gold fish (dropsy)

Dropsy (arrow) in cultivated mrigal

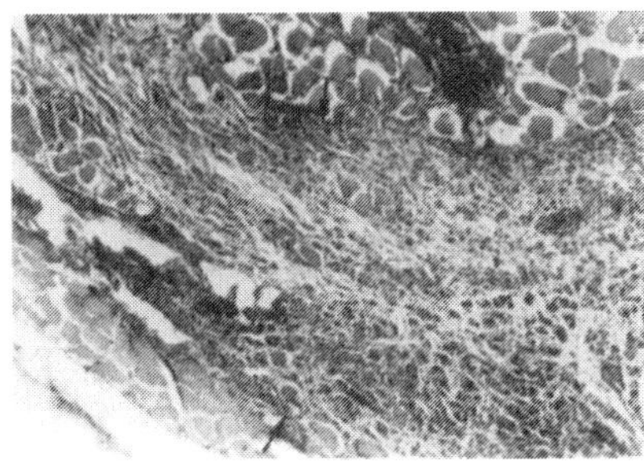

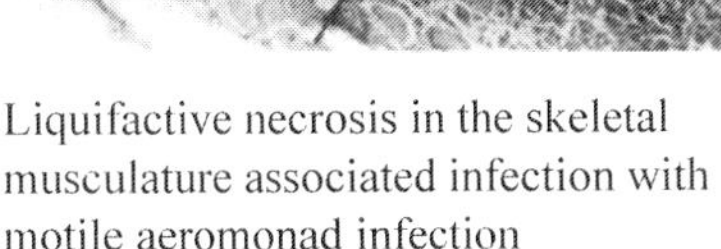

Liquifactive necrosis in the skeletal musculature associated infection with motile aeromonad infection

Surface ulcer associated with aeromonad

Plate 4: Bacterial Diseases of Fish (See colour version on page 372)

Suspected *A. hydrophil* infection of skin

Suspected *A. hydrophil* infection of gills

Plate 5: Bacterial diseases affecting Carps (See colour version on page 372)

Motile Aeromonad Septicemia (MAS)

Motile aeromonads causing septicemia are free living, mesophilic ubiquitous in soil, water and occur as normal flora of fish. They also form 0.2 to 8 % of diseases as secondary pathogens in human, avian and bovine host. However, they are most commonly found in aquatic species either as normal flora or primary/secondary pathogen.

Classification of motile aeromonads (MA) is and has been controversial. Currently, Bergey's manual has three species, *Aeromonas hydrophila*, *A. sobria*, *A. caviea* which are characterised by biochemical, genetic and serological methods. Despite this, it has been difficult to classify the species and in overall, it is preferred to place all the species under *A. hydrophila* or *A. hydrophila* complex or simply motile aeromonads (MA).

Motile aeromonads cause both acute and chronic diseases in fish. However by virtue of their ubiquitous nature, it is difficult to establish experimental infection in fish. Acute infection results in exophthalmia, distension of abdomen, septicemia with no signs, internal hemorrhage, necrosis and edema. Chronic infection includes deep dermal ulcers with hemorrhage, inflammation and petechial hemorrhage on the body.

Studies employing cell culture have demonstrated that motile aeromonads are capable of binding to a variety of tissue proteins, including collagen types I and IV, fibrinogen, laminin and nidogen. Phenotypic variations among the strains to express adhering receptors depend on condition of growth. Motile aeromonads produce a number of extracellular proteins (enzymes) and toxins (cytotoxins) capable of killing phagocytes. MA with surface S layer have been found to be difficult for phagocytosis by neutrophil compared to strains without S layer. MA have ability to compete with host for nutrients by production of catecholate siderophores. Immune response to infection include non-specific response (a natural non immuno-globulin agglutinin in the serum of carp which could agglutinate heat inactivated cells) and production of specific antibodies. In addition, cell mediated immunity (CMI) has been demonstrated which is antibody dependent.

Mechanisms of Cell and Tissue Damage

Extensive tissue damage in both chronic and acute phase of the disease has been recorded. Motile aeromonads produce extra cellualr proteins (ECP), cytotoxins and proteases. Cytotoxins affect cell line adhesion, cause shrinkage and rounding of cells. Cytotoxin is a hemolysin having both reversible and irreversible effect on host cell. In addition, MA elicit intensive inflammatory response with massive infiltration of monocytes and granulocytes into the infected tissue. This protective process of the host cells involved produce a variety of substance such as lysosomal enzymes and toxic free oxygen radicals leading to tissue damage characteristic of MAS. Several physiological changes are associated with MAS such as a general decline in red blood cells, hematocrit, and hemoglobin and shift in white blood cell differential counts from a predominant lymphocytes to predominant neutrophils. Along with the above, decrease in blood glucose, albumin, globulin, total protein, cholesterol and increase in bilirubin and uric acid are some of the common features in MAS.

Relative Virulence

MAS is strongly stress associated. The first factor of MAS epizootic is stress mediated immunosuppression leading to infection. Second, MAS is related to virulence of *A. hydrophila* strains which is highly variable. Strain virulence is related to ECP, exotoxins and surface characters. Among the surface characters a crystalline surface layer or S-layer is important.

Management

MAS is diagnosed by isolation of the bacterium from kidney of affected fish on Rimler Schotts medium in which motile aeromonads form yellow colonies. Additionally, specific rabbit antiserum and monoclonal antibody based immuno-diagnosis such as ELISA are available. As MAS is a stress associated disease, it can be prevented by good husbandry practices such as improved water quality and reduced stress.

Vibriosis

Vibriosis cause high mortality in warm water marine and brackish water fish. *Vibrios* are ubiquitous constituting 10-50% of the heterotrophs in marine environment. Earlier, vibriosis referred to vibriosis caused by *V. anguillarum*. However, with expansion in aquaculture, vibriosis due to several species of *Vibrio* such as *V. alginolyticus*, *V. parahemolyticus*, *V. vulnificus*, *V. ordelli*, *V. damsella*, *V. carcharia*, *V. salmonen*, *V. splendence* have been recorded. Vibriosis affect fish of both cold and warm water. Although pathology varies depending on host and species of *Vibrio*, basically clinical signs are similar to MAS in freshwater fish. Clinical signs and pathology associated with vibriosis are hemorrhages, skin lesions, focal necrosis of liver with atrophied hepatocytes, necrosis of spleen arteries, necrosis of tubules and glomerules of kidney.

Pathogenesis

Transmission of *Vibrio* is through water, and contact in crowded condition. The pathogen enters through skin and intestinal tract. Secondary infection is often through damaged skin. Primarily pathogenic *Vibrios* colonise fish without prior lesion. Some *Vibrio* are specific to intestinal tract causing necrosis, degeneration of mucosa, affecting epithelial cells between microvilli. *Vibrio* isolates are resistant to bactericidal action of normal serum which help avoidance of host immune system, proliferation and spreading. *Vibrio* also produce high affinity iron sequestering siderophores to avail sequestered nutrients from host. Type of sideropores vary among the species.

Vibrios produce proteases and extracellular enzymes which varies among species in number, kind and quantity. However, enzymes are serologically related. Purified Vibrio proteinases are toxic to fish and shellfish. Hemolysin is also produced.

Management

Vibriosis is diagnosed by isolation of the bacterium in specific medium such as Thiosulfate citrate bile salt sucrose agar (TCBS). A host of antibody and nucleic acid based diagnostic tests are also available for specific and rapid diagnosis. As *Vibrios* are secondary pathogens, avoiding stress due to poor water quality, high stocking density is important to prevent the disease.

Pseudomonas Septicemia

There are four species of ubiquitous *Pseudomonas* causing disease in warm water fish among which *P.anguliseptica*, *P.flourescence* are of economic significance. *Pseudomonas* are gram –ve, motile organisms with a single polar flagella. They are opportunistic pathogens with optimum temperature range of 20-25°C affecting both fresh and marine species. Clinical signs are similar to MAS. Improper management such as high stocking, low dissolved oxygen, handling stress and poor nutrients favour pseudomonad infection. There are two forms of disease, acute and chronic, the former is similar to that of MAS. Chronic form includes skin lesions, peritonitis, and pathological changes in hematopoitic tissue, skin, spleen and kidney. In tilapia, granulomas in several organs have also been recorded.

Pathogenesis

The pathogen enters as an opportunistic invader through mouth, skin and gills after primary damage. Several species produce siderophores. The pathogen is resistant to normal serum of eels, bluegills but not to that of carp, gold fish, tilapia and rainbow trout. Phagocytic activity is important in host immune system. Mechanisms of cell and tissue damage is similar to that of MAS caused by extracellular enzymes, toxins, lipases and proteases.

Management

Management approaches are similar to that followed in the case of MAS and vibriosis.

Table 1: Types of bacterial fish diseases

Pathogen	Site of Infection	Major Pathology	Major Clinical Signs
I. **Surface Ulcerative type**			
Pseudomonas sp.	a) Body surface b) Fins, tail	a) Inflammation in epidermis and dermis. Hyperplasia at ulcer edges. Hemorrhagic ulceration. b) Progressive disintegration of the tail/fin tissue	a) Body reddening, skin lesions, swollen belly, protruding scales, sunken or protruded eyes, inflamed anus with discharge from the vent. b) Loss of natural colour beginning from the outer margin of tail/fins, fraying of the tail/fin, swimming near water surface, progressive disintegration of tail/fin tissue.
Aeromonas hydrophila	Body surface	Inflammation in epidermis and dermis. Hyperplasia at ulcer edges. Hemorrhagic ulceration.	Skin lesion clearly visible in water, blood in the lesions, sluggishness, anorexia. Shallow open sores, mostly with white rim, presence of liquid and or blood, eroded fins and mouth.
Vibrio sp.	Body surface, fins, tail	Inflammatory response in the integument, ulceration, necrosis of tail and fin tissue	Shallow to deep open hemorrhagic ulcers, fraying and progressive disinfection of membranous tissue of fins and tail.
II Acute Systemic Type			
Aeromonas salmonicida (Furunculosis)	Skin and internal organs	Inflammation, myocardial necrosis, liquefaction of kidney and spleen, anemia.	Boils (furuncles) and or ulcers may be present. Skin lesions with fluid and or blood, pale gills, sluggish movement, fraying of the fins,bleeding from the gills, acute mortalities.
Aeromonas hydrophila (Aeromonad septicaemia)	Skin and internal organs	Surface ulceration, necrosis of haematopoietic tissue, internal haemorrhages, inflammation, hemorrhagic septicaemia.	Dropsy condition, loose scales, scale protrusion, ulceration, haemorhages, low to high mortality.

Pathogen	Site of Infection	Major Pathology	Major Clinical Signs
Vibrio anguillarum (Vibriosis)	Internal organs	Necrosis, ulceration on the skin, enlargement and liquification of the kidney.Focal haemorrhages on the surface of the heart, deep, necrotic haemorrhages within the myotomes.	Anorexia, darkening of the body, periorbital and/or abdominal dropsy, skin lesions with blood coloured exudate, low to high mortality.
Pseudomonas fluorescens	Internal organs	Hemorrhagic septicemia, peritonitis, lesions in the kidney and spleen, rupture of melanomacrophage centres (MMCs), hyperaemia.	Large haemorrhagic skin lesions, heavy mortality very shortly after the advent of lesions.
Yersinia ruckeri (Enteric red mouth)	Internal organs	Haemorrhagic septicemia, necrosis of abdominal organs, swelling and reddeningof kidney and spleen, necrotic foci on spleen, infiltration of leucocytes to spleen,capillary congestion of brain and eye vessels.	Red mouth (congestion cf the vessels of theoral area) ulceration, hemorrhage, external ulcerative lesions.
Edwardsiella tarda	Internal organs	Small cutaneous lesions extending down into the musculature with progressive fibrinous peritonitis, necrosis of the hepaticand renal tissue, focal colonies of bacteria throughout the tissues	Abdominal distension, deep cutaneous lesions.
III. Chronic Granulomatous Type			
Renibacteriumsalmoninarum (Bacterial kidney disease)	Kidney	Whitish, nodular granulomatous lesions in the kidney, large cavitations in the skeletal muscle, granulomatous lesions in the liver, heart, spleen or gills, muscle, proliferationof macrophages around the bacteria.	Dark colouration, occasionally exophthalmia, small haemorrhages at the bases of the pectoralfins, small raised vesicles on the sides of the fish, emaciation
Mycobacterium marinum (Piscine tuberculosis)	Internal organs	Tubercles in virtually any organ especially liver, spleen and kidney.	Dark colouration, swelling in the abdomen.

Emerging Bacterial Pathogens of Fish

Edwardsiella tarda

Edwardsiella tarda is a Gram-negative intracellular pathogen accounting for enormous economic losses in cultured seawater and freshwater fish. Infection by *E. tarda* often leads to the development of a systematic disease called edwardsiellosis. *Edwardsiella tarda* has a broad host range and geographic distribution, and contains important virulence factors that enhance bacterial survival and pathogenesis in hosts. Host range includes channel catfish, eels, mullet, Chinook salmon, flounder, carp, tilapia, striped bass, yellowtail, turbot and red seabream

Edwardsiella tarda causes septicemia with extensive skin lesions, affecting internal organs and leads to extensive losses. Clinical signs include ascites, hernia, exophthalmia, and severe lesions of internal organs. Although edwardsiellosis has been studied for many years, the major virulence factors of *E. tarda* are still poorly understood. It has been reported in humans as the cause of gastroenteritis and generalized infections mainly among individuals with impaired immune systems.

Outbreaks of edwardsiellosis have a very short latent phase, before epidemics. Therefore, quick detection of *E. tarda* is crucial for controlling the disease. Conventional methods of diagnosis including isolation, and biochemical identification and detection by ELISA of antigens in spleen, kidney and liver. PCR based methods have been developed by using either the 16S rDNA sequence or the hemolysin gene.

Several antigen-preparation of *E tarda* have been evaluated for vaccination however, commercial vaccines are not yet available. And control of edwardsiellosis by antibiotics was not as effective as that of extracellular pathogens such as *Vibrio* spp.

Vibrio causing Big belly disesae*in Lates calcarifer*

'Big belly' disease is a chronic, granulomatous bacterial enteritis and peritonitis, first reported in 3- to 4-weeks-old Asian seabass. Affected fry are emaciated with a swollen abdomen and the condition is referred to as 'skinny pot-belly' or 'big belly' disease. PCR, *in-situ* hybridization, and phylogenetic analysis of 16s r RNA suggest that 'big belly' disease in *Lates calcarifer* is caused by a novel *Vibrio* species. Examinations of diseased fish from a batch of 2-month-old, 6- to 8-cm *L. calcarifer* fingerlings, kept in seawater recirculating aquaculture systems, showed histoppathology resembling 'big belly' disease. Ethanol-fixed tissues tested positive with specific PCR primers based on 16SrRNA genes. *In situ* hybridization using dioxygenin-labelled positive PCR products on formalin-fixed paraffin-embedded tissues showed positive reactions with intralesional clusters of the large, 'big belly' coccobacilli.

Streptococcus species

Streptococcus spp cause a world wide fish health problem. *Streptococcus iniae* and *S. agalactiae* are important emerging pathogens that affect many fish species worldwide, especially in warm-water regions. Infection results in septicemia and neurotropic disease, with cumulative per cent mortality between 30 to 50. *Streptococcus* infections in fish can cause high mortality rates (> 50%) over a period of 3 to 7 days. Some outbreaks, however, are more chronic in nature and mortalities may extend over a period of several weeks, with only a few fish dying each day. Seven different Streptococcus species, *Lactococcus garvieae, Lactococcus piscium, Streptococcus iniae, Streptococcus agalactiae, Streptococcus parauberis* and *Vagococcus salmoninarum* are considered potential fish pathogens. *Lactococcus garvieae* infection is considered the most serious disease affecting farmed yellowtail *Seriola quinqueradiata* and amberjack *Seriola dumerili* in Japan.

Fish streptococcal disease in fish was first reported during 1957, affecting cultured rainbow trout in Japan. The pathogen has a wide host range, including salmon, mullet, golden shiner, eel, sea trout, tilapia, sturgeon, and striped bass. *Streptococcus* has also been isolated from a variety of ornamental fish, including rainbow sharks, red-tailed black, rosy barbs, danios, some cichlids including Venustus (*Nimbochromis ("Haplochromis") venustus*) and *Pelvicachromis* sp., and several species of tetras. Several species of Streptococcus are pathogens in fish and mammals with zoonotic potential

Streptococcus infection should be suspected, anytime fish are observed behaving in an unusual manner, with abnormal swimming behavior(spinning or spiraling). More often stress and poor immune status favour streptococcus infection as mere presence of the pathogen is inadequate to cause a disease outbreak. Clinical signs usually include hemorrhage, pop-eye, spinning, and rapidly progressing mortalities. Internal examination may reveal the presence of blood tinged fluid in the body cavity, an enlarged reddened spleen, pale liver, as well as inflammation around the heart and kidney. Many *Streptococcus* infect the brain and nervous system of fish, explaining the erratic swimming frequently observed in infected fish. Streptococcal disease is very difficult to control because these pathogens are resistant or become resistant to most chemicals and drugs. The successful application of commercial *L. garvieae* vaccine has resulted in remarkable decline in fish losses

Streptococcus dysgalactiae**:** An emerging pathogen of fishes and mammals

Streptococcus dysgalactiae subsp. *dysgalactiae* (GCSD) is of special interest to aquatic health experts due to its interesting veterinary and public health importance. Increasing records of GCSD infections in farmed fishes have been documented through diverse worldwide aquatic habitats in Japan, China, Malaysia, Indonesia, Taiwan and Brazil. GCSD causes a syndrome that is characterized by systemic multifocal inflammatory reaction, micro-abscessiations, severe septicemia, and high mortality rates with pathognomonic necrotic ulcers at the caudal peduncle region.The genetic basis of its virulence remains unknown despite the intraspecies/interspecies dynamic spread of fish GCSD, GCSD are identified as fish specific pathogen based on sequence analysis of16S rDNA, *sodA* gene and *tuf* gene. The biochemical characteristics (a-hemolytic and streptokinase activities) of the fish isolates are different from those of the typical mammalian strains *S. dysgalactiae* subsp. *dysgalactiae* and *S. dysgalactiae* subsp. *equisimilis*. Furthermore, the fish specific isolates of α-hemolytic GCS. *S. dysgalactiae* subsp. *dysgalactiae* [GCSD] are Gram positive cocci arranged in long chains, catalase negative, oxidase positive and auto-aggregated in saline , while the mammalian isolates of α-hemolytic GCS *S. dysgalactiae* subsp. *dysgalactiae* are Gram-positive cocci or oval cells arranged in short- to medium-length chains, catalase negative, and oxidase positive. The optimal temperature for growth of fish and mammalian isolates is 37 °C. Growth does not occur at 10 or 45 °C or in 6.5% NaCl or in 40% bile, bacitracin, or at pH 9.6. Fish specific isolates exhibit α-hemolysis on cattle and sheep blood agar, however the hemolysis pattern on sheep blood agar is mostly changed to β-hemolysis type after prolonged incubation or incubation at 4 °C. Epidemiologically, GCSD has been reported to be the main cause behind several mammalian infections such as streptococcal mastitis/endometritis in domestic mammals with some records of skin lesions, meningitis and bacteraemia in humans

Streptococcus iniae

Streptococcus iniae is globally distributed throughout warm water finfish aquaculture is a Gram-positive, sphere-shaped bacterium. The first report on the isolation of *S. iniae* was during 1976 from a diseased Amazon freshwater dolphin (*Inia geoffrensis*). *Streptococcus iniae* causes high mortality in cultured and wild fish stocks globally.

S. iniae infections have been reported in at least 27 species of cultured or wild fish from around the world. Freshwater and saltwater fish including tilapia, red drum, hybrid striped bass, and rainbow trout are among those susceptible to infection by *S. iniae*. Streptococcosis has been reported in the golden shiner (*Notemigonus crysoleucas*)) and striped mullet (*Mugil cephalus*), menhaden (*Brevoortia patronus*), sea catfish (*Arius felis*), pinkfish (*Lagodon rhomboides*) Atlantic croaker (*Micropogon undulatus*), spot (*Leiostomusxanthurus*) stingray (*Dasyatis* sp.) and silver trout (*Cynoscionnothus*)). Rainbow trout (*Oncorhynchus mykiss*), coho salmon (*Oncorhynchus kisutch*), Japanese amberjack (*Seriola quinqueradiata*), red drum (*Sciaenops ocellatus*), and barramundi (*Lates calcarifer*). Common carp (*Cyprinus carpio*), channel catfish (*Ictalurus punctatus*), and goldfish (*Carassius auratus*) appear to be resistant. Large scale epizootics *Streptococcus* infections were reported in Japan, Italy, Spain, Israel, U.S.A., Korea, Bahrain, South Africa and Australia.

S. iniae has occasionally produced infection in humans in the West, especially fish handlers of Asian descent. Human infections include sepsis, toxic shock syndrome, and inflammation of the skin, intervertebral discs, or inner layer of the heart.

Streptococcus iniae infection appears in fish raised in intensive aquaculture and subject to stressors such as suboptimal temperature, poor water quality crowding and handling. Wild fish populations located both near and far from[1] aquaculture operations have also proven susceptible to *S. iniae* infection. There is possible transmission of *Streptococcus iniae* from wild fish to cultured marine fish as recorded in gilthead sea bream and European sea bass. Species-specific PCR and ribotyping confirmed that wild and cultured fish got infected by a single *S. iniae* clone. Wild fish are therefore potential amplifiers of pathogenic *S. iniae* strains

Clinical signs and the gross pathology of streptococcosis caused by *S. iniae* are very similar to those for enterococcal infection. Streptococcosis is an infectious systemic scepticemic disease characterized by meningoencephalitis and skin leisons. Clinical signs include bilateral exophthalmia, listless swimming lethargia, anorexia, dark coloration of the skin and petechiae on the internal side of the opercula. The gross anatomy includes ascites, hepatomegalia splenomegalia and/or congestion of liver, spleen, kidney, brain, pectoral and caudal fins and mouth. Panophtalmitis and meningitis/meningoencephalitis are recorded in all infected fish species examined.

Clinical signs of *Streptococcus iniae* infection in freshwater ornamental fish also are similar to those from marine fish such as darkening of the skin and lethargy as first signs followed with hemorrhages on body and low level exophthalmia in a low percentage of fish (~ 10%). Moribund fish demonstrated the characteristic spinning, swimming pattern during streptococcal infections.

Histological analysis revealed leukocyte infiltration in the intestinal area, spleen, posterior kidney and brain. Necrosis and tissue degeneration were observed in the same organs in addition to degeneration of the renal tubules.

The site of *S. iniae* infection and its clinical presentation vary from species to species. In tilapia, *S. iniae* causes meningoencephalitis, with clinical signs including lethargy, dorsal rigidity, and erratic swimming behavior with death following in a matter of days. In rainbow trout, it is typically associated with septicemia and central nervous system damage. Symptoms are consistent with septicemia, and include lethargy and loss of orientation (as in tilapia) exophthalmia corneal opacity, and external and internal bleeding.

Tenacibaculum maritimum

Tenacibaculum maritimum is the etiological agent of an ulcerative disease known as tenacibaculosis, which affects a large number of marine fish species in the world and is of considerable economic significance to aquaculture. Problems associated with epizootics include high mortality rates, increased susceptibility to other pathogens, high labour costs of treatment and enormous expenditures on chemotherapy.

Tenacibaculum maritimum, a Gram-negative and with a wide host range infecting several species around the world after its first record of isolation from an outbreak with mortality in marine fish in Japan. Countries where Tenacibaculosis is recorded are Scotland, France, Spain, Portugal Italy, Australia, USA, Chile and Canada. Host range include black sea bream *Acanthopagrus schlegeli*, Japanese flounder, yellowtail *Seriola quinqueradiata*, Dover sole *Soleasolea*, seabass *Dicentrarchuslabrax*, turbot *Scophthalmus maximus*, Atlantic salmon *Salmo salar*, sole *Solea senegalensis* and *S. solea*, gilthead seabream *Sparus aurata*, tub gurnard *Chelidonichthys lucerna* ,rainbow trout *Oncorhynchus mykiss*, striped trumpeter *Latris lineata*, greenback flounder *Rhombosoleatapiriña*, yelllow-eye mullet *Aldrichetta forsteri* and black bream *Acanthopagrus butcheri*, white sea bass *Atractoscion nobilis*, Pacific sardine *Cardinops sagax*, northern anchovy *Engraulis mordax* Chinook salmon *Oncorhynchus tschawytscha*, Atlantic salmon smolts from British Columbia, Canada, turbots *Psetta maxima. Tenacibaculum maritimum* may infect other host, in either cultured and/or wildfish, however, due to its demanding growth requirements and correct identification for its diagnosis has been difficult.

The pathology of the organism is characteristic gross lesions on the body surface of fish such as ulcers, necrosis, eroded mouth, frayed fins and tail rots and sometimes necrosis on the gills and eyes. However, due to some variation in external pathological signs of the disease, depending on the species and age of fish, different names have been used to designate this ulcerative condition. The pathogen infects through skin-subcutaneous route, but intraperitoneal experimental infection also demonstrated. However, subcutaneous inoculation route reproduces the disease in a faster and more reliable way than the intraperitoneal route.

Francisella sp

Francisella sp. is a serious emergent bacterial pathogen known to cause acute to chronic disease in fresh and marine, warm and cold water cultured and wild fish species and molluscs world wide. *Francisella* have "emerged" as serious pathogens of various fish and molluscan species, both farmed and wild, from various geographical regions worldwide. The pathogen has been reported from Costa Rica (Latin America), USA, Indonesia, England, Japan, and China Francisellosis similar in different host species and is commonly characterised by the presence of multi-organ granuloma and high morbidity, with varying mortality levels. Etiological agents of Francisellosis identified only recently, as these bacteria can not be easily cultured and detected by histology. In most cases the diseases caused by this group of bacteria remain under-diagnosed. The disease is highly infectious and often prevalent in affected stocks. Most, if not all strains isolated from teleost fish belong to either *F. noatunensis sub*sp. *orientalis* in warm water fish species or *Francisella noatunensis* subsp. *noatunensis* in coldwater fish species. The "modern" emergence of francisellosis started with the identification of a *Rickettsia*-like organism (RLO) in diseased tilapia farmed in fresh and saltwater in Taiwan. Other species *F. tularensis* and *F. philomiragia*, the latter closely related to fish pathogenic species and widely recognised as virulent zoonotic agent in immunocompromised individuals. However, no evidence exists for zoonotic potential amongst the fish pathogenic *Francisella*

The organism infects a number of fish species such as Atlantic cod, *Gadus morhua*; tilapia, *Oreochromis* sp; Atlantic salmon, *Salmo salar*; hybrid striped bass, *Morone chrysops* × *M. saxatilis* and three-lined grunt, *Parapristipomatrilinineatum*. Ornamental blue-eyed plecostamus, *Panaquesuttoni* and dragonet, *Callionymus lyra and* Nile tilapia *Oreochromis niloticus*.

Francisellosis in fish has common clinical signs such as chronic, granulomatous infections resulting in varying degrees of mortality. Extensive occurrence of white, partly protruding nodules (granuloma) of various size in the spleen, kidney and liver are the usual clinical signs. Other organs include virtually any tissue type such as gill , heart, testes, musculature, brain and eye. Extensive chronic granulomatous inflammation with multiple granuloma in all have few to numerous, small Gram-negative bacteria (pleomorphic coccobacilli). In general, in macroscopic and microscopic examination, several internal organs (mainly spleen and kidney) are enlarged with white nodules. *Francisella* are "transmitted by direct contact with infected animals, through contaminated water or food, or by vectors such as biting insects.

Francisellosis is quite readily diagnosed by histology. Identification by culture on cysteine rich media or PCR with primers to the *Francisella* genus confirm the preliminary diagnoses. Although several vaccines have been developed and tried against *Francisella*in tilapia and cod, commercial vaccines are not currently available.

Aeromonas hydrophila causing Red spot disease (RSD) in grass carp

A disease, known in Vietnam as "red spot disease" (RSD) in grass carp with clinical signs similar to that of epizootic ulcerative syndrome, is causing significant economic loss in freshwater aquaculture in Vietnam. Clinical signs of RSD are haemorrhage, red spots on the body, scale loss, swollen vent and darkened skin. Grass carp is most frequently reported to be affected by RSD, although in some farms mrigal and rohu are also infected. RSD appears to have a seasonal pattern, occurring mainly in March-April and October-November. *Aeromonas* spp. suggested as the causative agent of red spot disease in grass carp *Ctenopharyngodon idella*. Of the several bacterial isolates isolated from RSD of grass carp, only *A hydrophila* at 10^{5}CFU/ml by injection or immersion could cause RSD

The Ricketsia, An Emerging Group of Pathogens in Fish

With expanding aquaculture world over, there has been increase in incidences of diseases due to ricketsia and ricketsial like organisms(RLO). Riketsiosis in some cases caused mortlity up to 60-90 %. Piscirickettsiosis and piscirickettsiosis-like diseases have affected aquaculture productivity and profitability. These fastidious intracellular bacteria are increasing and has been recorded in diverse group of fishes from different geographical regions and aquatic environments. However, the source, reservoir, and method of transmission as well as appropriate methods of disease prevention and control of these emerging disease remain to a large extent to be established. Earliest report of a ricketsial organism in fish occurred during 1939 in *Tetradon fahaka* from the Nile river in Egypt. Later during 1975 ricketsia like organism(RLO) was isolated from rainbow trout and by 1989 when ricketsia was isolated in CHSE 214 cells, the role of these pathogens in fish disease became apparent

Piscirickettsia salmonis was the first "rickettsia-like" bacteria to be recognized as a pathogenic agent of salmon in Chile during 1980s. Since then *Piscirickettsia*-like bacteria have been recognized with increasing frequency in a variety of fish and shellfish species, from both fresh and salt waters around the world. *Piscirickettsia*-like bacteria are now being frequently associated with disease syndromes in non-salmonid fish such as white seabass (*Atactoscionnoblis*), black seabass (*Dicentrarchus* sp.), tilapia (*Oreochromis, Tilapia and Sarotherodon* spp.), blue-eyed plecostomus (*Panaquesuttoni*), Asian Seabass and *Ophiocephalus argus*.Ricketsial disease has been recorded from a wide geographical region- Taiwan , China, Israel and Portugal.

The diseased fish showed an enlarged abdomen, the millet-like nodules in internal organs, and the swollen kidney which was composed of 5–10 sarcoma-like bodies in cream or gray-white colour or ulcerated into beandregs-like substance. Clinical signs include moribund fish,, loss of appetite, pale gills, hemorrage and leisons on skin to begin with and changes in overall internal organs of fish. The bacterium found in within macrophages and the cytoplasm of infected cells. The mode of transmission not confirmed nor demonstrated.

Management of Riketsial Diseases in Aquaculture

Histopathology and electron microscopy have been employed for detection of ricketsial organisms. Cell culture has been found to be very sensitive method for isolation. Also PCR has been developed. Treatment with antibiotics although suggested but remained ineffective because of its insufficient level in cytoplasm to kill these intracellular organisms. Prevention is the best method.

References

Alicia E Toranzo,2005, a review of main bacterial fish diseases in mariculture system, aquaculture 246

Agnieszka Pekala Safinska 2018, Contemporary threats of bacterial infection in fresh water fish, Journal of Veterinary Research, 6293, 261-267

M Abdeisalam , 2023, A review of bacterial fish diseases, Aquaculture International 31,417-437

5

Bacterial and Other Diseases of Significance of Shellfishes

Naveen Kumar B.T. and K.M. Shankar

A. Bacterial Diseases in Shrimp

Bacterial diseases in hatcheries and farms are being increasingly recognised as major hurdles to successful shrimp farming (Plate 6). Among the several bacterial diseases, diseases due to *Vibrio* sp. (vibriosis) are common. As majority of the *Vibrio* are secondary opportunistic pathogens, problems related to vibriosis can be traced to stress, poor water quality and bad management. It is widely accepted that the ubiquitous opportunistic vibrios are always responsible for causing mortalities in shrimp which are already stressed by poor water quality, external fouling and primary pathogens like viruses.

Vibrios are gram negative, oxidase positive, motile rods. Most frequently reported species in hatchery situations include *Vibrio harveyi*, *V. vulnificus*, *V. parahaemolyticus and V. alginolyticus.* All species of cultured shrimp are susceptible to vibriosis under stressful condition.

The common clinical signs associated with vibriosis are: 1) high mortalities, particularly in PL and young juveniles shrimp, 2) Moribund shrimp appearing hypoxic and often coming to pond surface and edges, 3) Reddening of shrimp, 4) Shell and appendage necrosis with blackening, 5) Presence of luminiscence in affected shrimp in tanks and ponds.

Presumptive diagnosis of vibriosis is performed based on clinical signs and presence of large numbers of the bacterium in the hemolymph as determined by wet mounts, smears or direct microscopic examination of larvae in wet mounts. Tissue level pathological changes and associated inflammatory response caused due to vibriosis can also be used for presumptive diagnosis by histological methods. Definitive confirmatory diagnosis is performed following isolation (from shrimp tissue or haemolymph), purification and identification using classical and rapid molecular methods.

Vibriosis in shrimps may be broadly classified as cuticular vibriosis, enteric vibriosis and systemic vibriosis. However, there are umpteen number of clinical descriptions of vibriosis in literature (shell necrosis, appendage necrosis, luminous vibriosis, red disease). Vibriosis has been recorded in all stages of shrimp.

Vibriosis of Larvae and Post Larvae

Larval vibriosis is caused by luminescent forms of *Vibrios* (*V. harveyi, V. vulnificus and V. parahaemolyticus) and related non-luminescent forms (V. alginolyticus*).

Appendage and Cuticular Vibriosis: In this condition appendage tips get colonised by vibrios, often resulting in necrosis of the underlying cuticular epithelium, consequent haemocytic inflammation and melanisation.

Oral and Enteric Vibriosis: Affected larvae typically show heavy bacterial colonisation visible as basophilic plaques on the cuticle of mouth parts, or the cuticular lining of the oesophagous and foregut. Accompanying this cuticular colonisation is the rounding up and sloughing of hepatopancreatic tubule and mid-gut epithelial cells into their lumens leading to typical enteritis. Following cuticular colonisation, bacterial invasion of the midgut and hepatopancreas may occur. Haemocytic inflammation is very commonly seen.

Systemic Vibriosis: This condition may occur in individuals with severe cuticular colonisation and midgut and hepatopancreas invasion.

Vibriosis of Juvenile and Adult Shrimp

Vibrio infection in juveniles and adults are normally the result of trauma, environmental extremes, secondary to other pathogen, or due to highly virulent forms.

Localised Cuticular Vibriosis: (Wounds, shell disease, haemocytic enteritis). These vibrio infections are typically well circumscribed by hemocytes forming capsules which are melanised. Bacteria are visible within an or adjacent to such lesions.

Systemic Vibriosis

V. parahaemolyticus, V.harveyi and V.vulnificus are believed to be involved in systemic vibriosis. This disease is mainly characterised by hemocytic nodules with septic centres in the lymphoid organ, heart, gills, hemocoel spaces and in the loose connective tissue. Such multifocal nodules may be melanised and or non-melanised.

Enteric Vibriosis (Septic Hepato-Pancreatitis)

Septic hepatopancreatitis syndrome (SHPS) is a major disease of cultured *P. monodon*. It is often referred to as "red disease" due to expansion of subcuticular red chromatophores and contraction of other pigments and is an expression of severe stress shown by some shrimp in terminal phase of SHPS. Several reports link SHPS to rancid feed and afflatoxins. SHPS has a septic phase from which *Vibrio* spp. are the dominant isolates. Generalised necrosis, a haemocytic inflammation, melanisation and basophilic masses of bacteria colonising tissue debris in the hepatopancreatic tubule lumens are the characteristic features.

Emerging Bacterial Diseases of Shrimp

a Acute Hepatopancreatic Necrosis Disease(AHPND)

Acute Hepatopancreatic Necrosis Disease (AHPND), also known as Early mortality syndrome (EMS) is a recently emergent shrimp bacterial disease that has resulted in significant financial losses since 2009. The disease is spread across much of the shrimp regions in Asia and Northern Hemisphere, devastating the global shrimp industry. In Asia the disease has been reported from China, Vietnam, Malaysia, Thailand, Mexico and the Philippines.

AHPND is caused by infectious strains of *Vibrio parahaemolytics*(V p AHPND) that has acquired a ~70 kbp "shellfish plasmid" with the genes that encode homologues of Photorhabdus insect related (Pir) binary toxins, $PirA^{vp}/PirB^{vp}$. The plasmid pVA1 carries a collection of genes associated with the conjugative transfer, which makes it potentially transferable to other. VpAHPND is expected surviving up to 9 to 18 days in estuarine and seawater at an ambient temperature of 28 ± 2°C. There are two main phases to the AHPND pathology, the acute and the terminal. In the acute phase, the infected hepatopancreas exhibits tubule epithelial degeneration in the absence of bacterial cells and detachment of tubule epithelial cells from the basement membrane. The hepatopancreas exhibits severe intratubular hemocytic infiltration and the emergence of a major secondary bacterial infection. Susceptible shrimp to AHPND are *Penaeus monodon* and whiteleg shrimp, *Litopenaeus vannamei*). Mortalities occur as early as 10 days or within 30-35 days of stocking shrimp post larvae (PL) or juveniles in ponds.In some cases, mortalities were observed as late as 46 to 96 days post stocking in the pond.

AHPND is an enteric vibriosis involving the target organs of gut-associated tissues and organs. The disease is transmitted horizontally via oral routes and co-habitation. Evidence suggests that AHPND is almost always prevalent in areas where farmed shrimp are enzootic. Characterized by sudden mass mortalities, the farmer may observe up to 10% mortality usually within 3- to 35 days of stocking post-larvae or juveniles in the pond Older juveniles may also be affected. The disease has two distinct phases:

i. **The acute phase** is characterized by a massive and progressive degeneration of the hepatopancreas tubules from proximal to distal side, with significant rounding and sloughing of the cells into HP tubules, HP collecting ducts and posterior stomach in the absence of bacterial cells

ii. **The terminal phase** is characterized by marked intra-tubular hemocytic inflammation and the development of massive secondary bacterial infections that occur in association with the necrotic and sloughed HP tubule cells

The incidence of the disease seems to reduce in water sources with low salinity (<20 ppt). Peak occurrence seems to occur during hot and dry season from April to July. Overfeeding, poor seed, water and feed quality, algal blooms or crashes may also lead to occurrence of AHPND in endemic areas.

AHPND is diagnosed by several methods methods

1. **Field diagnosis:** The onset of clinical signs can start as soon as 10 days after stocking. Clinical signs include pale-to-whitish hepatopancreas, significant atrophy, discontinuous or no content in gut, soft shell, and black streaks or spots within the hepatopancreas due to melanized tubules. In addition, the hepatopancreas cannot be squashed easily between thumb and forefinger, probably due to the increased content of fibrous connective tissue and hemocytes

2. **Isolation and identification of the pathogen:** Any suitable bacteriological medium (tryptic-soy broth or alkaline peptone water containing 2.5% NaCl supplement) can be used to perform preliminary enrichment of the pathogen VpAHPND from the diseased shrimp, sub-clinically infected shrimp, or environmental samples. After incubation overnight at 30°C the bacteria in the culture broth are centrifuged into pellets. The pellet can be used for isolation of pure culture using microbiological media for isolation of *Vibrio* sp. DNA also can be isolated from the pellet for PCR examination. Identification of the pathogen may be carried out using 16S rRNA PCR, toxR-targeted PCR, sequencing, and bioassay.

3. **Development of PCR for detection:** PCR tests have been developed to detect VP AHPND, pVA1plasmid and toxic genes. Various PCR

tests include One-step PCR detection of pVA1plasmid, One-step PCR detection of PirA/PirB toxin genes, AP4 Nested PCR detection of VpAHPND, Specific real-time PCR protocol for AHPND.

Prevention and Control

a. Good Biosecurity practices in hatchery such as, screening PL before stocking, good brood stock management, use of high-quality PL and overall good shrimp farm management can prevent AHPND.

b. Bacteriophage therapy, is a promising method for the prevention and treatment of vibriosis in aquaculture. Administering phage treatments at the right time or frequently enough to be effective would be the best course of action.

c. Employing probiotics is another strategy to reduce AHPND as they maintain the biological balance between the bacteria and algae in ponds and the digestive tracts of shrimp. Several crucial processes, including the direct and indirect immune response and food absorption, are influenced by these microenvironments. *S. cerevisiae* as a probiotic improved shrimp survival while preserving the microbial flora in the gastrointestinal tract.

d. Prophylactic measures such as immune priming and passive immunity with IgY have been tried. Passive immunization showed that the particular IgY significantly reduced the growth of *V. parahaemolyticus* and gave shrimps passive immunity. However, products for commercial applications against AHPND are not yet available.

Bacterial White Spot Syndrome (BWSS)

During 1999, a disease syndrome showing similar gross clinical signs of white spots, was reported as "bacterial white spot syndrome" (BWSS) in *P.monodon* from Malysia and Thailand. The clinical effects of BWSS, appear far less significant than that of the white spot virus infection, but severe infections could reduce moulting and growth. The syndrome is included in the Asia Diagnostic Guide due to the possibility of diagnostic confusion with viral White Spot Disease (WSD).

The syndrome was recorded with the presence of white spots on shrimp cuticles without significant mortality. Dull white spots are seen on the carapace and all over the body but are more noticeable when the cuticle is peeled away from the body. The white spots are rounded and not as dense as those seen in white spot viral disease. Wet mount microscopy reveals the spots as opaque brownish lichen-like lesions with a crenellated margin similar to spots in the early stages of WSD. The spot center is often eroded and even perforated. During the early stage of syndrome, shrimp are still active, feeding and able to moult - at which point the white spots may be lost. However, in severe syndrome there was delayed moulting, reduced growth and low mortalities.

Bacillus subtilis has been suggested as the possible causative agent of BWSS due to its association with the white spots but neither causal relationship nor experimental infection has been demonstrated. *Vibrio cholerae* is also often isolated in significant numbers with similar white spots in farmed shrimp in Thailand suspected due to high pH and alkalinity in ponds. However, in the absence of the White Spot virus or bacterial colonisation of the spots, indicated that the bacterial involvement may be secondary. The syndrome can be diagnosed by several methods. In wet mounts which show an opaque brownish lichen-like appearance with a crenallated margin and the center shows signs of erosion and/or perforation, along with extensive bacterial involvement. In Polymerase Chain Reaction (PCR) samples with gross clinical signs will be negative for white spot virus suggestive of the alternate etiology of BWSS. Histological examinations of the soft-tissues associated with the cuticular lesions do not show signs of the WSDV characteristic endodermal and mesodermal intranuclear inclusion bodies. In scanning electron microscopy(SEM), presence of spot lesions together with numerous bacteria will confirm BWSS.

Necrotising Hepatopancreatitis (NHP)

NHP is also known as Texas necrotizing hepatopancreatitis (TNHP), Texas Pond Mortality Syndrome (TPMS) and Peru necrotizing hepatopancreatitis (PNHP) is an emerging bacterial disease causing mortality. Elevated mortality rates reaching over 90 % can occur within 30 days of onset of clinical signs if not treated. The NHP bacterium occupies a new genus in the alpha Proteobacteria, and is closely related to other bacterial endosymbionts of protozoans. The bacterium causing Necrotising Hepatopancreatitis (NHP) is relatively small, highly pleomorphic, Gram negative and an apparent obligate intracellular pathogen. The NHP bacterium has two morphologically different forms: one is a small pleomorphic rod and lacks flagella; while the other is a longer helical rod possessing eight flagella on the basal apex of the bacterium and an additional flagellum (or possibly two) on the crest of the helix.

NHP was first described in Texas in 1985 followed by reports of outbreaks from most Latin American countries on both the Pacific and Atlantic Ocean coasts, including Brazil, Costa Rica, Ecuador, Mexico, Panama, Peru and

Venezuela. Host range of NHP includes *Penaeus vannamei* and *P. stylirostris* with higher mortalities in the former. NHP has also been reported in *P. aztecus*, *P. californiensis* and *P. setiferus*. Molecular testing of PL from infected broodstock indicates that vertical transmission does not occur

A wide range of gross signs which include: lethargy, reduced food intake, higher food conversion ratios, anorexia and empty guts, noticeable reduced growth and poor length weight ratios ("thin tails"); soft shells and flaccid bodies; black or darkened gills; heavy surface fouling by epicommensal organisms; bacterial shell disease, including ulcerative cuticle lesions or melanized appendage erosion; and expanded chromatophores resulting in the appearance of darkened edges in uropods and pleopods. The hepatopancreas may be atrophied with any of the following characteristics: soft and watery, fluid filled center; paled with black stripes (melanized tubules); pale center instead of the normal tan to orange coloration. The NHP bacterium apparently infects only the epithelial cells lining the hepatopancreatic tubules. The hepatopancreas is a critical organ involved in food digestion, nutrient absorption and storage, and any infection has obvious and serious consequences, from reduced growth to death. Environmental factors, the most prominent water salinity over 16 ppt and water temperature of 26°C or higher appear to be important for the onset of NHP clinical signs

Early diagnosis/detection is advocated to avoid spreading of the bcterium by cannibalism. In wet mounts hepatopancreas show reduced or absence of lipid droplets and/or melanized hepatopancreas tubules. In histopathology NHP is characterised by an atrophied hepatopancreas showing extreme atrophy of the tubule mucosa and the presence of the bacterial forms through histological preparations. Principal histopathological changes due to NHP include hemocytic inflammation of the intertubular spaces in response to necrosis, cytolysis, and sloughing of hepatopancreas tubule epithelial cells. The hepatopancreas tubule epithelium is markedly atropied, resulting in the formation of large edematous (fluid filled or "watery") areas in the hepatopancreas. Tubule epithelial cells within granulomatous lesions are typically atrophied and reduced from simple columnar to cuboidal in morphology. They contain little or no stored lipid vacuoles and markedly reduced or no secretary vacuoles. In transmission electron microscopy (TEM) two distinct versions of the NHP bacterium occur in infected hepatopancreatic cells. The first is a rod-shaped rickettsial-like form measuring 0.3 µm x 9 µm which lacks flagella. The second is a helical form measuring 0.2 µm x 2.6-2.9 µm which has eight periplasmic flagella at the basal apex of the bacterium and an additional 1-2 flagella on the crest of the helix.

Diagnosis also can also be confirmed employing commercial Dot Blot kit and *In situ* Hybridization kit for screening Asymptomatic Animals. Polymerase Chain Reaction (PCR) test has been developed and is recommended for early detection of clinical NHP which is important for successful treatment and curtailing potential for cannibalism which amplify and transmit the disease.

The disease is controlled by early detection of clinical NHP to avoid spreading by cannibalism. Periodic population sampling and examination (through histopathology, TEM or commercial gene probe) are highly recommended. The use of oxytetracycline (OTC) in medicated feeds is probably the best NHP treatment.There is also some evidence that deeper production ponds (2 m) and the use of hydrated lime (Ca(OH)2) to treat pond bottoms during pond preparation before stocking can help reduce NHP incidence. Preventive measures can include raking, tilling and removing pond bottom sediments, prolonged sun drying of ponds and water distribution canals for several weeks, disinfection of fishing gear and other farm equipment using calcium hypochlorite and drying and extensive liming of ponds.

Ricketsial Diseases in Shellfishes

a. Shrimp

Several rickettsial diseases have been recorded recently in shrimp culture associated with mortalites. Rickettsia or rickettsial like organisms (RLO) are very small in size and these agents have not been isolated, cultured or characterised from shrimp disease syndromes. Shrimps lightly infected with RLO are asymptomatic, while heavily infected ones become lethargic, go off-feed, and show pale colouration and atrophy of hepatopancreas. Rickettsial Infection in Penaeid shrimp recorded in *Penaeus monodon*, *Penaeus merguiensis*, *Penaeus marginatus* and *Penaeus stylirostris*. Co-infections in *P. monodon* occurred concurrently with other biotic agents including Gram negative bacterial septicemia, *Monodon baculovirus (MBV)* and Reo-like virus. The rickettsia-like bacterium (RLB), causing severe mortalities of commercially farmed *P monodon* in the southwest region of Madagascar, could also experimentally infect *P. vannamei*. Rickettsial infections are diagnosed by the demonstration of RLO in the cytoplasm of infected cells. Rickettsial microcolonies or inclusions are intracytoplasmic, membrane bound, basophilic and gram negative. Infections with RLO evoke marked systemic haemocytic response. There are no known treatments for this disease.

b. Lobster and crayfish

Rickettsial pathogens have been recorded in Spiny lobsters (*Panulirus homarus and Panulirus ornatus*)in farms of Indonesia. Ricketsial disease such as Milky haemolymph disease (MHD) is often seen as a symptom to mass mortality. Rickettsial infection in hepatopancreas and systemic type has also been recorded.

Hepatopancreatic rickettsia-like organism has been recorded in the redclaw crayfish *Cherax quadricarinatus* from Queensland, Australia. The tissue tropism of the organism is distinct from the systemic rickettsia-like organism previously described from *C. quadricarinatus*.

c. Molluscs

Rickettsia-like organism (RLO) for the first time recorded in the cultured tropical marine pearl oyster *Pinctada maxima* with mass mortality in China. This organism parasitizes the cytoplasm of host cells and forms intracytoplasmic eosinophilic inclusions.

Other Disease of Significance in Shrimp Black Gill Disease

The disease is also known as black gill syndrome, burned gills, black spot disease, and branchiostegal melanization. It is a condition or syndrome, that is most appropriately described as branchiostegite melanization. Melanization is referred to as the increase in the dark brown pigment melanin present in the tissue. Melanization is an immune response that is triggered locally as a response to an injury in some tissues. Branchiostegite melanization is a natural response to gill fouling or parasitic infestations of the gills in shrimp and occurs due to melanization at sites of tissue necrosis, or dying tissue. In other words, the shrimp's natural immune response to protect the gills where the protozoan parasites (ciliates) have damaged the gill tissue is by encasing the damaged area in protective tissue containing a dark pigment. However, this natural response can prove problematic. The protective tissue encasing the gill also blocks oxygen transfer and waste excretion through the gill. Over time, if enough of the gill area becomes encased in protective tissue, the shrimp cannot breathe or secrete waste efficiently. In severe cases, the condition can eventually result in the death of the shrimp.

It is an ubiquitous disease. The disease is associated with poor growth. It can spread and can cause high cumulative mortality of prawns, leading to enormous economic loss in aquaculture. The disease may be due to a variety of biotic and abiotic agents. It can be caused by many things in densely cultured marine shrimp populations including ciliated protozoa in the Genus *Hyalophysa* or gill fouling organisms such as bacteria or fungi (ambient, not infectious), and in instances of poor management decomposing organic materials. Even water conditions where there are high concentrations of suspended particles such as sand have been shown to cause a darkening of the gills due to irritation.

The symptoms include destructed and malfunctioning gill processes along with secondary infections, multifocal black or brown spots in/or general discoloration of the gills because of the melanization at the sites of tissue necrosis. Gill melanization may be visible through the side of the carapace. Histology of the affected shrimp shows massive hemocyte accumulation, tissue necrosis, and melanin deposition in the affected areas of the gills. Secondary infections of bacteria, protozoa, and fungi may occur. Prevention of the disease or control may be feasible if the cause can be identified.

Red Discoloration

The disease is also known as red disease. Although the etiological agent is unknown, a Grampositive coccus has been observed consistently in hepatopancreas cells and tubules, but microbial toxins from rancid feed has been implicated. The disease has been observed in Philippines, Taiwan, and Malaysia. *Peneaus monodon*, possibly *Peneaus stylirostris* and other penaeids are the host species for this disease.

Initially a yellowish-green discoloration of the body of the shrimp which eventually becomes distinctly red and often results in death. The red discoloration that characterizes this disease is from the normal hepatopancreatic carotenoid pigments that are distributed throughout the body by the haemolymph at the time of necrosis and atrophy of the hepatopancreas. Because necrosis of the hepatopancreas from any cause (such as Taura syndrome and Reo-like virus disease) may have a similar effect, red disease may be a collection of several disease syndromes, each with a different etiology, but with similar gross signs and histopathology. In advanced cases, the hepatopancreas may be pale or yellowish and atrophied. In histology, hepatopancreas show massive inflammation as well as necrosis and Gram-positive cocci (1.0-1.5 μm) infecting the hepatopancreas epithelium.

Aspergillus sp. is the common prawn feed contaminating agent leads to aflatoxin poisoning, resulting in red discoloration of the prawn. Slow growth and the changing of body color from yellowish to reddish with short streaks on gills are other symptoms of the disease. The disease can lead to gradual mortality which can result in loss of about 98% or more of the stock in three months.

B. Bacterial Diseases of Giant Freshwater Prawn *Macrobrachium rosenbergii*

Macrobrachium rosenbergii was considered relatively less susceptible to diseases. Nevertheless, with increased intensification of culture, the freshwater prawn culture has noticed an increase in disease outbreaks in both larval stages and adults caused by viruses, bacteria, fungi and parasites.

Several bacterial diseases are frequently reported in hatcheries and grow-out ponds. The level and type of bacterial populations associated with farmed

M. rosenbergii are useful indicators of quality and safety of prawns. Due to high stocking density, long rearing period, and organic load accumulation at the pond bottom, larval stages are most susceptible to bacterial diseases The bacterial infections caused by *Vibrio* species are specifically the more important cause of mortalities.

Vibriosis

Vibriosis is caused by *Vibrio* sp. which are ubiquitous in freshwater, estuarine, and marine aquatic systems, worldwide. The *Vibrio* sp. infecting scampi are *V. cholerae, V. nereis, V. algniolyticus, V. vulnificus, V. mediterranei, V. parahemolyticus, V. splendid, V. fluvialis, V. mimicus, and V. harveyi*etc. Among various prawn pathogenic Vibrio species, *V. parahemolytics*, is an important pathogen of *M. rosenbergii*, which causes 'Vibriosis', with high mortality. In general, *Vibrio* are prevalent in eggs, larvae, and post-larvae of freshwater prawns and their increase during the culture period is a serious concern. Luminescent *V. harveyi* causes massive losses of scampi larvae in hatcheries. The luminescence of affected larvae at night is a unique clinical sign of the disease. The diagnosis of *Vibrio* infection in prawn can be based on the combination of clinical signs, histological examination and bacterial culture since the palpable lesions of *Vibrio* infection are not pathognomonic.

Chitinolytic Bacterial Diseases

Shell disease is the most common bacterial disease that affects the crustaceans. Two types of chitinolytic bacterial diseases are known depending on the infection of developmental stage of *M rosenbergii*.

Type I: Chitinolytic bacterial disease also known as shell disease, brown spot disease, black spot disease, burned spot disease and rust disease. The juveniles and adults are reported to be the affected life stages of freshwater prawns is caused by different bacterial species like *Vibrio* sp. *Aeromonas* sp., *Flavobacterium* sp., *Pseudomonas* sp., *Bacillus* sp., *Benekia* sp. and *Spirillum* sp. The disease is ubiquitous, common in prawns that have been cultured/held in captivity for a long term. The affected prawn exhibits dark or brown discoloration as the exoskeleton becomes eroded, pitted, and melanized at the site of infection. All the life stages of *M. rosenbergii* can get affected by this disease. There is no treatment known for the disease. It can be managed in captive and cultured populations by following proper husbandry practices, maintaining hygiene, and reducing stocking density.

Type II :Bacterial Necrosis. Bacterial necrosis has variously been termed as 'black spot', 'brown spot', 'shell disease' or chitinolytic bacterial disease. Bacterial necrosis is restricted to larval phase only. Larval stage IV and V are generally more susceptible showing the necrosis and melanization of body appendages.

It is caused by the invasion of chitinolytic bacteria, which break down the chitin of the exoskeleton, leading to erosion and melanization at the site of infection. The bacteria can be localized but if the infection becomes systemic, it will be fatal. *Aeromonas hydrophila*, *Aeromonas caviea*, *A. sorbia* and *Aeromonas* sp. are bacterial flora isolated from bacterial necrosis. The clinical signs of bacterial necrosis are bluish color or discoloration, an empty gut, bluish color of affected larvae, gill obstruction, weak larvae falling to the bottom of the tank, and brown spots on antennae and newly formed appendages. Such larvae gradually fall to the bottom of the tank. Development of bacterial necrosis is rapid and if left untreated, it can lead to mass mortality of larvae within a short span of time. The disease can be controlled by avoiding overcrowding preventing sudden temperature changes; reducing stress, and antibiotic treatment.

Diseases of Significance of Unknown Etiology in *Mrosenbergii*

Larval Mid Cycle Disease (MCD)

Larval Mid-cycle disease (MCD) or MCD-like disease syndromes of larval *Macrobrachium rosenbergii* has reduced survival of larval cultures from a normal 50–70% to 5 to 10%. The main etiological agent remains unknown, but *Enterobacter aerogenes* has been associated with this disease. The disease has been observed in Thailand, Mauritius, Hawaii, northern Australia, Philippines and Malaysia. The larval stages 4-11 are the most affected by this disease. Observations revealed a characteristic mortality pattern, where high larval mortality occurs during the second third of the larval cycle with a peak in daily mortality occurring near the middle of the rearing cycle. Mid-cycle disease (MCD) usually affects the larval stages of *M. rosenbergii*, specifically between 15-22 days of production. The clinical signs of the disease include signs of lethargy, weak spiraling erratic swimming activity, reduced growth and feeding, bluish-grey discoloration of the body, possible cannibalism, and epibiotic fouling. Atrophy of hepatopancreas epithelium causes death Histology shows hepatopancreas atrophy, and coccobacilli within the digestive tract.

Establishment of standard hatchery sanitation procedures (i.e., routine disinfection between larval rearing cycles, separation of rearing tanks, use of individual equipment for each larval rearing units helps prevention of MCD. After MCD occurs, reducing stocking density, improving pond husbandry and sanitation, and providing adequate nutrition through good quality *Artemia*. Advanced epizootics are not treatable.

Idiopathic Muscle Necrosis (IMN)

Idiopathic muscle necrosis (IMN) is known by various names such as white muscle disease, muscle necrosis, spontaneous muscle necrosis, muscle opacity, or milky prawn disease. The disease has been observed in ponds with high stocking density and stressful environmental conditions such as low dissolved oxygen levels, salinity, and temperature fluctuations. It causes massive larval mortalities in hatcheries but can also affect all the life stages, i.e., larvae postlarvae, juveniles, and subadults. The symptoms include the whitish color of muscular tissue becoming necrotic with a reddish appearance. The disease shows focal and multifocal diffused opacity of striated muscles. At the chronic stage of the disease, the necrotic affected area may increase in size and acquire a reddish color, similar to cooked shrimp, due to the decomposition of muscle tissue. Striated muscle necrosis with cytoplasmic and acidophilic inclusions in muscular and connective cells can be clearly observed in the histology of the organs. Diagnosis of IMN in *M. rosenbergii* larvae can be made by gross demonstration of the opacity of the abdominal musculature. Histology is suggested for confirmation of the characteristic focal degenerative and/or regenerative change in the musculature. Improving pond management practices to avoid IMN-inducing stressors is the only way to prevent the disease.

Exuvia Entrapment Disease (EED)

EED is also known as metamorphosis moult mortality syndrome and moult death syndrome (MDS). This occurs primarily during the early and late larval stages and most commonly at the metamorphic moult stages from stage IX to PL. The diseased prawn shows the presence of localized deformities and fails to complete molting. The affected prawn are not able to free the exuvia from the appendages, eyes, or rostrum during and after ecdysis in which they get entrapped. Mortality occurs due to unsuccessful attempts by larvae to shed the exuvia which may reach up to 80% during metamorphosis.

The precise etiological agent of the disease is unknown, but it is thought that poor water quality, and inadequate nutrition may be responsible for this dreadful disease. Improvement in management in the hatchery particularly the water quality and nutrition helps in preventing the outbreak of the disease.

White Prawn Disease (WPD)

White prawn disease (WPD) is a disease of adult *Macrobrachium*, mainly in females. The subcuticular tissues appeared milky but the muscles were normal. No micro-organisms were demonstrated. It was characterized by a dense opaque white colour with a soft skeleton. A reddish abdominal discoloration affecting adult prawns is often observed in rearing ponds, but the etiology is unknown. The major symptoms are white opaque patches in the carapace (thus named White Patch Disease), necrosis, whitish blue coloration, loss of appetite and pale white muscles. Infected shrimp subjected Gram staining found that the rod-shaped bacteria was the pathogen. Genomic identification also confirmed that the causative pathogen is *Bacillus cereus*. Too much light diet, and stress have been considered to be responsible for this abnormality

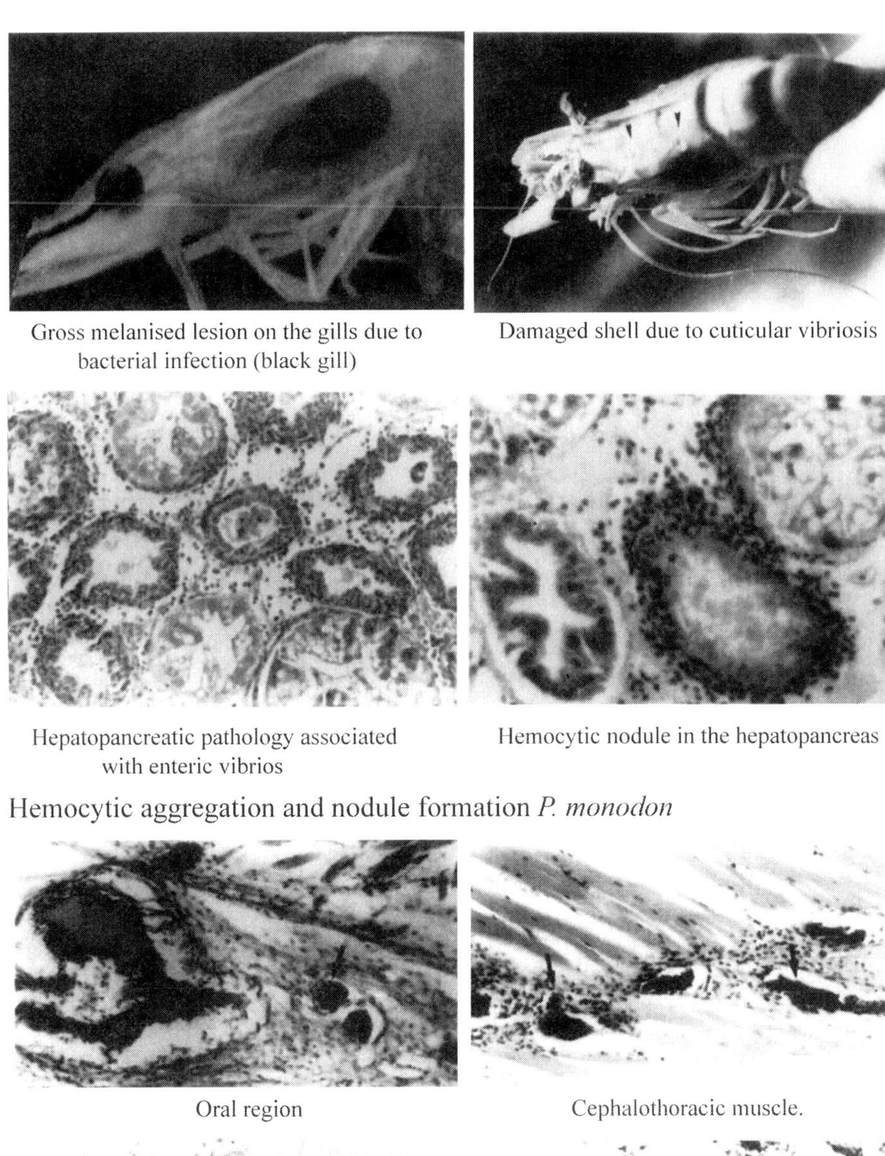

Gross melanised lesion on the gills due to bacterial infection (black gill)

Damaged shell due to cuticular vibriosis

Hepatopancreatic pathology associated with enteric vibrios

Hemocytic nodule in the hepatopancreas

Hemocytic aggregation and nodule formation *P. monodon*

Oral region

Cephalothoracic muscle.

Abdominal muscle

Antennal gland

Plate 6: Bacterial Diseases of Shrimp (See colour version on page 373)

References

Carry O Cunningham 2002 Molecular diagnosis of fish and shellfish diseases : present status and potential use in disease control. Aquaculture Vol 206 (1 and 2)

Devika Pillai, Jean Robert Bonami 2012A review on the diseases of freshwater prawns with special focus on white tail disease of *Macrobrachium rosenbergii* vol 43 issue 7 p 1029-1037,First published: 12 June 2012, https://doi.org/10.1111/j.1365-2109.2011.03061.

FAO (2013). Report of the FAO/MARD Technical Workshop on Early Mortality Syndrome (EMS) or Acute Hepatopancreatic Necrosis Syndrome (AHPNS) of Cultured Shrimp (under TCP/VIE/3304), 2013. Hanoi, Viet Nam, 25–27 June 2013. FAO Fisheries and Aquaculture Report No. 1053. Rome, Italy, 54 p.

Ramya Kumar, Tze Hann Ng, Han-Ching Wang 2020, Acute hepatopancreatic necrosis disease in penaeid shrimp, Reviews in Aquaculture Vol 12(3) p1867-1880

S.E. McGladdery. Shellfish diseases (viral, bacterial and fungal). Fish diseases and disorders. Volume 3: Viral, bacterial and fungal infections https://doi.org/10.1079/9781845935542.0748

6

Viral Diseases of Fishes

K.M. Shankar

Viruses are submicroscopic particles made up of nucleic acid covered with a protein coat. The nucleic acid is either RNA or DNA and not both. Protein coat is often covered with an envelope and depending on the presence or absence of an envelope, virus can be an enveloped or non enveloped virus. Size of the virus vary from 20-200 nm. Shape of the virus varies from a simple icosahedra to bullet shape to complex brick shape.

Virus are classified based on their nucleic acid (DNAor RNA) and protein profile. However, additional details on type of disease, host range, and geographic distribution of the virus are also used in classification. The classification of the virus is decided by the International committee on taxonomy of virus (ICTV).

Viruses are obligatory intracellular parasites requiring a living cell to replicate. As the virion is lacking its own metabolic machinery, it depends fully on the living host cell for machinery and raw materials for its metabolism and replication. This complex biology of the virus has made it difficult to clearly identify a viral process to design drugs for selective killing without damaging the host cell. Any attempt to kill the intracellular virus affects the host cell.

Replication of Virus in a Cell

There are two kinds of entry into and exit of virus from a cell: a) receptor mediated endocytosis (RME), b) direct entry by fusion. Non-enveloped virus enter the cell by RME (Fig. 1). RME of virus is similar to regular entry of other macromolecules into the cell. With this receptor mediated specified process, virus enter a cell in a vesicle which later fuses with lysosome. Inside the lysosome, viral particles undergo partial changes in protein coat due to digestion by lysosomal enzymes and move to cytoplasm. Finally, this uncoated nucleic acid replicates followed by synthesis of proteins and assembly of viral capsid proteins and genome to full blown virus particles. The site of assembly of the virus may be in the cytoplasm or nucleus of a cell which is referred to as inclusion body. Inclusion bodies which varies in size, shape, location and staining reaction can be used as a diagnostic feature for some viruses. Non-enveloped virus exit from a cell by damaging plasma membrane and hence clearly show cytopathic effect(CPE) in cell culture.

Enveloped viruses are known to use both RME and direct fusion for entry into and exit from a cell. The virus envelope is a phospholipid bylayer derived from host which is attached with viral derived protein. The protein is used for attachment, while the envelope involves in fusion with the plasma membrane. Enveloped virus exit out of the cell by budding, without damaging the host plasma membrane and hence CPE is usually not clear with an enveloped virus.

Aquatic viruses are transmitted mainly by two routes; horizontal and vertical. In the vertical route, virus is transmitted from parents to offsprings through sex products. Depending on whether virus is present inside or on sex products, it could be true or false vertical transmission. More commonly, virus is present on sex products and in plasma fluid. In horizontal transmission, virus is transmitted from one fish to another through water, feed, contact with asymptomatic host and other carriers. The mode of transmission has a key role in determining severity and rapidity of virus infection and disease manifestation, spreading and persistence of the virus in an area. Overall, the amount of virus (infectious dose) in brood, seed, water or other infected material determines severity of virus infection in a population. Normally, virus enters fish through skin, lateral line, gills and gut and reach target organs through blood.

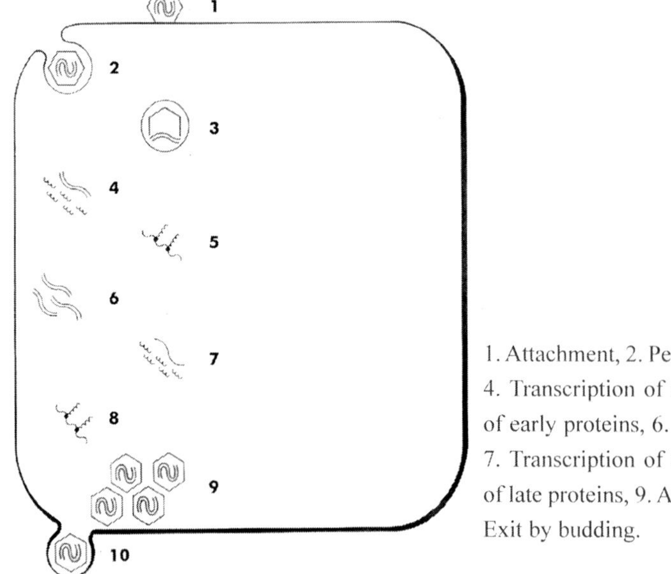

1. Attachment, 2. Penetration, 3. Uncoating 4. Transcription of mRNA, 5. Translation of early proteins, 6. Viral DNA replication 7. Transcription of mRNA, 8. Translation of late proteins, 9. Assembly of virions, 10. Exit by budding.

Fig. 1: Entry, replication and exit of a virus from a cell.

Virus Pathogenicity

Virus is a highly obligatory intracellular parasite. Virus replication is invariably at the cost of the host cell, either the cell is affected partly or fully (Plate 7). At host level, virus affect target organ partially damaging and impairing its function or fully destroying it leading to morbidity and death of the host.

Outcome of disease in a fish population due to virus infection is complex and not straight forward which depends on several factors. Among the various factors, immune status of individuals and infectious dose of virus (quantity of virus in water, feed, infected material) are very important in the outcome of disease. Depending on these variables, outcome of virus infection are of several types; virus infects host but get aborted in some individuals; virus infects and cause disease but host overcomes; virus infects cause disease and death; virus infects and host survives infection and remain as carrier. Hence, clinical signs and mortality pattern of viral infection in a population differs between individuals.

In a population viral infection need not always cause 100% mortality due to differences in the individual host immune status and infectious dose. In some cases, virus remain at a low level of infection establishing a delicate balance with the host. In addition, there are carriers which are survivors of a mass scale infection and mortality. Carriers will be spreading the virus life long to the environment and are potential risk and concern in viral disease management. Other type of carriers are non target hosts which are not affected at all but become asymptomatic carriers. Latent infection of virus at very low level which express late in the life of an host is also risky. Usually it is difficult to detect virus in carrier or latent infection stage.

Virus Variants

Replication of virus is very rapid. In general, virus requires 6-10 h for one complete cycle which is faster in RNA virus compared to that in DNA virus. Because virus replicate rapidly, they also undergo mutation very often. Mutation in RNA virus is higher than in DNA virus. Intensive aquaculture measures such as high stocking density, extensive culture area, stress to animals and high traffic provide ideal conditions for outbreak of viral disease and resulting mutation leading to several strains. The new strains of virus may be virulent or avirulent which differ in pathogenicity and mortality pattern in hosts. Due to rapid evolution of virus, designing vaccines or standardising diagnosis becomes difficult.

Viral Disease Diagnosis

Diagnostic methods include both traditional and modern molecular biological tools. Traditional methods such as histopathology, electron microscopy, cell culture assay, virus neutralisation tests have their own importance. Electron microscopy provides vital useful information on size, shape and site of replication of the virus inside host cell. However, the method is cumbersome, time consuming and hence cannot be recommended for routine diagnosis. Histopathology gives idea about target organs and cells involved. The replication sites are identified by the presence of inclusion bodies. Greater specific details on viral pathology can be obtained by immunohistochemistry or DNA based *in situ* hybridisation.

Diagnosis by cell culture is the first step in identification and study of viruses. Virus replicates only in living cell and hence cell lines are essential for study, diagnosis, purification and vaccine development. Cell lines of many cultivable species of fish are available now. Since list of cultivable species is growing, there is emergence of new viral disease, and hence there is a need for development of more cell lines. Further, cell lines from various organs of fish such as gonad, gill, spleen, liver, skin and heart are required for efficient culture and detection of virus. Virus enters cells, replicates and damage the cell monolayer in a culture. The damage is termed as cytopathic effect (CPE). In some cases virus besides destroying the cells change the morphology of cells. Nature of CPE and inclusion body give preliminary idea about the involvement of a virus and to some extent type of the virus. Quantity of the virus in the infected fish can be estimated by cell culture by two methods; tissue culture infectious dose 50% ($TCID_{50}$) assay and plaque assay. The $TCID_{50}$ is a quantal assay which indirectly gives virus concentration in terms of the dose causing 50% cell death. The plaque assay gives actual number of infectious particles present in the fish. The $TCID_{50}$ assay is ideal for mass scale quantitation of virus, while plaque assay is ideal for small scale studies. Once virus is identified by histopathology, cell culture CPE, further specific tests are required to confirm the type of the virus. This involves specific serological tests such as neutralisation test (NT) employing rabbit antisera or monoclonal antibodies. A suspected virus should be treated with available antisera against established viruses and the neutralisation of the virus is detected by inoculating the reactants to cell culture. Depending on the occurrence of CPE in the neutralisation test, virus can be identified. A new virus cannot be neutralised by the antisera and hence cause CPE in cell monolayer. Neutralisation test is the most specific and invariably required in virological investigations. In addition to NT, other serological tests available for fish viruses are hemagglutination hemeadsorption and complement fixation.

Antibody based tests such as ELISA, immunodot, Western blot are commonly used in diagnosis of viral disease. Other antibody based *in situ* tests used such as immunoperoxidase (IPO), immunofluorescence (IF) give more specific information about the target organs and cells involved in pathogenesis, besides diagnosis. In addition, field or farmer level antibody based test kits are available for detection of many aquatic viruses. These tests were developed using either antisera or highly specific monoclonal antibodies. Most commonly available test kits are lateral flow test and flow through test, which are rapid, low cost providing sensitivity matching I step PCR. DNA based methods such as DNA hybridisation and Polymerase chain reaction (PCR) are also becoming popular in diagnosis. Several simplified versions of these tests are available as kits for field diagnosis in developed countries.

Investigations on New Fish Diseases Suspected to be of Viral Etiology

In recent years, several new diseases suspected to be of viral etiology have emerged in tropical aquaculture. Such new diseases could also be due to bacteria. Therefore, it is always customary to look for primary etiology by differentiating virus from the bacteria step by step. First step involves filtering the infected tissue homogenate through 0.2 *u* filter and performing Koch's postulate using the bacteria-free filtrate. Next, the bacteria free filtrate is inoculated to cell culture in the presence of antibiotics to find out whether it causes CPE. As bacteria get separated in 0.2 *u* filter and hence cannot cause CPE in the presence of antibiotics, while the filtrate cause CPE indicating involvement of a virus. This result can be further confirmed by using the filtrate with available antiserum of known viruses and checking for CPE. Simultaneously, bacteria-free filtrate producing clinical disease and mortality will indicate viral involvement. Further tests such as electron microscopy and virus neutralisation will positively confirm the viral etiology.

Common Viral Diseases of Fishes in the Cold Countries

In some countries of Europe and North America, scientific fish culture particularly salmonid culture started about a century ago. In salmonid culture where intensive hatcheries are common, several viral diseases of very serious nature have been recorded. Economically important fish viral diseases of cold countries are given in Table 1.

Among the various cold water fish viruses, infectious pancreatic necrosis virus (IPNV) which was reported back in 1940 in salmonid hatcheries is very well studied. Information on this virus will be useful in fish and shellfish viral disease management of warm water aquaculture. IPN virus is an RNA virus affecting young ones of salmonids in hatcheries causing mortality ranging from 60-100%. The virus cause disease by damaging pancreas and intestine of young fish and hence the name infectious pancreatic necrosis. Juveniles and adults are refractory to infection. Young ones which survive the infection will become life long carriers spreading the virus. In addition to salmonids, the virus has been isolated from non-salmonids such as tilapia, carp with little or no mortality. The virus has also been recorded in molluscs and rotifers. The virus which originated in North America and Europe gradually spread to some of the Asian countries through transport of salmonid eggs and juveniles. Feral population of salmonids in remote areas of North America acting as carriers of IPN virus is also on record. e.g. presence of IPN virus in feral lake trouts of Sub Arctic region of Canada. There are several serotypes of IPNV-VR299, SP and Ab being the major ones. Also several variants of the IPN virus differentiated using gene and monoclonal antibody probes have been reported. Hence, vaccine development is a difficult task. As such, no effective drug is available. Chemicals such as vescodyne is available for washing trout egg which is not 100% effective. Since no effective drug or vaccine is available preventing the entry of virus to the culture system has been the best option. Proper screening and certification of seed, brood and water has been the major strategy for control of the virus. If disease is rampant crop holiday is the suggested measure to reduce the viral load in the culture system.

Fish Viral Diseases in the Asia Pacific Countries

Compared to viral diseases in North America and Europe, fish viral diseases recorded are few in Asia Pacific countries. This could be due to several reasons. Intensive fish culture started in these countries only recently. Immune system of warm water fishes is well developed compared to that of coldwater species. Cold water species are more prone to viral diseases compared to that of warm water species. Furthermore, in some of the Asian countries virological studies are not conducted systematically due to lack of trained persons, facilities and cell lines. Nevertheless, new emerging diseases, suspected to be of viral etiology in cyprinids, perches, groupers and host of aquarium fishes have been recorded. In EUS, primary viral involvement is still suspected.

Table 1: Viral pathogens of economic importance in cold water aquaculture

Virus	Host	Target Tissue	Major Pathology	Major Clinical signs	Vaccine Under Development.
a. Infectious Pancreatic Necrosis virus (IPNV)	Salmon, Trout	Pancreas and liver	Necrosis of acinar pancreatic tissue, catarrhal exudate	Distended abdomen, blackening of body, swirling movement, exophthalmia	Killed, live attenuated, subunit
b. Infectious Haematopoietic Necrosis virus (IHNV)	Salmon, Trout	Kidney, spleen and liver	Necrosis of haematopoietic tissue,anaemia, internal haemorrhages	(Non-specific)	Subunit, live attenuated
c. Viral Haemorrhagic septicaemia (VHS)	Salmon, Trout	Vascular system, kidney	Haemorrhagic anemia, Necrosis and exudate formation	Dropsy (Non-specific)	Subunit, live attenuated
d. Rhabdovirus carpio	Carp, Pike	Swim bladder	Haemorrhage in the swim bladder	Non-specific	Killed, live attenuated, subunit
e. Channel catfish virus (CCV)	Channel Catfish	Pancreas	Necrosis, spongiosis	(Non-specific)	Live attenuated
f. Lymphocystis virus	Walley	Lymphocytes, Epidermal cells	Hypertrophy of lymphocytes andepidermal cells	(Non-specific)	None

Emerging Viral Pathogens of Fish in the Asia Pacific Region

Koi Herpes Virus (KHV)

Koi Herpes Virus (KHV), also known as cyprinid herpes virus-3 (CyHV-3) causes a serious emerging disease of ornamental fish. CyHV-3 is highly contagious pathogen, resulting up to 100% mortality in Koi. KHV is a member of the genus *Cyprinivirus* in the family Alloherpesviridae measuring 100-110 nm with a ds DNA of 250-300Kb. It is a divergent herpes virus unclassified as it differs in number of polypeptides from other fish herpes viruses of catfish, eel, shellfish and oysters.

First identified in Israel during 1999,it is now widespread in Japan, USA, Europe, Asia and South Africa. In later years KHV affected the carp (*Cyprinus carpio*) resulting in heavy losses. In view of its significance on fish trade, KHV is designated as OIE listed pathogen of finfish.

The virus causes mass mortality in Common carp and Koi in 2-3 weeks is species specific infecting Koi of all age and younger stage of common carp. Permissible temperature for infection is 18-28 ^{0}c while incidences are less at temperature more than 30° C. Clinical signs include high mortality, skin discoloration, gill necrosis, swollen kidney and spleen. Intranuclear inclusions can be found in cells of target organs gill, skin, liver, kidney and gas bladder. The virus causes interstitial nephritis and gill necrosis in carps, and it is also termed as carp interstitial nephritis and gill necrosis virus. Virus spreads by both vertical and horizontal transmission. Experimental infection has been standardised. The virus can be isolated from dead fish using KF cell lines with 7-10 days incubation at 20° C. The virus is difficult to detect from carrier live fish. Electron microscopy and PCR including real time PCR have been developed for diagnosis. An unique method of vaccinating fish developed where in fishes are first experimentally infected with the virus at the favaourable temperature 18-28 ^{o}C followed by raising temperature to more than 30 ^{o} C, increased temperature inactivates the virus in the fish serving as vaccine to offer protection.

Although reported in common carp (*Cyprinus carpio*) in other South East Asian countries, KHV has not been reported in common carp or Indian major carps in India. Indian Major Carps (IMC), appears not to be susceptible to KHV. As there is every possibility that unregulated import of ornamental fish can lead to KHV host range expansion, including other carp species a strict surveillance and monitoring of KHV has been taken up in India.

Nervous Necrosis Virus (NNV)

Nervous Necrosis Virus (NNV) causes Viral Encephalopathy and Retinopathy (VER) with a wide host range in a number of cultured marine fishes all over the world. It is a piscine noda virus, non enveloped, icosahedra 25 nm, with ss + RNA, RNA 1 for structural protein, RNA 2 for coat proteins. The virus belongs to the genus *Betanodavirus* of the family Nodaviridae. There are four major genotypes/serotypes of NNV each with different host specificity.

First reported during 1996 the virus has wide geographical range- Japan, Norway, Taiwan, Italy, China, India and Singapore. The virus also has a wide host range; Groupers, barramundi, European sea bass, striped jackand rarely in some freshwater ornamental fish in India. The virus cause high mortality in larvae and juveniles (60 to 90%) but less serious in adults/brood serving as carriers. Clinical signs include lack of appetite, altered coloration, inflammed gas bladder, hyperactivity and abnormal swimming behaviour. Both experimental and natural infection have been demonstrated. The disease affects the neuronal tissues of brain, spinal cord and eye. Pathological changes include cellular vacuolation, neuron degeneration in central nervous system(CNS), retina, and ganglia of peripheral nervous system. Virus can be detected by MAb based imunohistochemistry of retina, brain and spinal chord. NNV although found in CNS of larvae mainly, can hide in different organs of adult carriers such as gonads, intestine, stomach, kidney and liver which can be detected. The virus can be isolated easily in preferred cell lines such as grouper cell line. Virus can also replicate in snakehead cell line. Histopathology provides presumptive diagnosis, which can be confirmed by antibody based immunofluorescence test. Vaccine mostly a recombinant DNA type evaluated elicits immune response in halibut and turbots.

Irridoviruses

Irridoviruses cause Epizootic Hematopoitic Necrosis Disease(EHND**)**. The irridoviruses shave been reported from both feral and culture including ornamental fishes from fresh and marine environment all over the world. The virus first recorded during 1998 caused epizootic hematopoitic necrosis/ viral erythrocyte necrosis, and sleepy grouper disease caused upto 100% mortality in fry and 30 % in adults. It is an enveloped icosahedra ds DNA virus, measuring 120-130 nm, with several isolates, genotypes but all with same pathogenecity. In antibody test such as ELISA irridovirus of fish and frog cross react.

Irridoviruses have wide host and geographical distribution: groupers, red seabream, dwarf gouramy, striped beak perch, Red drum in china, sword tail, mandarin fish, pike, perch ,white sturgeon in USA, and redfin perch (*Perca fluviatilis*). Several species of fish present in irridoviruses endemic regions such as *Gambusia affinis*, holbrooki and native Australian fish *Macquaria australasica*, some European fishes, *Ameiurus melas*, *Esox lucius* and *Sander lucioperca* are susceptible to the virus. The potential host range of irridoviruses is greater than that which has been observed in natural disease outbreaks. The viruses also found in frogs and toads. The viruses have been recorded from China, Japan, Taiwan, USA, and Australia

Infected fish have hemorrhage at the base of the fins, peritoneal effusion, enlargement of the spleen and kidney with pale foci on the liver. Histopathology includes enlargement, necrosis of renal/spleenal hematopoitic tissue. Experimental infections indicate an incubation period of 10–28 days with a brief period of anorexia, darkened coloration and abnormal swimming preceding death.

Irridoviruses are detected by isolation in cell line (blue gill fry cell line-BF2) and polymerase chain reaction (PCR). The World Organization for Animal Health (OIE) lists infection with irridoviruses to facilitate biosecurity measures (OIE, 2015).

Tilapia Lake Virus (TiLV)

Tilapia tilapinevirus, known as tilapia lake virus, belongs to genus Tilapinevirus. This orthomyxo-like virus possesses 10 negative-sense RNA segments enclosed within a membrane-bound nucleocapsid. The virus particle is roughly circular with a diameter 55 to 100 nm. The virus was initially categorized under Orthomyxo-like virus but later on classed into family Amnoonviridae and genus Tilapinevirus.

First scientific reporting of TiLV was from Israel and Equador in the year 2013. However, the genomic epidemiology study suggests the origin of TiLV 5–10 years before the first report of the virus in Israel. TiLV is widely distributed in Asian countries *viz*. Thailand, Chinese Taipei, India, Malaysia, Indonesia, Philippines and Bangladesh. Incidences of the virus in Colombia, Peru, USA and , Mexico, and African countries also reported. Genome sequence analysis of the viral isolates from different countries shows 97–99% nucleotide sequence identity. More than 45 countries are believed to be at high risk of TiLV, including different southeast Asian countries and Sub-Saharan African countries. The widespread occurrence of this virus across the tropical region of the world from America to Asia and Africa poses a significant threat of serious pandemic in tilapia farming. A number of tilapia species, namely Nile tilapia (*O. niloticus*), red tilapia (*Oreochromis* sp.), and hybrid tilapia (*O. niloticus*× *O. aureus*) have been reported to be affected by TiLV, causing mortality up to

90%. TiLV has been identified from various species of wild tilapias as well. All the life stages of Tilapia viz., fertilized eggs, yolk-sac larvae, fries, fingerlings and adults are susceptible to TiLV.

The virus can spread horizontally and can survive in both freshwater and brackish water (OIE 2018b). The disease can occur across a range of temperatures (15–30°C); however, a water temperature of 25°C is most commonly associated with spreading through a population. All age groups are susceptible and cumulative mortality ranging from 10–90% has been reported. Tilapia infected with TiLV often exhibit non-specific clinical signs including lethargy, loss of appetite, and decreased schooling behavior. Diseased fish may also present with "pop eye" (i.e., bulging eyes or exophthalmia), darkening of the skin and ulcerated or bleeding skin. Additional signs include pale gills, a swollen celome due to accumulated fluid internally and scale loss. Damage to the skin often leads to secondary bacterial infection (e.g., *Aeromonas* spp.).There are also reports of co-infection with different bacterial pathogens in the TiLV infected fishes. Different species of *Aeromonas viz A. veronii*, *A. ichthiosmia*, *A. enteropaelogenes*, *A. hydrophilia*, *Flavobacterium* spp. and *Streptococcus* spp, were detected from TiLV-infected fishes. TiLV with *Aeromonas* spp. was more frequent than with other bacterial species.

Indian Major Carps (*Labeo rohita*, *Catla catla*and *Cirrhinus mrigala*), milkfish (*Chanos chanos*), pearl spot (*Etroplus suratensis*), *Mugil cephalus*, *Cyprinus carpio* and *Liza ramada* cultured along with Tilapia did not show any mortality during the outbreak of the viral disease. Furthermore, species like *Trichogaster pectoralis*, *Pangasianodon hypophtthalmus*, *Clarius macrocephalus*, *Channa striata*, *Anabas testudineus*, *C. carpio*, *Barbodes gonionotus Lates calcarifer were* found to be resistant to TiLV infection. Other than tilapia, the river barb (*Barbonymus schwanenfeldii*) and giant gourami (*Osphronemus goramy*) were found to be infected by TiLV.

In experimental infection the TiLV can be transmitted from one fish to another. In the co-habitation study, cumulative mortality of 50% with the clinical disease was observed. Adult Tilapia may have asymptomatic infections as the apparently healthy adults were also found positive for TiLV. The vertical transfer potential of this virus has been demonstrated by several studies. It is believed that intragastric route is the prime route of infection due to the aggressive feeding habit of tilapia. TiLV observed in vital organs like brain and liver. Detection of virus in the mucus of Tilapia suggests mucus as a possible mode of horizontal transmission through co-habitation. A co-habitation study revealed that healthy Tilapia might acquire infection within 1–2 days after exposure to infected fish. The transfer of the virus may be directly correlated with the stress status of the animal, as stress is the most crucial risk factor for the disease.

Along with many generalized clinical signs, high mortality associated with corneal opacity may be one of the overt clinical signs during outbreaks. Clinical signs like anorexia, poor body condition, severe anemia, bilateral exophthalmia, skin abrasion and congestion, scale protrusion, and abdominal swelling were observed in fishes in both natural and cultured conditions. Histopathological examination of brains of TiLV infected fishes revealed multifocal haemorrhages, edema, proliferative glial cells and capillary congestion. Syncytial cell formation and massive hepatocellular necrosis with pyknotic and karyolitic nuclei most commonly observed in the liver of infected fish. In addition, eosinophilic intracytoplasmic inclusion bodies were present in the liver and spleenic cells of infected fish. Multiple necrotic foci were observed in the anterior kidney with increased melanomacrophage centers (MMCs), and in spleen with dispersion of melanin granules. Regardless of the route of transmission, Tilapia developed identical clinical and histopathological lesions, when challenged with both intracoelomic and intragastric pathway.

Spring Viraemia of Carp Virus (SVCV)

Spring viraemia of carp (SVC) is caused by the spring viraemia carp virus (SVCV), which is also known as Rhabdovirus carpio. Isolates can be divided into four genetic groups; Genogroup Ia from Asia, Genogroups Ib and Ic from Russia, Moldova and Ukraine and. Genogroup Id mainly from the U.K. SVCV also reported from cultured carps in China and USA. Although disease is endemic to Europe for the last 50 years, it is an emerging diseasein other countries in the last two decades. SVC is a contagious viral disease mainly in farmed carp and related species causing substantial economic losses. SVC can be highly fatal in young fish, with mortality rates up to 90%. In Europe, where this disease has been endemic for at least fifty years, 10-15% of one-year-old carp are lost to SVC each year.

SVCV primarily infects carp and other species in the family Cyprinidae. It is also found in a few species from other fish families. Infections have been reported in common carp (*Cyprinus carpio*), koi carp (*Cyprinus carpio koi*), grass carp/ white amur (*Ctenopharyngodonidella*), silver carp (*Hypophthalamicthys molitrix*), bighead carp (*Aristicthys nobilis*), crucian carp (*Carassius carassius*), goldfish (*Carassius auratus*), tench (*Tinca tinca*) orfe (*Leuciscus idus*) and sheatfish/ European catfish/ wels (*Silurusglanis*). Common carp is the most susceptible species and considered as principal host. Very young fish of various pond species, including pike and perch are also susceptible. Experimental infections have been reported in roach (*Rutilus rutilus*), zebrafish (*Danio rerio*), guppies (*Lebistesreticulates*), northern pike (*Esox lucius*), golden shiners (*Notemigonus crysoleucas*) and pumpkinseed (*Lepomis gibbosus*). Interestingly, a SVCV–like virus has also been found in diseased cultured shrimp (*Penaeus stylirostris* and *P. vannamei).*

Fish can carry SVCV with or without clinical signs. Fish up to a year old are most likely to be affected, but illness also occurs in older animals. The clinical signs are nonspecific. In carp, the most common clinical signs include abdominal distension, exophthalmia, inflammation or edema of the vent (often with trailing mucoid fecal casts), and petechial hemorrhages of the skin, gills and eyes. The body is often darkened with pale gills. Diseased fish tend to gather at the water inlet or sides of the pond, swim and breathe more slowly than normal, and react sluggishly to stimuli. Loss of equilibrium, with resting and leaning, are seen in the late stages. Concurrent bacterial infections (carp-dropsy complex) or parasitic infections influence the clinical signs and mortality rate. SVC outbreaks are most common in farmed carp, but can also occur in wild fish. The morbidity and mortality rates vary with stress and population density, species, age and condition of the fish. Incubation periods from 7 to 15 days have been reported in experimental infections. Water temperature affects the development of disease, with clinical signs most common at 17°C (63°F) or below. Mortality rates up to 70% have been reported in young carp during outbreaks while up to 90%. in experimental infection.

Both clinically ill and asymptomatic fish carry the virus. This virus is shed in the feces ,urine, gill and skin mucus of infected fish. It is also found in the exudate of skin blisters and edematous scale pockets. Transmission is by direct contact or through the water. The virus enters most often through the gills. SVCV has been found in ovarian fluids and "egg-associated" (vertical) transmission is suspected. SVCV can also spread by fomites and invertebrate vectors. Known vectors include carp louse *Argulus foliaceus* and leech *Piscicola geometra,* including aquatic arthropods and fish-eating birds. Infectious virus can persist in 10°C water for more than four weeks and in 4°C mud for at least six weeks. It is difficult to erradicate the virus once it is established in a pond.

SVC should be suspected in cyprinid fish with signs of a systemic infection and an increased mortality, when water temperatures are below 20°C. A particularly high incidence of disease in common carp, with reduced susceptibility in carp hybrids and lower disease prevalence in other cyprinids, indicates the viral involvement. SVCV is diagnosed by virus isolation in cell lines such as EPC (*Epithelioma papulosum cyprini*), FHM (fathead minnow) and carp leukocyte cultures. SVCV can also replicate in some mammalian cell lines. The identity of the virus is confirmed by virus neutralization or polymerase chain reaction (PCR) and nucleotide sequence analysis of PCR products. Immunofluorescence or ELISA can be used for rapid presumptive identification of the virus in cultures.

Viral Disease Management

Specific drugs for viral disease treatment are not available or difficult to develop since virus is host cell dependent for all its metabolic activities. However, virucidal chemicals capable of killing virus outside the host are available. Chlorine, iodine, ozone and UV rays are some of the commonly employed virucidal agents in aquaculture. In recent years, though several nucleic acid analogue based drugs have been developed, they are found to be cytotoxic and as such they are not successful. Vaccines are the most effective preventive strategies in mammals. However, vaccines in general are not found to be effective in fish viral disease management. Nevertheless, killed, attenuated and recombinant DNA vaccines have been developed. Protection from vaccines against viral disease in fish is found to be for short period with variable results. Poor immune system of the fish and young age at infection are some of the responsible factors for susceptibility to disease. Further, there are several variants of the virus. All these factors contribute to the poor performance of viral vaccines. In the absence of successful drug or vaccine, avoidance of the virus in culture system is the best strategy. This can be achieved by proper biosecurity measures such as sensitive screening of brood, seed and certification programmes. Crop holiday is one of the best strategies to prevent viral disease in aquaculture and is still practised in salmonid culture to prevent IPN disease. During crop holiday, infectious dose in water and carriers gradually comes down due to reduction in host and carrier availability in the system.

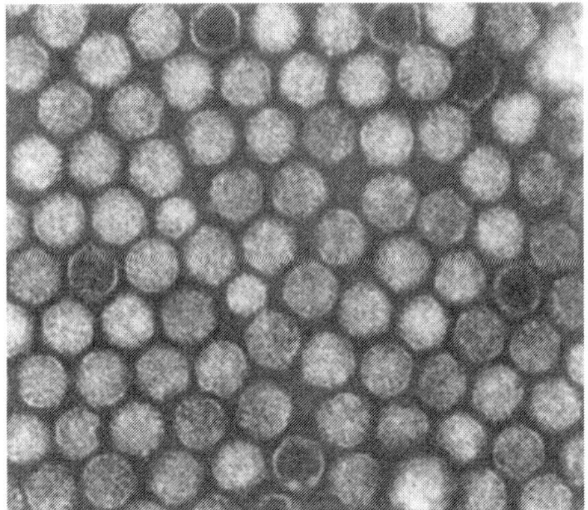

Electron micrograph of purified IPN virus (icosahedral shape)

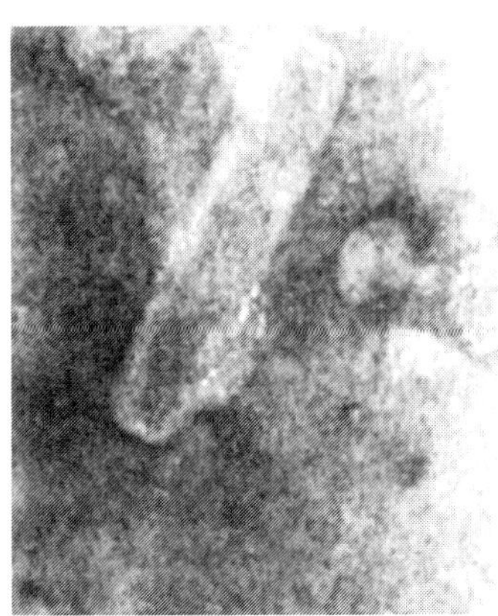

Electron micrograph of a bullet shaped virus

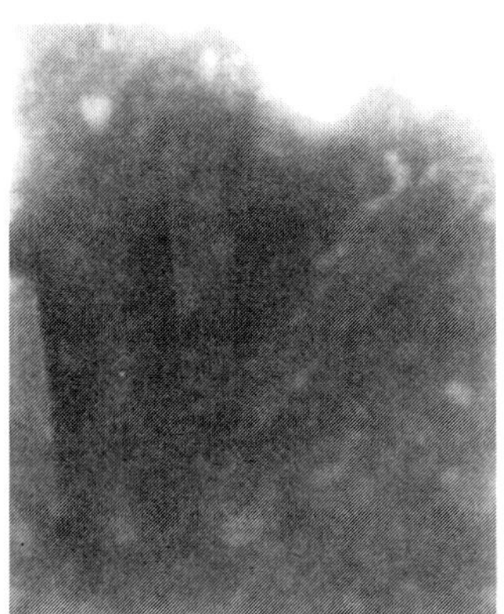

Electron micrograph of nucleocap-sids of a bullet shaped virus

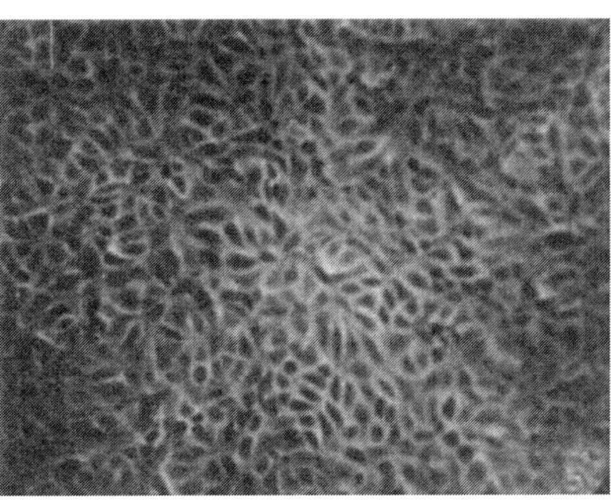

Normal healthy monlayer of a fish cell line

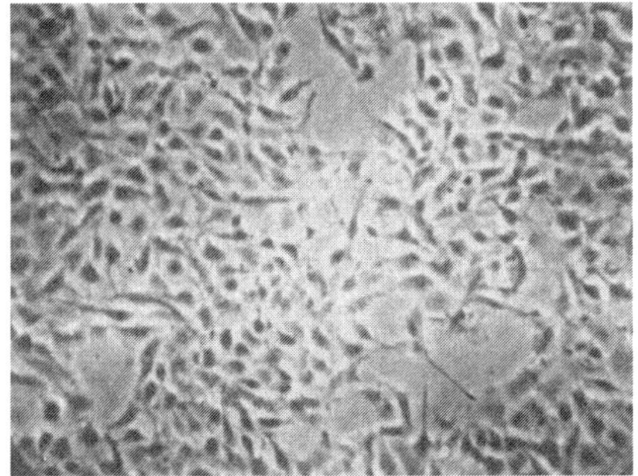

Cytopathic effect (CPE) of IPN virus

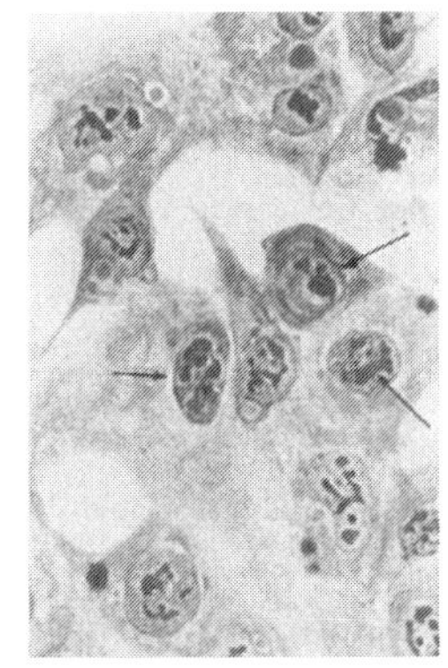

Basophilic intranuclear inclusions in viral infected cells

Plate 7: Virology of Fish

References

Gaurav Rathore, Gokhlesh Kumar, T. Raja Swaminathan, and P. Swain. 2012,Koi Herpes Virus: A Review and Risk Assessment of Indian Aquaculture, Indian J Virol. 23(2): 124–133. Published online 2012 Sep 6. doi: 10.1007/s13337-012-0101-4

Japhette E Kembou-Ringert , Dieter Steinhagen John Readman, Janet M Daly, Mikolaj Adamek. 2023 Tilapia Lake Virus Vaccine Development: A Review on the Recent Advances Vaccines ;11(2):251. doi: 10.3390/vaccines11020251.

K. M. Shankar and C.V Mohan,2002, Fish and Shellfish Health Management , ISBN 81-7525-328-2, Fish Pathology and Biotechnology Laboratory, Dept of Aquaculture, UAS, College of Fisheries, Mangalore-575002

Ken Wolf 1988. Fish viruses and Fish viral diseases, Cornel University Press Pages 480.

Mark Crane and Alex Hyatt 2011. Viruses of fish: An overview of significant pathogens Viruses, 3 (11) 2025-2046. doi: 10.3390/v3112025

7

Viral Diseases of Shellfishes

K.M. Shankar and Abhiman

A. Viral Diseases of Shrimp

Shrimp farming which has grown to an industry in the last three decades is young compared to fish farming in the world. Yet the shrimp farming is facing serious problems with several types of viral diseases. Intensive culture in hatcheries and grow outs along with movement of stock across national and international borders of these animals with poor immune system has encouraged emergence of large number of viruses. During a short span of 30 years, 15 viruses of economic concern have been reported in shrimp culture of the Asia Pacific region. Shrimp viruses (Plate 8) can be broadly categorised into two groups:

1. Viruses infecting ectodermal (epidermis, hypodermal epithelium of fore and hindgut, nerve cord and nerve ganglia) and mesodermal (hematopoitic organs, antennal gland, gonads, lymphoid organ, connective tissue and striated muscle) origin tissues and replicating within the cytoplasm (YHV, TSV) and within the nucleus (IHHNV, WSSV), 2. viruses like MBV, HPV, BMNV, and BP infecting endodermal origin tissue and replicating within the nucleus of hepatopancreatic and mid gut epithelial cells. Lack of established shrimp cell line is the major drawback in study of shrimp viruses. Following are some of the common viruses causing diseases in shrimp hatcheries and grow out ponds in the Asia Pacific region.

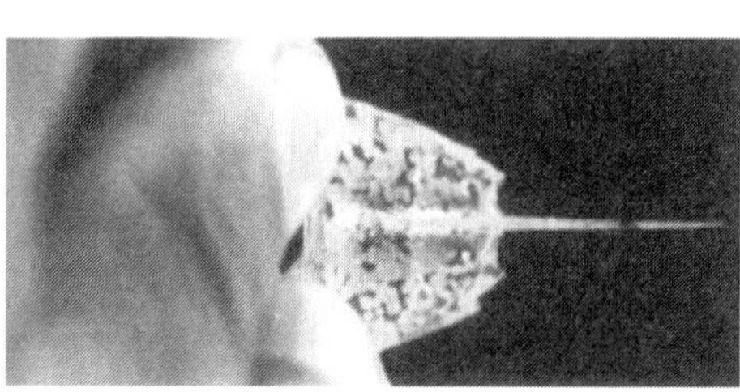

Clinical white spots associated with WSS Vinfection in *P. monodon.*

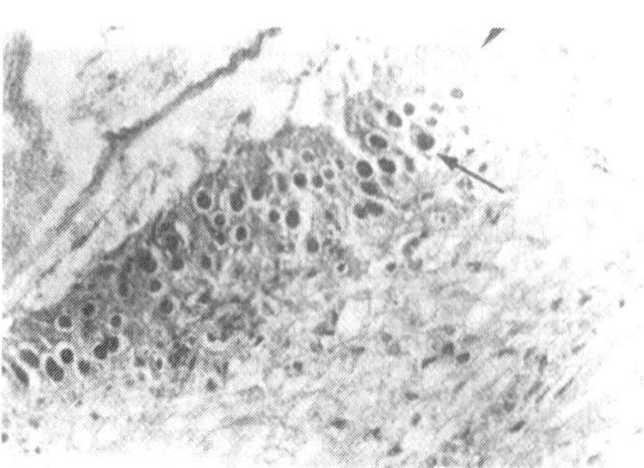

Basophilic intranuclear inclusions of WSSV in foregut cuticular epithelium

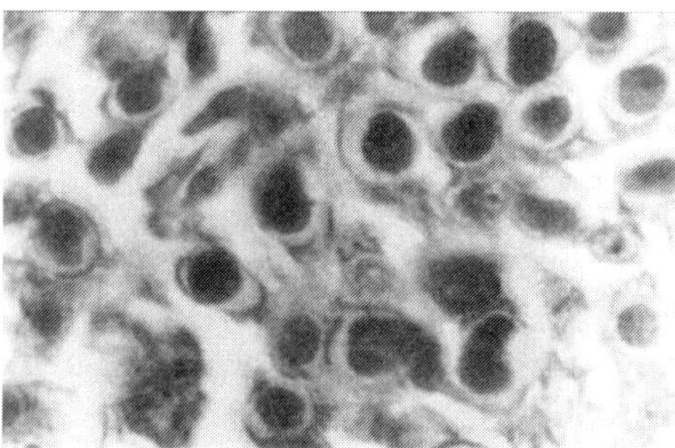

WSSV inclusions in connective tissue

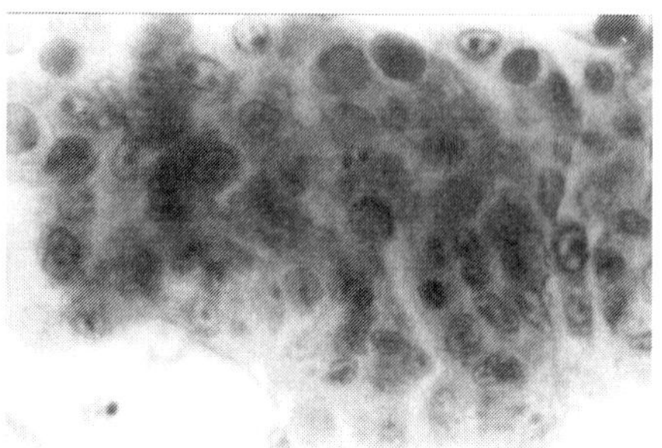

Dual infection in *P. monodon* with WSSV (intranuclear inclusion) and YHV (dense intracytoplasmic inclusion)

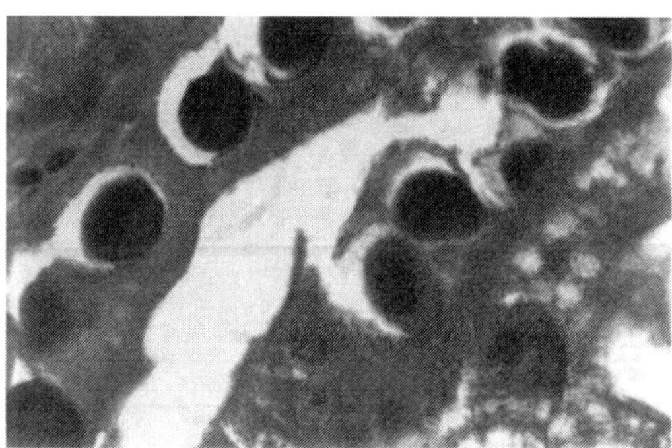

Prominent intranuclear basophilic inclusions of HPV in the hepatopancreas

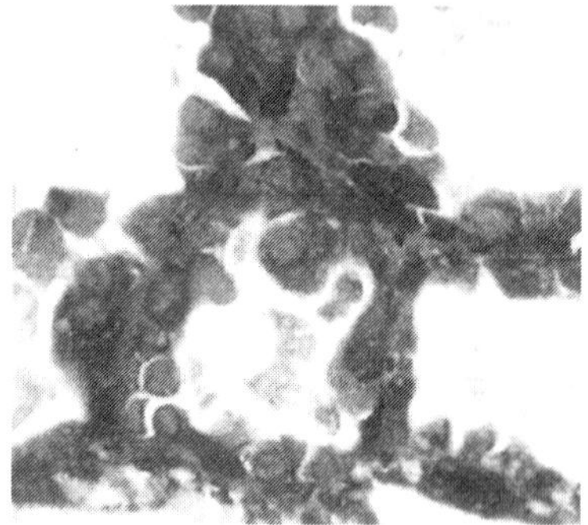

Occlusion bodies of MBV in the hepatopancreas

Plate 8: Shrimp Viral Diseases (See colour version on page 374)

1. White Spot Syndrome Virus (WSSV)

White spot disease is caused by a systemic ectodermal and mesodermal non-occluded DNA virus (121 x 276 nm), now commonly known as white spot syndrome virus (WSSV). It is the single most destructive virus very prevalent in shrimp culture industry causing billions of dollar loss world wide and hence it is described here in greater detail.The overall epidemiology of the WSSS is first affected shrimp show rapid reduction in food consumption, become lethargic with reddening and loose cuticle. Characteristic white spot of 0.5 to 2 mm dia at inner side of carapace is common in moribund and dead shrimp. Mortality rates reach upto 100% in 3-10 days. The virus infects all sizes and species of penaeid shrimp in all types of culture systems. In addition to the Asian countries listed above, farmed shrimp exhibiting the gross signs and histology of WSD have been reported in the USA and Latin America.The virus is widespread, though named differently in different parts of the world (SEMBV, WSSV, RV-PJ, WSBV, SEED).

Hosts and vectors of WSSV: White spot syndrome virus is a generalist virus mainly infecting decapod crustaceans with an extremely wide host range. The virus can infect a wide range of wild and cultured aquatic crustaceans including marine, brackish and freshwater penaeids, crabs, lobsters, crayfish, squilla, and gaint freshwater prawn. Non crustacean hosts and vectors/carriers include green algae, rotifers, polychetes, copepods, oysters, clam, snail, aquatic insects and birds. Host and vector species of WSSV belong to 50 families including 11 families of non-crustacean hosts. Molecular technique such as PCR has been used to confirm infection of non-penaeid carriers of WSV.

WSSV in widely cultured freshwater prawn is a major concern. Infection of freshwater prawn with WSSV has been detected by PCR. Larvae, PL, juveniles and adults have been found to be PCR positive. Amplified PCR product has been found to be similar to that of natural infected penaeid WSSV. Freshwater prawn is susceptible to WSSV, hence great care should be taken to prevent epizootics of WSSV in freshwater prawn culture system.

The virus can be detected by light microscopy of fresh stained samples, histopathology, gene probes, antibody based methods and EM. Early stages of inclusion bodies are eosinophilic, centronuclear and resemble that of IHHNV. However, presence of larger, more fully developed basophilic inclusion bodies of WSSV can be differentiated easily at later stages.Since 1993, white spot disease (WSD) has been causing significant production losses in cultured shrimp all over south east and south Asia. In order to overcome the losses, several farmers have switched over from open to closed or semiclosed culture systems. Despite this, WSD outbreaks continue to be a problem. WSD still occurs even in post larvae certified as virus free not withstanding PCR checks which have become a frequently adopted practice. Everyone in the shrimp culture industry has now realised that there is no 'quick-fix' solution to WSD.

Since 1994, scientific work has yielded several new research findings, about WSD. Many of these new findings, if analysed and understood properly, can be of immense help in minimising the impact of WSD in shrimp culture.

The WSSV is transmitted by vertical and horizontal means. Histopathological and PCR evidences have demonstrated that mother shrimp caught from the wild are positive for WSSV. Recent evidence has also shown that WSSV is transmitted vertically from mother to larvae. In addition, horizontal transmission can take place from contaminated water, through cohabitation with infected shrimp, carrier organisms and feeding of infected shrimp.

Three-Stage Sequence of WSD Development:The sequence of WSD development in shrimp has been divided into three stages, namely latent transition and patent. Shrimps in latent stage are apparently healthy, active and do not have any white spots. These are histopathologically negative, and surprisingly, negative, even by 1-step PCR. However, they may be positive by 2-step PCR. Shrimp in the transition stage are active, may have very tiny white spots and there may not be any mortality. These are positive by 1-step PCR and may be positive or negative by histopathology. Shrimp in the patent stage are weak, refuse feed, have prominent white spots and test positive by I step PCR and histopathology. Shrimp in the patent stage tend to come to pond margin display characteristic clinical sign and die.

Latent stage infection in shrimp might persist for months, but progression to transition and patent stage can happen within a few hours under stressful conditions. The time taken for progression from latent to patent stage within a shrimp is dependent on several factors. Shrimp size and stress appear to be very important. It has been convincingly demonstrated that, a 2-step PCR negative latent shrimp can become 1-step PCR positive in less than 24 h under conditions of acute stress. The rapidity with which latent infection can become patent under stress, highlights the need for stress management. This is especially true in endemic areas, where most of the cultured shrimp would be having the latent type of infection. Shrimp in a culture pond would have become latently infected either by vertical or horizontal means. Through better and rational stress management it should be possible to prevent latent stage turning to patent stage leading to mortality.

Basis for WSD Outbreaks: Significant mortalities due to WSSV infection taking place in a short period of time can be regarded as an outbreak. WSSV has a tropism for all cells of ectodermal and mesodermal origin and does not infect the cells of endodermal origin (midgut and midgut gland). Since, the endodermal origin tissues are negative for WSSV, shedding of virus through the feces is highly unlikely. Looking at the tissue tropism of WSSV, it becomes very clear that a live infected shrimp in any stage (latent, transition or patent) may not contribute substantial load of the virus to the system for horizontal transmission. An outbreak is largely due to the death and disintegration of a few patent shrimp. The rapidity with which latent infection becomes patent and the rapidity with which patently infected shrimp die and contribute to the viral load for horizontal transmission (through cannibalism and water) determines the nature of outbreak.

Intervention strategies aimed at preventing outbreaks should target the transition and patent stage shrimps. Patently infected shrimp should be prevented from dying and disintegrating in the pond. Regular pond monitoring, removal of moribund and dead shrimp should become important. Selective removal of shrimp in transition and patent stages through formalin stressing has given encouraging results in some countries.

Do All WSSV Infected Shrimp Produce White Spots?: This question has been bothering shrimp farmers for sometime. The answer is no. Now it is very clear that WSSV infection can produce three types of outbreaks with different clinical manifestations in an endemic area. In type I outbreak, mortalities take place within 2-3 days, major portion of the affected population display massive pinkish coloration and have a very high tissue severity index. There may not be any recognisable white spots. In type II outbreak, mortalities take place within 3-10 days with white spot as the dominant clinical sign and the tissue level severity is medium to high. In type III outbreak, major portion of the population appears healthy and the severity is low to medium. Mortality is slow and gradual.

These findings have far-reaching management implications. As of now, most of the farmers rely on white sports for making decisions on emergency harvests and pond disinfection. Massive pinkish colouration in shrimp in an endemic area may be regarded as bacterial red disease and farmers may resort to antibacterial therapy without any success. In addition, farmers may not disinfect the water before releasing it to the environment. Similarly, low level mortalities taking place in an endemic area, without white spots (type III outbreak), may be neglected and this may progress to become an type I outbreak under conditions of stress.

Types of White Spots: All white spots are not caused due to WSSV infection. Three different types of white spots have been recognised. In Type 1 white spots, the shrimp refuse feed, come to pond margins, and die in large numbers. These shrimp test positive to WSSV infection by PCR and other WSSV specific methods. Nothing can be done to save the shrimp at this stage, for, by this time, most of the shrimp would have picked up the infection by feeding on the moribund and dead infected shrimp. In addition, the water will have a high viral load as a result of disintegration of WSSV infected shrimp. The only thing that can be done here is emergency harvest and pond disinfection.

In Type 2 white spots, shrimps are active, feed normally and there is no mortality. These are negative for WSSV by PCR. This is thought to be due to prolonged exposure to high pH in water. This will cause calcium deposits on the shrimp shell giving the white spot appearance. Maintaining the pH between 7.5 and 8 should overcome this problem. The white spots will disappear following molting. In this case, there is no need to panic. The shrimps will grow normally and reach harvestable size. There is no need for any prophylactic or therapeutic intervention. Such an intervention will only stress the shrimp and may lead to other complications. Assessing the efficacies of different types of prophylactic and therapeutic measures in this type of white spot scenario will give misleading inferences.

In type 3 white spots, shrimps would be active, feed consumption may be slightly lower, gills will be dark brown or dirty. There will of course be white spots. The level of mortality may be even low and are negative for WSSV by PCR. However, these shrimp often show bacterial infection in a number of organs. Removal of unhealthy shrimp and improving pond water quality conditions will help the shrimps to recover. Intervention strategies aimed at managing bacterial diseases might also show promising results. Recently, reports of a new type of bacterial white spot syndrome (BWSS) associated with increased use of probiotics have been reported from some countries.

Integrated Risk Reduction Approaches for Management of White Spot Disease: WSSV is a generalist virus with wide host range including crustaceans and non crustaceans. The virus can easily spread beyond shrimp farms and farm environment. Hence virus transmission is complex and its containment in farms and natural environment is challenging.It is well known that it is impossible to eradicate certain pathogens that have become endemic and established in natural populations. Integrated disease control and risk reduction programs have the objective of managing health problems in a practical way at the farm level. Such programs aim to reduce health related losses to acceptable levels, restore confidence and assist the industry to keep going. These programs integrate broadly targeted disease control measures, using basic principles of pathogen elimination, pathogen exclusion and regular health monitoring using available knowledge of particular diseases. The option for interventions to reduce risk should be examined at each stage of the culture operation such as brood and seed selection, pre-stocking and growout management.

Selection of Brood: Viral infection status of spawners is considered very important in integrated risk reduction programs. Reliable identification of infected brood stock and their elimination is the ideal approach. Such a strategy would eliminate the need for PCR screening of PL. Practical considerations are plenty. Strategies should be put in place to reduce vertical and false vertical transmission of WSSV. Egg washings with disinfectants may help to substantially reduce viral infection in seed resulting from false vertical transmissions. It is important to consider the option of buying PL coming from batches where egg washing regimes have been rigidly followed.

Wild caught brood stock with high viral load (1 step PCR +ve) normally fail to spawn and should be rejected. It is well known that spawning stress induces viral replication. Therefore, brood stock screening should be done following spawning. Progeny of brood stock testing positive by 1-step PCR after spawning should be rejected. On the other hand, the progeny of brood which test positive or negative by nested PCR following spawning can be used.

Selection of Seed: Good quality seed is normally sought after by all shrimp farmers. PCR screening for WSSV is a very good strategy to minimise the entry of virus to the pond through seed. However, there are several practical problems with PCR screening of PL. The validity, sensitivity and reliability of the PCR results needs to be properly addressed. Sampling error has been recognised as a serious issue. A batch of PL usually derived from a single hatchery tank, may number up to 500,000 and represent the progeny of several female spawners. Typically, only a relatively small number of these PL will be included in the sample to be tested for PCR. Consequently there is high probability of sampling error and of consequent false negative results. Choosing appropriate number of PL for screening a batch should be based on statistical protocol. The protocol recommends to screen appropriate sample size depending on the suspected rate of infection say 5,10 or 15 % and hatchery lot size in 1000, lakh and crores. This is important and often not followed in hatcheries either due to cost and time, leading to error. Furthermore, there are no programs in place to destroy the infected PL at the hatchery and hence, invariably, infected PL enter culture systems. For PL screening to work as an elimination strategy, there is need for checks and counter checks. Lack of consistent associations with outcomes (disease/no disease) and the PCR status of PL (negative/positive) has added further problems to this seed screening strategy. PL batches with high load of virus (1 step PCR +ve) and high prevalence should not be used for stocking.

Exposure of PL to 100 ppm formalin for up to 30-40 min prior to stocking is believed to eliminate weak PL. Removal of all the weak PL at the end of the treatment and stocking of only the strong PL ensures higher survival. Survival rates of such PL in the pond has been shown to exceed 90%. It has been shown by PCR that WSSV infection is significantly lower in PL remaining after pre-stocking formalin exposure treatment compared with prevalence in the batch before treatment and in the removed weak PL. Pond trails have supported this hypothesis.

Pre Stocking Strategies: Interventions should be targeted at pathogen entry points into the pond to minimise the risks. Pond drying and liming are essential components of pond preparation. Treatment of water with 30 ppm of 60% calcium hypochlorite for 3 days prior to use in ponds has been widely recommended. Holding the water in the culture pond for up to 5-7 days without any animal is another option. Using fine mesh at pumping points (40-100 um mesh size bolting silk) would prevent the entry of potential virus carriers. Keeping the crabs out by crab fencing is also an option worth considering.

Management During Grow Out: Knowing the PL survival during the first month post stocking is very essential for implementation of integrated health management. There is a need to accurately estimate the survival up to one month. Later survival figures can be obtained from feed tray estimates. If low survival can be determined within the first two weeks of stocking, farmers may be able to restock the pond quickly, thereby avoiding subsequent problems with size differences in the population.

Using nursery before stocking is being seriously considered and being implemented successfully in several countries. Nursery area with in the grow out pond is not ideal. Separate nursery ponds for nursing the PL for 15-30 days is considered ideal. Transfer of survivors from nursery to grow out ponds has given good results in some countries.

Monitoring and decision making is very important in integrated health management. Regular monitoring and looking for predictors like size differences, clinical signs, nature and cause of mortality, disease history of the area, etc. are very useful.

Interventions for preventing outbreaks should aim at controlling events in the pond. Closed and partially closed systems are at less risk. Reduced water exchange in the beginning and avoiding exchange during periods of outbreaks in the area is strongly recommended. Stress management to prevent latent carriers from becoming patent (clinically sick) animals is important but not always possible. Preventing patent or sick animals from dying and getting eaten by other shrimp is the next important step. Healthy shrimp feeding on sick shrimp could trigger outbreaks in the pond. Hand picking of dead and moribund shrimp and destroying them is an useful approach. Using fishes like seabass to remove the weak and dead shrimp is also being considered as an option.

Using formalin in ponds to selectively stress and remove weak ones from the ponds is also followed in some countries.

In summary, by exercising proper intervention strategies at every step in the culture operation, it should be possible to take successful harvests in endemic areas. Selection of PCR negative brood and seed will go a long way in minimising the impact of the disease. Measures aimed to minimise viral load in the water should concentrate on reservoir concept, closed system, and elimination of carriers. Stress management to prevent latent infection turning to patent infection in a shrimp is of crucial importance. Most important of all, is to remove patently infected shrimp before they become moribund and die.

There are no known treatments for shrimp infected with WSV, however, a number of preventative measures are recommended to reduce spread. Screening of brood for WSD by nested PCR, screening of large number of PL before stocking by nested PCR. Further during cultivation, avoiding rapid changes in water temperature, hardness and salinity, or reduced oxygen levels (<2 ppm) for extended periods, to prevent WSD outbreaks. Avoiding fresh or fresh-frozen feeds of aquatic animal origin in the grow out ponds, maturation units and hatchery facilities is a good measure. Disinfecting infected ponds with 30 ppm chlorine to kill the infected shrimp and any potential carriers. avoiding equipment, clothes etc from infected farms also helps preventing the virus.

2. Yellowhead Virus (YHV)

Yellowhead disease of *P. monodon* is caused by YHV, a rod shaped virus measuring 44x173 nm. Lymphoid organ virus (LOV) and gill associated virus (GAV) of *P monodon* in Australia are related to the YHV complex viruses, although, of the two, only GAV is known to cause mortality. Characteristic clinical signs are typically displayed by tiger shrimp. Natural infections occur in *P monodon*, but experimental infections have been shown in *P. japonicus*, *P. vannamei*, *P. setiferus*, *P. aztecus*, *P. duorarum and P. stylirostris* and *Penaeus merguiensis*. These shrimp appear to be resistant to disease but not necessarily to infection. *Palaemon styliferus* has been shown to be a carrier of viable virus. *Euphausia* spp. (krill), *Acetes* spp. and other small shrimp are also reported to carry YHV.

Mortality occur within 2 to 4 days of occurrence of gross signs of disease following an interval of exceptionally high feeding activity that ends in abrupt cessation of feeding. Mortalities can reach 100% within 3-5 days. Diseased shrimp aggregate at the edges of the ponds or near the surface. The hepatopancreas becomes discoloured which gives the cephalothorax a yellowish appearance, hence the name of the disease. The overall appearance of the shrimp is abnormally pale. Post-larvae (PL) at 20-25 days and older shrimp appear particularly susceptible, while PL<15 appear resistant.The virus can cause mortalities in the absence of the classic yellowish appearance of the cephalothorax. Clinical signs are not always present. Therefore, confirmatory diagnosis including a minimum of whole, stained gill mounts and haemolymph smears should be carried outto find out YHV involvement in any cases of rapid unexplained mortality In histopathology, YHD virions are found generally in tissues of ectodermal and mesodermal embryonic origin, including: interstitial tissues of the hepatopancreas, systemic blood cells and developing blood cells in the haematopoietic tissues and fixed phagocytes in the heart, the lymphoid (Oka) organ, gill epithelial and pillar cells, connective and spongiform tissues, sub-cuticular epidermis, striated and cardiac muscles, ovary capsules, nervous tissue, neurosecretary and ganglial cells, stomach, mid-gut and midgut caecal walls. The epithelial cells of hepatopancreatic tubules, mid gut and midgut caeca (endodermal origin) are not infected with YHV although underlying muscle and connective tissues are. The Oka organ, gill, heart and sub-cuticular tissues, including those of the stomach epithelium, contain the highest levels of YHV. Infected cells show nuclear pyknosis and karyorrhexis which are apparently signs of viral triggered apoptosis

Only juvenile to subadults (50-70 days in grow out) are affected. Initially feeding rate increases abruptly which then decreases totally and within a day, a few moribund shrimp appear near margin of pond. The animals have a light yellow cephalothorax. Within the second day, the number of similar affected shrimp increases dramatically, and by third day the shrimp stock in entire pond is lost. Affected shrimp show generalised body pallor, with a yellowish often swollen cephalothorax, frequently whitish or pale yellowish to brown gills and often a pale yellow hepatopancreas. The YHD has been confirmed only in Thailand and Taiwan, even though reports on YHD like diseases have been recorded from Indonesia, Malaysia, China, Philippines, Texas and India. The collapse of the Taiwanese shrimp culture industry in 1980's is believed to be primarily due to yellowhead syndrome.

The disease is diagnosed by hemocyte staining, histopathology, gene probes,Reverse Transcriptase-Polymerase Chain Reaction Assay and TEM.

In histopathology generalised multifocal to diffuse severe necrosis, with prominent nuclear pyknosis and karyorrhexis are characteristic features. Basophilic usually spherical, perinuclear cytoplasmic inclusion occur in affected cells especially in hemocytes, lymphoid organ, hematopoitic tissue, gill pillar and epithelial cells in the lamellae. The simplest bioassay method is to allow shrimp(+10 g wet weight) to feed on carcasses of suspect shrimp or alternatively exposing homogenates of gill tissues from suspect shrimp. Infected shrimp show clinical signs within 24-72 hours and 100% mortality within 3-5 days. Infections should be confirmed by histology of gills and haemolymph.

YHV infections are generally believed to be horizontally transmitted. Survivors of YHD infection, however, maintain chronic sub-clinical infections and vertical transmission is suspected. There are a number of known or suspected carrier crustaceans including the brackish water shrimp, *Pelaemon styliferus* and *Acetes* sp., which can potentially transmit YHD to farmed shrimp. There are no known treatments for shrimp infected with YHV. However, a number of preventative measures such as screening broodstock for YHV, destruction of infected individuals and their offspring in a sanitary manner, disinfection of associated equipment and rearing water, using pre-stocking PL screens to exclude YHD carriers, avoidance of rapid changes in pH or pro-longed periods of low (<2ppm) dissolved oxygen which can trigger sub-lethal out breaks of YHD. pH should not vary more than 0.5 pH units daily and avoiding water pH levels>9, avoiding fresh aquatic feeds are recommended to reduce spread.

3. Taura Syndrome Virus (TSV)

TSV is a picornavirus (SS RNA) measuring 30-32 nm replicating in cytoplasm. The virus causes `red-tail' disease or Taura syndrome in *Penaeus vannamei*, with 80-95% mortality. A number of penaeid species are suceptible, the most susceptible species is the Pacific white shrimp *Penaeus vannamei*, although

P. stylirostris, and *P. setiferus* can also be infected. Post-larvae and juvenile *P. schmitti*, *P. aztecus*, *P. duorarum*, *P. chinensis*, *P. monodon*, and *Marsupenaeus (Penaeus) japonicus* could be infected experimentally.

The virus distribution though originally limited to the Americas, has now extended to other countries in Asia such as Taiwan, China through seed and brood of *P. vannamei*. TSV infection causes disease in shrimp between 14-40 days of stocking PL (0.05 to 0.59) in grow outs. Larger shrimp may also get infected. The disease has peracute and recovery (chronic) phase which are grossly indistinguishable. In per acute phase, moribund shrimps exhibit expansion of red chromatophores, overall reddish coloration with distinct red color in tail fan and pleopod (hence the name 'red tail' disease). Usually focal epithelial necrosis is noticed in the thin appendages (uropods and pleopods). Soft shell and empty guts are other clinical signs. Shrimp die during ecdysis suggesting molting is important in pathogenesis of Taura syndrome. In chronic phase, fair to moderate number of shrimp show multifocal melanised cuticular lesions. These are the survivors of acute infection, which usually look normal. Survivors of acute TSV infection pass through a brief transition phase to enter the chronic phase which may persist for the rest of their lives. This sub-clinical phase of infection is believed to have contributed to the spread of the disease via carriage of viable TSV.

In histology pyknotic and karyorrhectic nuclei and the generally spherical cytoplasmic inclusions give a "peppered or buckshot-riddled" appearance. Chronic stages of infection are characterised by the presence of spherical accumulations of cells in the lymphoid organ, referred to as 'lymphoid organ spheroids' (LOS). These masses are composed of presumed phagocytic hemocytes, which have sequestered TSV and aggregate within intertubular spaces of the lymphoid organs.

The disease is diagnosed by histopathology, immunoassays, Rt-PCR, gene probes- *in situ* hybridisation, TEM and bioassay studies. In bioassay, specific Pathogen Free (SPF) juvenile *Penaeus vannamei* can be used to test suspect TSV-infected shrimp by exposing i. chopped pieces ii. inoculation with homogenate iii hemolymph samples. If the suspect shrimp were positive for TSV, gross signs and histopathological lesions should become evident within 3-4 days of initial exposure. Significant mortalities usually occur by 3-8 days post-exposure. The control shrimp should stay healthy and show no gross or histological signs of TS.

Shrimps that have survived the acute and transitional phases of TS can maintain chronic sub-clinical infections within the lymphoid organ, for the remainder of their lives. These shrimp may transmit the virus horizontally to other susceptible shrimp. Vertical transmission is suspected, but this has yet to be conclusively demonstrated. In addition to movement of sub-clinical carriers of TSV, aquatic insects and sea birds have been implicated in transmission of the disease. Viable TSV has also been found in frozen shrimp products. The virus is managed in farms by avoidance Employing seed from wild brood has shown promising results. Another management strategy has been doubling post-larval stocking densities in semi-intensive pond culture to compensate heavy losses due to TS early in the production cycle , survivors are TS tolerant. Development of TSV resistant stocks of *P. vannamei* and *P. stylirostris* show a 20-40% improvement in survival.

4. Monodon Slow Growth Syndrome (MSGS)

Monodon slow growth syndrome (MSGS) is a major concern in South and Southeast Asia with significant economic loss. MSGS is characterized by abnormally slow growth with coefficient of size variation of >35 %. A new virus, Laem-Singh virus (LSNV) first detected in Laem-Singh district of Thailand was identified to be associated with MSGS affected shrimp. LSNV a RNA virus of about 25 nm diameter is related to the insect-borne viruses in the families Barnaviridae, Tymoviridae and Sobemoviridae. LSNV is considered a necessary but insufficient cause of MSGS, requiring yet unknown pathogen or environmental factors—because it is found in both healthy and growth-retarded shrimp. A key component is the positive detection of Laem-Singh virus (LSNV) by RT-PCR, considered a necessary but insufficient component cause for MSGS. Known pathogens are unlikely to be the cause of MSGS and previous trials have indicated that a filterable infectious agent is involved. *P monodon* appears to be primary host while other penaeids may be susceptible.

Abnormally slow growth, resulting in irregularly sized shrimps due to low average daily weight gain of less than 0.1 g/day at 4 months of age. There are no definitive microscopic pathological signs. The complete aetiology for MSGS is uncertain and there isn't a clear case definition for this syndrome. The LSNV infections have been confirmed in various organs of infected shrimp such as lymphoid organ, gills and nervous tissues by various diagnostics-*in situ* hybridization, RT-PCR, quantitative real-time RT-PCR and reverse transcription loop-mediated isothermal amplification combined with a lateral flow dipstick (RT-LAMP-LFD).

The transmission of LSNV through horizontal and vertical routes has been experimentally demonstrated. Despite the fact that the actual etiology of MSGS is unknown, it is recommended that farmers avoid stocking ponds with LSNV-infected post larvae. Further, *L. vannamei* and *P. monodon* should be reared separately, particularly at the maturation and hatchery phases and LSNV should be added to the list of pathogens for exclusion.

5. Loose Shell Syndrome(LSS)

Loose shell syndrome first reported during 1998 is one of the economically important chronic diseases of shrimp in Asia. LSS is a slowly progressive disease of penaeid shrimps which is characterized by a flaccid spongy abdomen from muscular dystrophy and the development of a loose exo-skeletal covering over the abdominal musculature. Loose shell syndrome causes low-level mortality. The affected shrimps can be identified with paper-like carapace, soft muscle with a gap between muscle and exoskeleton and pigmented hepatopancreas.

Muscle and gills are usually damaged and fouling in the gill leads to respiratory problems. Behavioral changes include off-feeding, loss of weight, reduction in daily growth rate, lethargy, impaired molting with progressive mortality. Usually small shrimp of less than 20 g are affected within about 50 days of culture. Shrimp populations have low-level, progressive mortality which go unnoticed initially. Progression of LSS is gradual, leading to low-level progressive mortalities.

LSS manifestation suspected due to several conditions such as poor water quality, nutritional deficiency, low-quality feed, poor pond bottom condition and exposure to certain pesticides. LSS can also be caused by fungi, bacteria, protozoa in which deposition of the suspended particles on the shrimp appendages and exoskeleton makes them more vulnerable. LSS could be induced in healthy tiger shrimp by challenge studies using membrane-filtered LSS-affected shrimp tissues, suggesting involvement of a filterable infectious age

The clinical signs and gross pathology of LSS closely resemble those of Necrotizing Hepatopancreatitis (NHP) reported in Pacific white shrimp (*Litopenaeus vannamei)* from the Americas. Marked general atrophy of the hepatopancreas tubules, intertubular edema, nodulation, and low levels of lipid storage in hepatopancreas cells were observed in LSS-affected shrimp. Histological examination of the lymphoid organs of these animals revealed lumen enlargement, diffused necrosis with reduction in the number of stromal matrix cells and separation of tubules. In addition, histopathological investigations showed shrinkage of extensor and flexor muscles with occasional hemocytic infiltration. The hepatopancreas showed inflammation of hepatopancreatic tubules with enlargement of intertubular spaces, hemocytic infiltration, and low levels of lipid reserves in the R cells. In advanced stages of LSS, many tubules were in highly necrotic condition with a sloughed epithelium, reflecting the dysfunction of the digestive gland. Use of good quality feed, low stocking density, maintaining good water quality by frequent water exchange helps to reduce the occurrence of LSS.

6. Gill Associated Virus (GAV)

GAV has only been recorded from Queensland , Australia and is endemic to *P. monodon* in this region. GAV is a single-stranded RNA virus of the family *Coronaviridae* closely related to yellow head virus and is regarded as a member of the yellow head complex. GAV can occur in healthy or diseased shrimp and was previously called lymphoid organ virus (LOV) when observed in healthy shrimp.Natural infection with GAV has only been reported in *Penaeus monodon* but experimental infection has caused mortalities in *P. esculentus*, *P merguiensis* and *P. japonicus*. An age or size related resistance to disease was observed in *P. Japonicus*. There are no reports of GAV from other countries

Shrimp with an acute GAV infection demonstrate lethargy, lack of appetite and swim on the surface or around the edge of ponds. The body may develop a dark red colour particularly on the appendages, tail fan and mouth parts; gills tend to be yellow to pink in colour. Fouling is common usually. However, the gross signs of acute GAV infection are variable and hence are not reliable, even for preliminary diagnosis. GAV is predominantly found in the gill and lymphoid organ but has also been observed in haemocytes. During acute infections, there is a rapid loss of haemocytes, the lymphoid organs appear disorganised and devoid of normal tubule structure and the virus can be detected in the connective tissues of all major organs.

The virus is detected by histopathology and Electron microscopy. In histopathology, lymphoid organs from diseased shrimp display loss of the normal tubule structure, where tubule structure is disrupted, with no obvious cellular or nuclear hypertrophy, pyknotic nuclei or vacuolization. Foci of abnormal cells are observed within the lymphoid organ and these may be darkly eosinophilic. The gills of diseased shrimp display structural damage including fusion of gill filament tips, general necrosis and loss of cuticle from primary and secondary lamellae. The cytology of the gills appears normal apart from small basophilic foci of necrotic cells. In Electron Microscopy,the cytoplasm of lymphoid organ cells contains both rod shaped enveloped virus particles (183-200 *nm* long and 34-42 *nm* wide) and viral nucleocapsids(166-435 *nm* in length and 16-18 *nm* in width). Enveloped virions are less common, occurring in about 20% of cells within the disrupted areas of the lymphoid organs and gills. Sensitive RT PCR has been developed

The most effective form of horizontal transmission of GAV is direct cannibalism but transmission can also be water-borne. GAV is also transmitted vertically from either or both parents. As GAV is endemic it is unclear whether the onset of disease results from environmental stress or whether the disease arises from a new infection with a pathogenic strain of GAV

7. Infectious Hypodermal Hematopoietic Necrosis Virus (IHHNV)

The infectious hypodermal and hematopoietic necrosis (IHHN) is caused by IHHNV, a small (22 nm), ss DNA parvovirus. The virus causes acute disease with very high mortalities in juveniles after approximately PL 35 or older stages.

Infected adults seldom show signs of disease or mortalities. Reduction in food consumption, followed by changes in behaviour and appearance are the early clinical signs. Affected shrimps rise slowly to water surface, become motionless, then roll-over slowly to sink to bottom. Later, they become weak unable to move. At this stage individuals often have white or buff-coloured spots in the cuticular epidermis. In some species, such as *P. vannamei*, the virus causes a chronic disease called Runt deformity syndrome (RDS). Juveniles with RDS display bent or deformed rostrum, wrinkled antennal flagella and cuticular deformities. Population of juvenile shrimp with RDS display a relatively wide distribution of sizes with many smaller than expected ("runted") shrimp. The virus infects all penaeid species, adults surviving infection become life long carriers. Virus is presumed to be enzootic in wild penaeids. The virus is widely distributed in America, Asia and Australia.

8. Baculoviral Midgut Gland Necrosis Virus (BMNV)

BMNV is a type C, non-occluded gut infecting baculovirus measuring 72x310 nm. The virus cause Baculoviral midgut gland necrosis (BMN), a serious epizootic in hatchery reared penaeids. It is characterized by sudden onset and high mortality rate. The disease first appear in protozoea and is serious up to PL-9 to PL-10, by which time cumulative mortality reaches up to 100%. The disease decreases by PL-20. A "white turbid" hepatopancreas (midgut gland) in larvae and post larvae is the first grossly visible sign of the disease, which becomes more clear as disease advances. Severely affected PL are easily distinguished, floating inactively on surface exhibiting a white midgut line through abdomen. All cultivable penaeids are susceptible. The virus has been reported from wild and cultured *Penaeus* sp in Japan, Korea, Australia, Indonesia and Philippines.

Infection with BMNV is diagnosed by clinical signs. Morbid or larvae heavily infected with BMNV shows a cloudy midgut gland, easily observable by the naked eye.Hypertrophied hepatopancreas (HP) cell nuclei in Giemsa stained smears and in histopathology,fluorescent antibody method and darkfield microscopy of wet mount of HP are typical of BMNV infection. Infection confirmed by Transmission Electron Microscopy (TEM)through demonstration of the rod-shaped enveloped virions.The oral route has been demonstrated to be the main infection pathway for BMNV infection. Viruses released with feces into the environmental water of intensive culture systems play an important role in spread of viruses and disease. Complete or partial eradication of viral infection may be accomplished by thorough washing of fertile eggs or nauplii using clean sea water to remove the adhering excreta. Disinfection of the culture facility and the avoidance of re-introduction of the virus are critical factors to control BMN disease.

9. Monodon Baculovirus (MBV)

MBV is a type A occluded baculovirus (42x246 nm) having ds DNA. MBV infections are diagnosed by the presence of a single or multiple, spherical, eosinophilic occlusion bodies in the nucleus of HP and midgut epithelial cells. Moderate to heavy infection of HP and anterior midgut of all life stages reported except nauplius and protozoea stage 1 and 2. Serious mortalities in larval, post larval and juvenile stages of *P. monodon* have been linked to MBV infection. MBV infection have also been found in healthy adult shrimps. Heavy infection with MBV leads to reduced feeding, decreased growth rate and increased surface and gill fouling which are the main clinical signs. The virus is widespread all over the world. MBV is enzootic in wild penaeid stocks. MBV infection is diagnosed by wet mount, histology, gene probes and EM.MBV infections appear as single or multiple spherical or sub-spherical inclusion bodies within enlarged nuclei of hepatopancreas or midgut epithelia. MBV occlusion bodies measure 0.1-20.0 μm in diameter. The occlusion bodies can be stained using a 0.05% aqueous solution of malachite green, which stains them more densely than the surrounding, similarly sized spherical bodies

MBV is transmitted orally via uptake of virus shed with the feces of infected shrimp, or cannibalism on dead and dying shrimp. Infected adults have also been shown to infect their offspring via fecal contamination of the spawned egg masses.Overcrowding, chemical and environmentally induced stress, have all been shown to increase the virulence of MBV infections in susceptible shrimp species under culture conditions. Exposure of stocks to infection can be avoided by pre-screening the potential broodstock and selecting adults shown to be free of fecal contamination by occlusion bodies of either baculovirus. Prevention of infections may also be achieved by surface disinfection of nauplii or fertilised eggs with formalin, lodophore and filtered clean seawater.Eradication of clinical outbreaks of MBV may be possible in certain aquaculture situations by removal and sterile disposal of infected stocks, disinfection of the culture facility, avoidance of re-introduction of the virus from other nearby culture facilities and wild shrimp.

10. Baculovirus Penaei (BP)

Baculovirus penaei (BP) is a type A baculovirus measuring 56x286 nm. BP causes serious epizootics in larval, post larval and juvenile stages of *Penaeus* sp, particularly in hatcheries with high mortality rates. Infection starts from protozoea stage 2 and becomes severe in mysis with cumulative mortality reaching more than 90%. Affected larvae exhibit a white midgut line visible through the abdomen. Reduced feeding, growth rate and increased fouling are the gross clinical signs. Infection is common under high density rearing in earthen ponds.BP is found throughout the Americas with multiple strains within this geographic range. The infection is diagnosed by the presence of tetrahedral intranuclear occlusion bodies in HP and midgut epithelial cells. Diagnosis is by wet mount, histology, gene and antibody probes and TEM.

BP is transmitted orally via uptake of virus shed with the feces of infected shrimp, or cannibalism on dead and dying shrimp. Infected adults have also been shown to infect their offspring via fecal contamination of the spawned egg masses.Overcrowding, chemical and environmentally induced stress, have all been shown to increase the virulence of BP infections in susceptible shrimp species under culture conditions.Exposure of stocks to infection can be avoided by pre-screening of potential broodstock and selecting adults shown to be free of fecal contamination by occlusion bodies of either baculovirus. Prevention of infections may also be achieved by surface disinfection of nauplii larvae or fertilised eggs with formalin, lodophore and filtered clean seawater.Eradication of clinical outbreaks of BP and MBV may be possible in certain aquaculture situations by removal and sterile disposal of infected stocks, disinfection of the culture facility, the avoidance of re-introduction of the virus (from other nearby culture facilities and wild shrimp.

11. Hepatopancreatic Parvo-like Virus (HPV)

HPV, causing disease and mortality in hatcheries, is a DNA virus which is distributed all over the world. It causes low scale mortality with secondary infection. HPV infection produces prominent basophilic intranuclear inclusions in the epithelial cells of HP and midgut. HPV infection is often associated with growth retardation in cultured shrimp. Currently, HPV is diagnosed by fresh smears, histopathology and gene probes.

12. Spawner Isolated Mortality Virus (SMV)

Spawner-isolated Mortality Virus (SMV) is associated with Spawner Mortality Syndrome (SMS) and Midcrop Mortality Syndrome (MCMS). It is one of the several viruses associated with mid-crop mortality syndrome (MCMS) which resulted in significant mortalities of juvenile and sub-adult *P. monodon*. The virus is a single-stranded icosahedral DNA virus measuring 20-25 nm most closely associated with the family Parvoviridae. The virus has been reported from Australia, Philippines and Sri Lanka.

There are no specific clinical signs known for SMV infection. SMV infect *Penaeus monodon*. *P. esculentus*, *P. japonicus*, *P. merguiensis* and *Metapenaeus ensis* with mortalities. In farmed freshwater crayfish (*Cherax quadricarinatus*) putative SMV infection demonstrated in moribund samples using DNA-probe. Cannibalism assumed to facilitate horizontal transmission

Histopathology associated with SMVD is not disease specific. In naturally infected juvenile *P. monodon*, haemocyte infiltration and cytolysis is focused around the enteric epithelial surfaces. In experimental infections, systemic infections is manifested by haemocytic infiltration, necrosis and sloughing of epithelial cells of the midgut and hepatopancreas. In Electron microscopythe icosahedral virions (20-25 *nm* in diameter) are found in the gut epithelial tissues.

Prevention of introduction of shrimp from SMV infected stock into historically uninfected areas is recommended. Daily removal of moribund animals from ponds, particularly early in production, has also been recommended. Stocking of ponds with progeny of spawners with SMV-negative fecal testing using PCR-probes has been shown to reduce mortality by 23%.

13. Covert Mortality Nodavirus (CMNV)

CMNV the cause of covert mortality disease of shrimp first recorded in china, has caused serious loss.The disease causes economic losses in hatcheries and farms with high mortality rates of up to 80% commonly found within 60–80 days post-stocking. CMNV is a new virus of the family Nodaviridae, genus Alphanodavirus. CMNV is different from other nodaviruses such as infectious myo-necrosis virus (IMNV), *Macrobrachium rosenbergii* nodavirus (MrNV) and *Penaeus vannamei* nodavirus (PvNV)

CMNV has a wide host range among cultured shrimp species, with a high prevalence and wide distribution in Southeast Asia and Latin American countries. CMNV was found in eleven species of invertebrates collected from ponds of cultured shrimp species, which may be vectors and reservoirs of the virus. CMNV has also naturally crossed the species barrier and infected several species of fish such as *Mugilo gobiusabei*, a common marine fish in shrimp farming ponds and coastal water in China and, farmed Japanese flounder (*Paralichthys olivaceus*). Shrimp infected with CMNV are commonly found in deep water on the bottom of the shrimp pond rather than swimming on the surface. CMNV cause hepatopancreatic atrophy and necrosis.

14. Shrimp Hemocyte Iridescent Virus (SHIV)

SHIV results in a high mortality in white leg shrimp (*Litopenaeus vannamei*). A new virus Xiaviridovirus of the family Iridoviridae associated with mortality has been recorded in China. SHIV besides *L. vannamei*, also detected in *Fenneropenaeus chinensis* and *Macrobrachium rosenbergii*.

15. Bamboo Back Disease in Tiger Shrimp, *Penaeus monodon*

Bamboo back disease affecting *Penaeus monodon* is reported from the Philippines. The cuticle of the abdominal segments of shrimp with bamboo back disease do not overlap properly which gives them a bamboo-like appearance. The appendages are shorter compared with normal shrimps. No bacteria were recovered from the hepatopancreas, lymphoid organ and hemolymph thus ruling out bacterial infection. Histopathology shows normal hepatopancreas, but the muscle fibers of the abdominal segments are fragmented and necrotic

B. Viral Disease of gaint fresh water prawn *Macrobrachiumrosenbergii*

In general very few diseases in *Macrobrachium rosenbergii* due to viruses have been reported. At the present a noda virus infecting and causing disease in *M. rosenbergii*is of considerable importance.

Macrobrachium rosenbergii nodavirus (*Mr*NV)

The giant freshwater prawn *M. rosenbergii* is widely cultivated in tropical and subtropical regions. The production of the prawn slumped significantly from more than 220,000 tons due to *Mr* NV during 2014. The first *Mr*NV outbreak was reported in Pointe Noire, Guadeloupe in 1997, followed by China, India Taiwan Thailand, Malaysia, Australia, and Indonesia. *Mr*NV, the causative agent of white-tail disease has severely affected production of the freshwater prawn *Macrobrachium rosenbergii* world wide. The virus infecting larvae and post larvae of the prawn in hatcheries often result in mortality up to 100%, leading to devastating economic losses. Experimentally transmitted to healthy animals, *Mr*NV caused 100% mortality in post-larvae but failed to cause mortality in adult prawns.

*Mr*NV has been classified within the Nodaviridae family of viruses. *Mr*NV is a small icosahedral non-enveloped RNA virus, 26–27 nm in diameter, replicating in the cytoplasm of connective tissue cells. One more virus named extra small virus (XSV), 15 nm has been reported associated with *Mr* NV with the white muscle disease in the giant freshwater prawn.Transmission electron microscopy of negative staining of diseased PL homogenates also showed the two types of viral particles, *Mr*NV and XSV. As XSV has always been found associated with *Mr*NVin muscle and connective tissue cells of diseased animals, it could be an autonomous virus, a helper-type virus or a satellite-like virus. This is also reflected inexperimental immersion challenge of PL with infectious clones of *Mr*NV and XSV where mortality accompanied by WTD lesions occurred with *Mr*NV alone or in combination with XSV but not with XSV alone, despite its replication. The RT-PCR assay revealed both viruses in moribund post-larvae in all the important organs tested except eyestalks or the hepatopancreas. The presence of these viruses in ovarian tissue indicates the possibility of vertical transmission. Pleopods have been found to be a suitable organ for detecting these viruses in brooders by the RT-PCR.Possibility of vertical transmission of *Mr*NV and XSV in *M. rosenbergii* and *Artemia* has been demonstrated. In oral or immersion challenged survivors without any clinical signs of WTD thc ovarian tissue and fertilized eggs were found to be positive for *Mr*NV and XSV. In *Artemia*, reproductive cysts and nauplii derived from challenged brooders were normal but found to be positive for *Mr*NV/XSV by RT-PCR.

A sensitive diagnosis is essential to detect presence of the virus in susceptible PL and asymptomatic adults. Three genome-based methods are available for the detection of *Mr*NV, i.e. dot-blot hybridization, *in situ* hybridization and reverse transcriptase-polymerase chain reaction (RT-PCR). *In situ* hybridization indicated that infection is confined to the striated muscle tissue. Sensitive RT-PCR can be used for in-depth investigations to examine the extent of the viral infection and establish the onset of infection in hatcheries. An one-step multiplex RT-PCR has also been developed to detect both *Mr* NV and XSV simultaneously in naturally and experimentally infected prawns. Infected prawns showed two prominent nucleic acid bands of 681 and 500 *Mr*NV and XSV respectively, in separate RT-PCR assays. Experimentally infected adult prawns showed two bands for these two viruses in all the organs, except hepatopancreas and eyestalk.

A rapid and sensitive detection of *Mr*NV and XSV, by loop-mediated isothermal amplification (LAMP) is also available. The LAMP reaction is suited for routine disease diagnosis as amplified DNA can be detected without the use of agarose gel electrophoresis, by the production of whitish precipitate of magnesium pyrophosphate as a by-product.

Appropriate biosecurity measures are required to prevent infection of susceptible post-larvae stage as adults *Macrobrachium rosenbergii* are immune to *Mr*NV infection. Currently, the most useful strategies for preventing an outbreak is physical biosecurity measures.Evidence of innate immune memory has been documented in several invertebrates including crustaceans, in which it confers long-term immune protection against specific pathogens independent of the adaptive immunity. The term "trained immunity" was coined to define the establishment of immune memory from the activated innate immune system. Hence there is hope and scope for exploring potential vaccine and immunostimulants for prophylaxis in culture of *M rosenbergii*

C. Emerging Viral Pathogens of Mollusks

Culture of molluscs- oysters, clams, mussels, abalon is gaining importance in the Asia pacific region. However, emerging viral diseases although in few numbers are posing serious problem.

Abalon herpesvirus/ Haliotid herpesvirus-1(HaHV-1)

Haliotid herpesvirus-1(HaHV-1) previously called abalone herpes virus causes Viral ganglioneuritis in abalon with extensive mortalities in both wild and farmed abalone *(Haliotisdiversicolor supertexta)* and significant economic losses in Asia and Australia since early 2000s.The viral disease is particularly serious in Taiwan and China with high mortality and loss. HaHV-1 is a spherical virus with an icosahedral core and an envelope of approximately 100 nm in diameter. Infection of abalone is characterized by acute mortality and necrotizing ganglioneuritis, therefore the disease is named abalone viral ganglioneuritis. In *H. supertexta*, lesions include necrosis of the cerebral ganglia and nerve bundles in the muscle of the foot as well as in the muscular layers beneath the visceral organs with associated hemocyte infiltrates. Experimental infection by immerse challenge has been standardised. Diagnostics such as histopathology and PCR have been developed.

References

Camilla Prester et al 2022. A review of viral diseases in cultured brachyuran crustaceans, Aquaculture International 31, 627-655

Dain Lee et al 2022. Viral shrimp diseases listed by OIE: A review, Viruses 14 (93)

Desrina et al 2022. White spot virus syndrome virus host range and impact on transmission. Aquaculture, 14 (94) p 1843-1860

F. Vega- villasante and M. E. Puete 1993, A review of viral diseases of cultured shrimp, Preventive Veterinary Medicine 17(3 and 4), p 271-282

K. M. Shankar and C.V. Mohan, 2002. Fish and Shellfish Health Management , ISBN 81-7525-328-2, Fish Pathology and Biotechnology Laboratory, Dept of Aquaculture, UAS, College of Fisheries, Mangalore-575002

8

Oomycete and Fungal Diseases in Finfishes

Pravata Kumar Pradhan and Neeraj Sood

Introduction

Oomycete and fungal diseases are considered as the second most economically impactful diseases in aquaculture, after bacterial diseases. However, among both the diseases, oomycetes are more common. The oomycetes are members of the Kingdom Chromista, which also includes diatoms, kelps, and brown algae. These group of pathogens form free-swimming zoospores and their cell walls are composed of cellulose and glycans rather than chitin. Fish-pathogenic oomycetes (*Saprolegnia*, *Achlya*, *Aphanomyces* and *Branchiomyces*), are commonly known as water moulds, are ubiquitously distributed in a variety of aquatic environments and ecological niches, and belong to the Order Saprolegniales. These pathogens have low host specificity and infect a diverse range of fishes. On the other hand, most of the fungi (*Fusarium* sp., *Aspergillus* sp., *Candida* sp., *Paecilomyces* sp., *Penicillium* sp. and *Cladosporium* sp.) responsible for infection in fish, form non-motile spores and their cell walls are made up of chitin. In addition, *Ichthyophonus* sp. an important pathogen which, superficially appears like fungus and produces spores and hyphae, belongs to a distinct clade of protistan parasites. Among the oomycete diseases, infection with *A. invadans* is considered as one of the most serious fish disease. Therefore, the disease is discussed elaborately.

Infection with *Aphanomyces invadans*: Infection with *Aphanomyces invadans* is popularly known as, Epizootic Ulcerative Syndrome (EUS). It has a complex infectious aetiology and is clinically characterised by the presence of invasive *Aphanomyces* infection and necrotising ulcerative lesions, typically leading to a granulomatous response (Plates 9 and 10). In an expert consultation held during the sidelines of the Fifth Symposium on Diseases in Asian Aquaculture (DAAV) at Gold Coast, Australia in November 2002, it was proposed to rename the disease as epizootic granulomatous aphanomycosis; however, the term EUS is still used popularly.

Geographical distribution: Infection with *A. invadans* was first reported in farmed freshwater ayu (*Plecoglossus altivelis*) in Japan in 1971. It was later reported in estuarine fish, particularly grey mullet (*Mugil cephalus*) in eastern Australia in 1972. The disease extended its range through Papua New Guinea into South-East and South Asia, and into West Asia, and reached India in 1988. Outbreaks of ulcerative disease in menhaden (*Brevoortia tyrannus*) in the United States of America (USA) had the same aetiological agent as the disease observed in Asia. The first confirmed outbreaks of infection with *A. invadans* on the African continent occurred in 2007 in Botswana, Namibia and Zambia, and were connected to the Zambezi-Chobe river system. During 2010 and 2011, infection with *A. invadans* was reported in wild freshwater fish in the Western Cape Province of South Africa, and in wild brown bullhead fish within Lake Ontario, situated in the Province of Ontario, Canada. Over the last 50 years, the disease has spread to the world's major continents, with the apparent exception of South and Central America. Considering the huge geographical spread and impacts, infection with *A. invadans* is listed by the World Organisation for Animal Health (WOAH), and is a disease of international concern.

Etiology: Since the disease occur in a wide geographical area and in diverse range of habitats, a diverse mix of microbial agents have been reported from the affected fish. So, importance of each agent in the disease outbreak is discussed in detail. When EUS first occurred in Southeast Asia, the explosive nature of the disease, the clinical signs and extent of losses led first to the conclusion that it was related to pollution and subsequently to the belief that it was due to the use of agricultural pesticides or other agro-chemicals. So, one of the first priorities for the 1985 FAO research team was to find relationship between pesticides or fertilizers and disease outbreaks. But the survey led to the conclusion that pesticide is an unlikely cause for occurrence of the disease. The spread of the disease in an explosive but gradually extending fashion strongly pointed to an infectious etiology. Therefore, the role of major group of pathogens in the disease development process are discussed below.

Open dermal ulcers in *Puntius* sp.

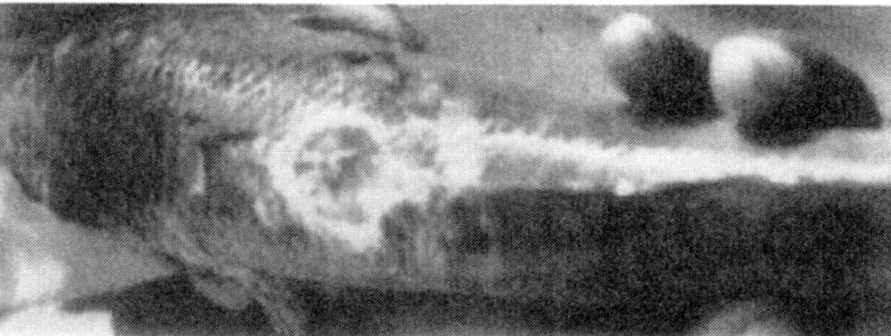

Dermal uclear in *Channa* sp.

Healing ulcer in *Channa* sp.

Plate 9: Gross Surface Lesions Associated with Epizootic Ulcerative Syndrome (EUS) (See colour version on page 375)

Oomycete: Since the initial outbreak in south-east Asia, oomycete has been known to be involved in the etiology of EUS. In 1985, FAO research team, in detailed histopathological studies have shown that EUS was a severe granulomatous condition characterised by presence of oomycete (Plate 11). Oomycete presence and mycotic granulomatosis conditions also have been consistently seen in EUS affected fishes, from different parts of the world. It is well established that the invading oomycete causes significant necrotic changes in the skin and muscle tissue and results in formation of dermal ulcers. It also causes ulcers in the contra lateral side and pathology in the internal organs. It has also been shown that the oomycete can penetrate the thick muscular wall of gizzard of mullets and spinal cord and bones of *Puntius* indicating its highly invasive nature. Prior to the initial occurrence of EUS in Southeast Asia, the pathogenic *Aphanomyces piscicida* had been isolated from fishes affected by mycotic granulomatosis (MG) in Japan. Subsequently, an *Aphanomyces* oomycete was also isolated from outbreaks of red spot disease (RSD) in Australia in 1989, as well as from EUS outbreaks in Thailand. These isolates were demonstrated to be capable of inducing typical EUS lesions when introduced beneath the dermis of susceptible fish. All of these pathogenic isolates associated with MG, RSD, and EUS were observed to have slow growth rates and were sensitive to changes in temperature in culture under *in-vitro* conditions. Comparison of pathogenic *Aphanomyces* isolates from different countries indicated that they are similar in terms of protein banding profiles, growth characteristics, and susceptibility to chemicals. Additionally DNA fingerprinting techniques demonstrated that these various isolates were genetically very similar.

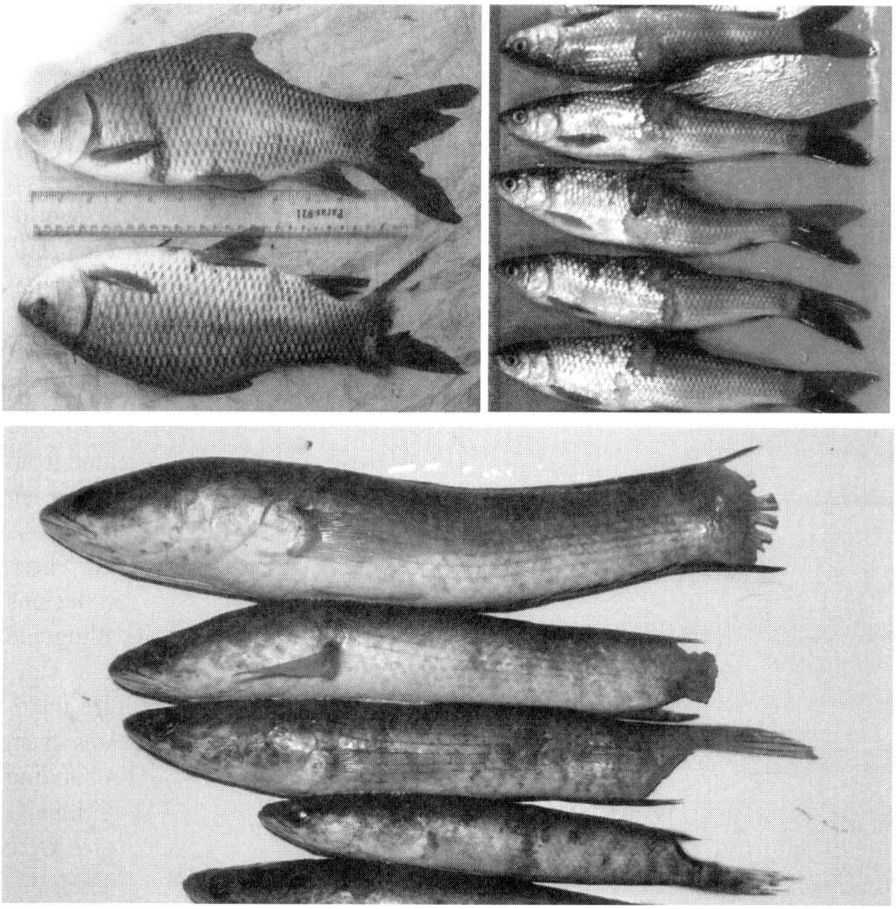

Plate 10: Gross surface lesions associated with Epizootic Ulcerative Syndrome (EUS) (See colour version on page 376)

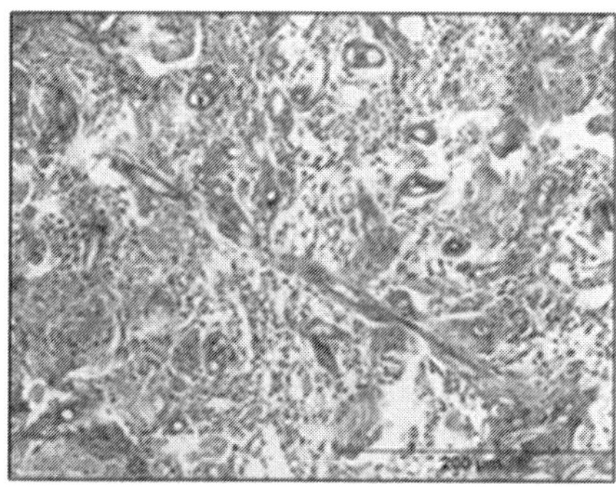

Histological section showing pathology in a susceptible fishinfected with *A. invadans*

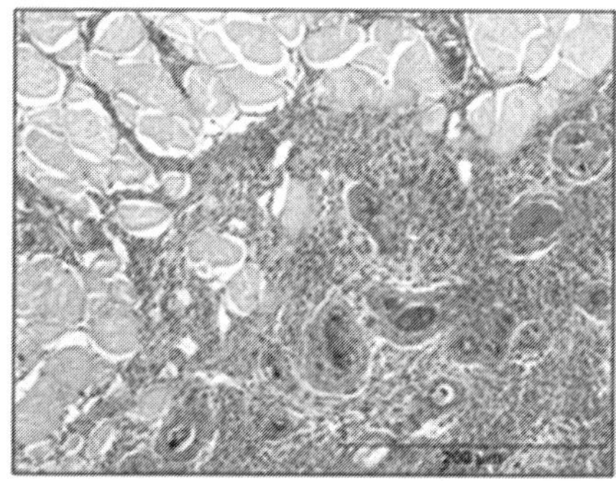

Histological section showing pathology in a fish which is able to resist the infection with *A. invadans*

Plate 11: Histopathology of EUS (See colour version on page 376)

Initial trials on isolation of the pathogenic oomycete from EUS lesions were difficult since attempts were made to isolate from the surface, where the mycelium inside the chronic inflammatory granuloma was dead and no longer viable. Later attempts of sterile microdissection down to the depth of the below lesion area and culture of minute viable hyphal tips, it was possible to isolate the pathogenic oomycete which was slow growing, thermo-labile and culturally distinct different compared tofrom other saprophytic fungi. In pathogenicity studies comparison of the *Aphanomyces* associated with EUS and other with other similar oomycetes, it was found that only the pathogenic *Aphanomyces* strains isolated from fish affected by EUS, RSD and MG were able to grow invasively through the fish muscle and produce the distinctive EUS gross and histological lesions. This provides further evidence to support the hypothesis that a single species of *Aphanomyces* is responsible for the oomycete related lesions in EUS, RSD and MG-affected fishes. On the other hand, saprophytic *Saprolegnia*, *Achlya* and *Aphanomyces* spp. isolated from EUS-affected areas were unable to reproduce typical pathology associated with EUS. Hence, it was suggested that all these pathogens are opportunistic in nature. Nonetheless, saprophytic *Saprolegnia*, *Achlya* and *Aphanomyces* spp. are commonly observed on the surface of EUS lesions and may contribute to the disease as opportunistic wound parasites. All these saprophytic isolates can be easily distinguished from pathogenic *Aphanomyces* oomycete in terms of pathogenicity and growth characteristics. As *A. invadans* is the only valid taxon name according to the International Code of Botanical Nomenclature (ICBN), *A. invadans* has been adopted to describe all the MG, RSD and EUS oomycete pathogens.

Viruses: Prior to findings on oomycete *Aphanomyces invadans*, viruses were thought to be as the most likely necessary infectious cause of EUS. Although several viral pathogens have been isolated from the EUS affected fishes, and varying interpretations have been made about their pathogenic significance, the involvement of viruses with EUS has been a highly debated and controversial issue for the following reasons; 1) to date no single type of virus has been consistently isolated, 2) all the recovered EUS associated snakehead rhabdovirus (SHRV), ulcerative disease rhabdovirus (UDRV) are not belonging to one group, 3) difficult to isolate virus at any one time from more than 5% of diseased EUS affected fish examined, and 4) not possible to reproduce the disease in bioassay and demonstrate any consistent lesion in pathogenicity studies using viruses alone. Hence, it was concluded that virus is not the causative agent. However, recent evidences suggest that rhabdoviruses can be more readily isolated from fish specimens collected during the early period of outbreaks. Therefore, if viruses have any role in the disease development process of EUS, they may be involved in causing skin lesions, thereby facilitating to allowthe entry of *A.invadans*.

Bacteria: Of all the infectious pathogens reported to be associated with EUS, bacteria had attracted much attention, and it was perceived as either a possible primary causative agent or a significant secondary opportunistic pathogen. However, it was not possible to isolate any single bacterial species consistently from EUS affected fishes. Further, *A. hydrophila* is recognised as a pathogen involved in bacterial haemorrhagic septicaemia in freshwater tropical fish. On the contrary, one of the pathognomonic feature of EUS is localised ulceration without significant bacterial involvement. All these studies clearly indicate that *A. hydrophila* is not the primary causative agent of EUS. Hence, it is suggested that bacteria may be playing roles at two stages in EUS pathogenesis but aren't essential. Firstly, they may be creating skin lesions which could facilitate oomycete entry, but there's no confirmed evidence for this. Secondly, opportunistic bacterial pathogens may be colonizing in the established ulcers, leading to lethal septicaemia.

Parasites: Several protozoan and metazoan parasites have been observed to be associated with EUS affected fishes. However, no parasite species is reported to be very intimately associated with EUS lesions. It may be possible that certain ectoparasites may induce skin lesions, thereby facilitating *A. invadans* to attach and invade the host dermis.

Environmental Factors: Since EUS outbreaks occur seasonally, it is suggested that a number of biotic and/or abiotic factors influenced by seasonal changes might be playing a role in lesion induction, and/or presence of infective forms of *A. invadans*. As it has been seen that low temperature is mostly associated with diseases outbreaks, it is suggested that it is an important determinant for most of the outbreaks. Low temperature might be crucial in suppression of immune response of fish thereby impairing the ability of the fish to contain and inactivate invasive oomycete. EUS outbreaks in estuaries often coincide with significant rainfall events. Hence, it is suggested that the inflow of freshwater into the estuary might be lowering the salinity to less than 1 ppt at outbreak sites, and creating conditions favourable for *A. invadans* sporulation. In addition, fish exposed to acidified runoff water from areas with acid sulphate soils may be resulting in epidermal necrosis, thereby facilitating the attachment and invasion of *A. invadans* zoospores and initiating EUS lesions. Importantly, floods are considered as one of the major risk factors which are believed to be responsible for spread of the disease by facilitating the movement of both infected fish and *A. invadans*. Therefore, to conclude, several biotic and abiotic factors including viruses, bacteria, ectoparasites, low pH, and low temperature, might acting as predisposing factors and might be responsible for inducing nonspecific skin lesions in fish, which subsequently might be getting colonized by *A. invadans* and initiating the EUS lesions.

Susceptible host species: *Aphanomyces invadans* causes disease and mortality in both farmed and wild fish, worldwide. Around 160 fish species representing 54 families and 16 orders have been reported to be affected by infection with *A. invadans*. It is to be noted that *A. invadans* has the highest number of documented aquatic animal host species those are reported to be susceptible to this disease. Such a wide host range indicate that many more species are likely to be susceptible to the disease those are yet to be reported which indicate the disease will continue to have impact on most inland fisheries in tropical and subtropical regions, including subsistence fisheries. Interestingly, a few of the commercially important fish species, such as common carp (*Cyprinus carpio*), Nile tilapia (*Oreochromis niloticus*) and milkfish (*Chanos chanos*) are considered naturally resistant to the disease.

Susceptible stages of the host: The susceptible life stages of the fish are usually juvenile and young adults. There is no report of infection with *A. invadans* being found in fish fry or fish larvae. Experimental infection studies with yearlings of Indian major carps against *A. invadans* revealed resistance to infection whereas same size/age groups of murrels are susceptible.

Target organs and infected tissue: The motile zoospores play an important role in the spread of the disease. Once the motile spores attach to the skin of the fish, the spores will germinate under suitable conditions and their resulting hyphae will invade the fish skin, muscular tissue and may reach the internal organs. Fish skeletal muscle is the main target organ.

Clinical Signs (Gross Pathology) and Histopathology: In the affected fish, red spots may be observed on the surface of the body surface, head, operculum or caudal peduncle at initial stages of infection which may progress to large red or grey shallow ulcers, often with a brown necrosis in the later stages. Most of the species other than striped snakeheads and mullet may die at this advanced stage of infection. In snakehead, usually the lesions are more extensive and can lead to complete erosion of the posterior part of the body or necrosis of both soft and hard tissues of the cranium. Gross appearance and histopathological lesions of EUS (Plate 11) have been characterised, classified into different types, based on studies conducted on EUS affected wild fish.

***Type I lesion (Erythematous Dermatitis)*:** The Type I lesion, also known as Erythematous Dermatitis, represents an early stage characterized by minute red spots measuring less than 1cm in diameter, scattered across the body surface. There are no prominent haemorrhages or ulcerations, and the epidermis remains intact at the lesion margins. Histologically, infiltration by mononuclear cells is observed in the epidermis. Other tissues surrounding the lesion appear normal. Occasionally, nodular structures may be observed within the epidermis. Hyphae or granulomas in the dermis and skeletal musculature may be present in minimal amounts.

***Type II lesion (Necrotising Dermatitis)*:** The Type II lesion, also known as Necrotising Dermatitis, represents moderately advanced lesions, and is characterized by circular discoloured areas ranging from 2 to 4 cm in diameter on the body surface. Histopathological examination of this type of lesions reveals a moderate to severe, locally extensive, necrotizing, granulomatous dermatitis response. Within the affected tissues, mycotic granulomas are prominently observed in the epidermis, dermis, and musculature, closely associated with numerous non-septate oomycete hyphae. Invasion of oomycete hyphae usually results in significant necrotizing dermatitis and myositis. Furthermore, severe floccular degeneration of muscle fibers is evident due to invasion of oomycete hyphae in the muscular tissue.

Type III lesion (Dermal Ulcer): Type III lesion also known as Dermal Ulcer (DU), is characterised by distinct circular or oval ulcers of approximately 1-4 cm in diameter. These open ulcers are usuallyhemorrhagic. Histologically, within the affected areas, the underlying musculature undergoes replacement by oomycete mycotic granulomas intermingled with host inflammatory tissue. A notable feature is the presence of considerable myofibrillar necrosis. Furthermore, oomycete hyphae extend spread in all directions from the centre of dermal ulcer and may invade internal organs (liver, kidney, viscera, spinal cord).

Type IV lesion (Healing Ulcer): In the Type IV lesion, also known as Healing Ulcer (HU), there is evidence of recovery in surviving fish. The lesion is characterized by circular to oval whitish or melanized healing areas on the skin. This phase is marked by increased macrophage activity, vascularization, fibrinolysis, and effective clearing of exudate. Within the granulomas, remnants of killed or crumbled oomycete hyphal material are observed. Additionally, epithelialization occurs, indicating the restoration of the skin's protective barrier. The presence of these regenerative features signifies the fish's resilience and the ongoing progression toward recovery from the previous infection.

Life cycle of *Aphanomyces invadans*: *A. invadans* has aseptate fungal-like mycelial structures and has two typical zoospore forms. The primary zoospores consists of round cells that are formed within the sporangium and are subsequently released to the tip of the sporangium, where it they forms aggregate as spore clusters. The primary zoospores rapidly undergo transformation into the secondary zoospores, which are characterized by their kidney-shaped morphology and two lateral flagella, enabling them to move freely in which is reniform with laterally biflagellate cells and can swim freely in the water. The duration of motility of secondary zoospores varies with environmental factors and the presence of a fish host or suitable substratum. Typically, after attaching to a suitable host, the secondary zoospores germinate to produce new hyphae. If the motile zoospores can't find suitable substrates, they will encyst. Many a times, further generations of zoospores are formed, a phenomenon known as polyplanetism. However, there is no suitable method to recover or isolate the encysted zoospore from affected fish ponds. How long the encysted spore can survive in water or on a non-fish substrate is still unclear. Under*in vitro* experimental conditions, the encysted zoospore has been reported to survive for at least 19 days. There is no evidence to suggest that fish can be lifelong carriers of *A. invadans*. Generally, most infected fish die during an outbreak and some of the fishes which survive with mild or moderate infections, they are most unlikely to act as carriers of *A. invadans*.

Pathogenicity mechanism of *A. invadans* infection: During the disease development process, the pathogen and host are involved in a constant molecular arms race, where the pathogen continuously develop virulence mechanisms to subvert host defences to their benefit and on the other hand, the host tries to resist the infection. Oomycetes are known to establish infections through the secretion of effector proteins. These proteins are released during the infection and they target host factors to overcome host defence mechanisms or they modify the host metabolism for the benefit. Recently, it has been reported that *A. invadans* genome has a good repertoire of effector proteins that are predicted to interact with various host processes , and it has been demonstrated that active proteases of *A.invadans* secreted into the culture filtrate are similar to the predicted proteases in the genome and the effector repertoire shows more similarities to the fish-pathogenic *Saprolegnia parasitica* than to the plant-pathogenic *Phytophthora infestans*.

Monitoring sequential changes of innate immunity parameters against any pathogen in susceptible and resistant hosts helps in revealing mechanisms of disease resistance.

Studies on sequential changes in various innate immune parameters in one of the susceptible host i.e. Indian major carp, *Labeo rohita* following experimental infection with *A. invadans* indicated that at early stages of infection, no significant changes in any of the studied innate immune parameters (respiratory burst activity, myeloperoxidase activity, alternative complement activity, total serum protein, albumin and globulin, lysozyme activity, total antiproteases activity, alpha-2 macroglobulin activity) were observed. However, at the advanced stages of infection, the respiratory burst and alternate complement activity were significantly higher whereas lysozyme, antiproteases and alpha-2 macroglobulin values were significantly lower than the control group and also from the infected group at earlier stages of infection. The findings indicated that the rohu immune system was strongly modulated by *A. invadans* infection. On the other hand, monitoring of sequential changes in various innate immune parameters in a resistant host i.e. common carp, *Cyprinus carpio* following experimental infection with *A. invadans* indicated that innate immunity parameters viz. respiratory burst, alternative complement and total anti-proteases activities of the infected common carp were higher compared to control fish, particularly at early stages of infection. Hence, the findings indicated that innate defense mechanisms of common carp are able to neutralize the virulence factors secreted by *A. invadans*, thereby, preventing its invasive spread and containing the infection.

Furthermore, comparison of RNA sequences from infected and uninfected samples gives insights into immune-related genes that are differentially expressed during an infection, and ultimately lead to a better understanding of molecular mechanisms involved in host immunity to a particular pathogen. To understand the molecular basis of susceptibility and resistance against *A. invadans*, the gene expression profile in head kidney of *A. invadans*-infected and control rohu and common carp was investigated using RNA sequencing. Time course analysis of RNA-Seq data of rohu indicated down-regulation of differentially expressed genes in some of the important pathways namely antigen processing and presentation (APP). In the APP pathway, genes belonging to both MHC I and MHC II were down regulated. As MHC I is generally activated in response to intracellular antigens and is the major route of presentation of viral antigens, down regulation of the genes in MHC I indicated both intracellular and extracellular mode of pathogenicity of *A. invadans*. In addition, many of the genes in Tolllike receptor signalling pathway were also down regulated. The observed down regulation could be a part of immune evasion strategy of the oomycete *A. invadans*. Interestingly, following infection of the resistant host i.e. common carp with *A. invadans*, majority of *MHC I* and *MHC II* genes in antigen processing and presentation pathway were upregulated. Further, genes in the several other pathways like Fc gamma receptor-mediated phagocytosis pathway, NOD signaling pathway, Chemokine signaling pathway were upregulated. Hence, it could be concluded that efficient processing and presentation of *A. invadans* antigens, enhanced phagocytosis, recognition of pathogen-associated molecular patterns, and increased recruitment of leukocytes to the sites of infection contribute to resistance of common carp against *A. invadans*.

Immunization of Fish Against *A. invadans*

To date, there is no protective vaccine available against infection with *A. invadans*. Immunization trials with *A. invadans* homogenate, filtrate, extract and extracellular products in conjunction with adjuvant revealed that immunization with oomycete extract mixed with adjuvant although resulted in significantly increase in antibody production but reduction in mortality in challenged fish was not statistically significant. In another study, immunization of fish with inactivated germinated zoospores of *A. invadans* in combination with adjuvant rendered short term protection. Further, sera of such immunized fish which survived the challenge, showed significant inhibition of germination of zoospores and germlings growth in comparison to control group. From these findings it was suggested that inactivated germinated zoospores of *A. invadans* could be a potential candidate for vaccine development. However, further studies are required to validate the duration of protection, and dose and duration time period of immunization.

Differential Diagnosis: Appropriate diagnosis of EUS is crucial to avoid confusion with other non-specific ulcerative conditions. It is to be noted that ulcer is a non-specific clinical lesion which may be caused due to many different other agents including mechanical damage, trauma, ectoparasites, surface ulcerative bacteria, saprophytic fungi, predators, etc. Importantly, such ulcerative conditions, are not associated with large-scale mortalities and may not be confined to any particular season. In addition, tissue sections of the lesion of those affected fishes will not show any mycotic granulomatous response. Therefore, for confirmatory diagnosis, the following aspects should be considered: outbreaks in known susceptible fish species should be associated with ulcerative lesions, (ii) the temperature during outbreaks should not exceed 35°C, (iii) septate oomycete hyphae should be observed in light microscopy in the muscle tissue underlying the lesion, (iv) mycotic granulomas and oomycete hyphae should be demonstrable in histological sections, (v) isolation of *A. invadans* from affected internal tissues and the demonstration of typical secondary cyst clusters characteristic of the genus

Aphanomyces should be possible in sporulating culture and the disease should be reproduced through bioassay using secondary zoospores, vi) detection of *A. invadans* through use of molecular techniques such as by PCR amplification and sequencing of the ITS region (Internal Transcribed Spacer)and/in-situ hybridisation.

Management of EUS: Research efforts have identified several causal (risk factors) associated with the EUS. Therefore, in the region or areas where EUS is regarded as endemic, management should involve eradication, exclusion and treatment. Removal of all susceptible fish from ponds, drying, and liming of ponds, disinfection equipments/nets should be considered for eradication programmes. Having eradicated the oomycete from a culture facility, the goal should be to prevent its re-introduction of the pathogen. As infection is spread by affected/carrier fish, contaminated water, equipment, etc., obtaining seed and water from *A. invadans* free zone, using screen to remove wild fish, treatment of pond water with disinfectants, etc. should be considered under exclusion strategies. Use of epidemiological evidences to devise intervention strategies and special management measures during periods of high risk can be beneficial. During EUS seasons, keeping low stocking densities, avoiding stress, preventing skin damage caused by cutaneous ectoparasites and bacteria, harvesting of susceptible fish before the high risk period, are some of rational approaches. Resorting to farming of EUS resistant species like tilapia, European carps, Chinese carps is also one of the options in high risk area. There is no published evidence for any effective treatment for EUS. However, application of agricultural lime to ponds during the culture period can decrease the severity of the disease. Furthermore, immersion treatment with broad spectrum antibacterial and antifungal chemicals including application of ICAR-CIFA product CIFAX might be effective in control of the disease. However, recently ICAR-NBFGR has developed a formulation named OoNIL for treatment of oomycete diseases. The formulation has been field validated in over 400 farms in a period of 4 years, and has been found to be very effective in controlling oomycete diseases (EUS and saprolegniosis) in fishes.

Other Important Oomycete and Fungal Fish Diseases

Majority of these pathogens are saprophytic and derive their nourishment by decomposing organic matter, and they can cause disease in fish either as a primary pathogens or as secondary invaders. Oomycete pathogens i.e. *Saprolegnia* sp. and *Achlya* sp. are opportunistic pathogens and are ubiquitous in nature. These group of pathogens can act as primary pathogens in physiologically debilitated fish characterized by a decline in mucus production and/or in immunologically compromised fish. On the other hand, in healthy fish, with optimum mucus production, the zoospores of these oomycete will not get an opportunity to settle, germinate and invade the underlying tissues. For the fish to get infection with these pathogens, primary epidermal damage acts as a predisposing factor. For the disease to develop, the zoospores settle on damaged skin, germinate and invade the underlying integument and superficial muscle. In severe and prolonged infections, the mycelia will penetrate beneath the dermal layer into the muscles, and many a times reach the inner organsin very small fish.

Saprolegniosis

In this disease, the invading hyphal tips induce massive necrotising lesions in the integument and muscle. The mycelia are often seen from outside as cotton-wool like growth. Therefore, the infection is also sometimes referred to as cotton-wool disease. Skin infection is easily detected by the appearance of patchy or extensive cotton wool cover (the oomycete mycelia) emerging usually from the haemorrhagic skin lesion. Infection generally leads to large wounds on the body surface leading to impaired osmoregulation and haemodilution. In addition, extensive lesions in the gills cause respiratory failure, and many a times, these large wounds and gill lesions, both can lead to mortality. Handling, bad husbandry practices and trauma predispose fish to such infections. *Saprolegnia* infection is also is a problem in hatcheries during incubation of eggs. The pathogen first establishes on dead and unfertilized eggs and gradually spread to healthy developing eggs destroying the entire batch of incubated eggs. Washing of eggs and removal of unfertilized and damaged/ dead eggs in hatcheries can help in minimising oomycete infection in hatcheries.

Gill Rot or Branchiomycosis

It is caused by an intravascular oomycete called *Branchiomyces* sp. The pathogen lives in the larger blood vessels of the gill arch and obstructs the flow of blood to the gill lamellae. Infection in the blood vessels of the gill results in blockage, haemostasis and thromboses which subsequently cause extensive necrosis of the gill filaments and the affected portions of the gill lamellae appear externally as white patches. Microscopic examination reveals branched non-septate hyphae containing numerous numbers of spores. *Branchiomyces sanguinis* and *B. demigrans* primarily affect the gills, while *B. sanguinis* infects the vascular system and *B. demigrans* grows in perivascular tissues in gills. *Branchiomyces* infections generally cause respiratory distress with associated high mortalities particularly above 20 °C.

Fungal Diseases: Fungal infections in fish are generally considered secondary to other pathogens or stresses associated with water or environmental changes including high organic load and high stocking density. This group of pathogens release digestive enzymes into their environment and absorb the digested simple biomolecules through their cell wall. Fungi produce non-motile spores with a chitinous cell wall. These spores can survive in unfavourable conditions, and the resistance of the spores is an important adaptation strategy to infect susceptible hosts. These spores play a crucial role in dispersal between hosts and dissemination within hosts. Some of the fungal pathogens which causes disease in fishes include *Fusarium*, *Candida*, *Aspergillus*, *Paecilomyces*, *Penicillium*, *Cladosporium* and *Ichthyophonus hoferi*.

Infection with *Fusarium* causes skin ulcers or can become systemic causing kidney and brain necrosis. *Candida albicans* has been reported to colonize the epithelial surface of fish, and invading tissues. During the invasion, morphogenesis of the pathogen from ovoid yeast to a filamentous hypha is important in causing tissue damage and mortality. In *Aspergillus* sp. infections, the spores get attached to the fish mucus and can invade the epidermis and internal organs. *Aspergillus* infection causes damage to gill lamellae with subsequent respiratory distress, but systematic infections from feed contaminated with *Aspergillus* sp. usually cause with high mortality. *Paecilomyces* sp. infections commence with ingestion of the fungal spores in the water by the fish. Some of the fungi like *Penicillium* sp. and *Cladosporium* sp. are present insoil andmay cause systemic infection.

Infection with *Ichthyophonus hoferi* is an economically important disease with broad host range. It is a systemic disease which particularly affects liver, kidneys, spleen and heart. The spores are believed to gain entry into the fish through the oral route. Infected marine trash fish used in fish meal is believed to be the source of infection. Taking advantage of the blood vascular system, the fungal spore reaches the target tissue where it starts to germinate, proliferate and invade the tissue. Infection causes enlargement of the liver and congestion of the kidneys, and raised nodules are observed mainly in liver, kidneys and spleen. The invasive and proliferative activity of this fungus stimulates inflammatory response in the host in the form of chronic granulomatous nodules. The extent of inflammatory activity can be severe and replace the functional cells of in the target tissue. Severe tissue damage can cause high mortality. Multinucleated schizont is the most common stage of *Ichthyophonus* in live fish is a multinucleated schizont. After the death of the fish, the schizonts fragment into multinucleate motile plasmodia which are probably the source of infection. The infection is transmitted largely by consumption of these plasmodia through consumption of diseased infected fish.

References

Baldock, F.C., Blazer, V., Callinan, R., Hatai, K., Karunasagar, I., Mohan, C.V. &Bondad-Reantaso, M.G. 2005. Outcomes of a short expert consultation on epizootic ulcerative syndrome (EUS): re-examination of causal factors, case definition and nomenclature. In: Diseases in Asian Aquaculture- Vol. V; Walker, P., Lester, R. &Bondad-Reantaso, M.G. (eds). Fish Health Section, Asian Fisheries Society, Manila, the Philippines, 555–585. Available at: www.researchgate.net/ publication/312928593_

Iberahim, N.A., Trusch, F. & van West, P. 2018. Aphanomyces invadans, the causal agent of Epizootic Ulcerative Syndrome, is a global threat to wild and farmed fish. Fungal Biology Reviews 32, 118e130.

Pradhan, P.K., Mohan, C.V., Shankar, K.M. & Kumar, M.B. 2008. Susceptibility of fingerlings of Indian major carps to Aphanomyces invadans. Asian Fisheries Science 21(4), 369-375.

Pradhan, P.K., Sood, N., Yadav, M.K., Arya, P., Kumar, U., Kumar, C.B., Swaminathan, T.R. & Rathore, G. 2018. Effect of immunization of rohu Labeorohita with inactivated germinated zoospores in providing protection against Aphanomyces invadans. Fish and Shellfish Immunology 78, 195-201.

Pradhan, P.K., Verma, D.K., Peruzza, L., Gupta, S., Haq, S.A., Shubin, S.V., Morgan, K.L., Trusch, F., Mohindra, V., Hauton, C., van West, P. &Sood, N. 2020. Molecular insights into the mechanisms of susceptibility of Labeorohita against oomycete Aphanomyces invadans. Scientific Reports 10(1), 19531.

Pradhan P.K., Mohan, C.V., Shankar K.M. & Kumar M.B. 2010. Injection of forty-five day old fingerlings of common carp with zoopspores of Aphanomyces invadans leads to histopathological lesions suggestive of Epizootic ulcerative syndrome (EUS). Asian Fisheries Science 23, 321-328

World Organisation for Animal Health (OIE), Infection with Aphanomyces invadans (Epizootic ulcerative syndrome) [Chapter 2.3.1]. In: Manual of Diagnostic Tests for Aquatic Animals. (Accessed December 10, 2023) Available from. https://www.woah.org/fileadmin/Home/eng/Health_standards/aahm/current/2.3.01_EUS.pdf

9

Non Infectious Diseases (NID) of Fishes

K.M.Shankar

In the past several Non infectious diseases (NIDs) of fish and shellfish have been observed in natural waters. However, with expansion in fresh, brackish and marine aquaculture including aquarium keeping world wide there is a phenomenonal increase in incidences of NIDs and economic loss. Common Non infectious diseases (NIDs) in fish are 1.Gas bubble disease(GBD), 2. Body deformities(BD), 3.Nutritional diseases(ND) and 4. Disease due to harmful algal bloom(DHAB)

1. Gas Bubble Disease (GBD)

Gas bubble disease (GBD), is caused by supersaturation of gases in water. The gas involved is nitrogen in most cases and rarelycarbon dioxide and oxygen can cause GBD as these two gases are used by fish tissues and readily processed.Oxygen can cause gas-bubble disease at about 350 per cent air saturation, but nitrogen can cause the disease even below 118 per cent. GBD becomes prominent whenever there is a change in temperature, pressure and turbulance in aquatic environment, leading to disturbance in metabolism of fishes. Bubbles of gas may form in the eyes, skin, gills, fins and within the internal organs where small bubbles may coalesce into larger bubbles.Fish absorb the excess gas, which forms bubbles in the small blood vessels, due to pressure changes inside their body. Gas bubble disease (GBD), is seen both in captive and wild fish. When bubbles affect internal organs, some fish may diesometimes without obvious clear clinical signswhile other fish may be chronically affected, and have a reduced appetite, activity and buoyancy. GBD can also cause secondary bacterial, viral or protozoan infections.All fish and shelfishspecies as well as amphibians and aquatic invertebrates are susceptible to GBD.Sensitivity of fish to GBD in terms of pathology and mortality varies with species and age.

Etiology of GBD

Gas bubble disease is caused by uncompensated hyperbaric pressure of total dissolved gases. Total dissolved gas (TDG) is more important than individual gases.Fishes are cold-blooded and their bloodstream can become supersaturated with gases when there is a sudden change in water temperature or pressure. Cold water and water under pressure can dissolve more air than at atmospheric pressure. Fish transported by air may also develop gas bubble disease.Aalgal blooms producing more oxygen during day than that can diffuse into the water also supersaturates the pond water causing GBD. When pressure compensation is inadequate, dissolved gases may form emboli (in blood) and emphysema (in tissues). The resulting abnormal presence of gases can block blood vessels (hemostasis) or tear tissues, and may result in death. The severity of the disease depends on the number of gas bubbles formed and the tissues affected.

Three stages of GBD have been identified. 1. a period of gas pressure equilibrium with increasing mortality, 2. a period of rapid and heavy mortality, 3. a period of protracted survival with dysfunction leading to death. Overal indicators of GBD are restless fishes with erratic movement, trembling fins and flaring opercula. Fish often bloat and float due to accumunlation of gases in gas bladder and peritoneal cavity. Hemorrhage of gills, skin, muscle, gonads, intestinal epithelium, eroded fins with whitened tips and exopthalmiaare commonly associated with GBD.

Factors and Practices Contributing to GBD

In supersaturation condition water hold more dissolved gases than it normally contains at a particular temperature and atmosspheric pressure. Therfore, a change in pressure, temperature or salinity cause supersaturation of dissolved gases leading to GBD.

Under high pressure or at low temperatures water can contain more gas. Gas supersaturation in water can occur by both natural and artificial causes. Photosynthesis in water with heavy aquatic plant growth or algal bloom can saturate water with oxygen and can become supersaturated upon warming. A temporary saturation of water with oxygen causes the fish's blood to become saturated with this gas. Other natural causes are ground water supersaturated with nitrogen or the supersaturation of water caused by cascades or waterfalls. Artificial causes such as Plungpools from dams, heated water from power plant, and air from pipes or pumps where pump pressure or gravity head forces gas into solution cause supersaturation.

Clinical Signs

Two forms of GBD 1. acute 2. chronic can be described In acute phase 100% mortality occurs in minutes with hyperinflation of swim bladder, pnemoperitonium, swollen gill lamellae and sweling of cranium. Exopthalmia,disorientaion, subcutaneous emphysema, swimming near water surface, darkened skin with hemorrages and heavy mortality are other clinical signs. Usually when total gas pressure (TGP) is above 110/115 %, GBD occus in acute form.

Chronic form of GBD is associated with low mortalirty often with hyper inflation of swim bladder and emboli in gastrointestinal ract. Secondary infection of emphysematous tissue with bacterial pathogens is common in chronic GBD which may lead to high mortality. Occurence of small gas bubbles under epithelium leading to gas bubbles on fins, gills and eyeballs is common sight with GBD. Excess amount of gas may cause floating of fishes. Eye is very susceptible to GBD because of bloood vessels network leading to exopthalmia and impair of vision leading blindness.

Diagnosis

First external clinical signs of GBD are viscible bubbles of gas on eyes, skin, gills and fins. In biopsy, presence of microscopic bubbles on gills, accumulation of gase in swim bladder and visceral peritonis leading gas embolism is clear. Other than clinical signs such as a testing of water by saturometer for concentration of gases by supersaturation confirms GBD. In histology oedima of secondary lamellae of gills and occlusion of large branchial vessels reveal GBD

Treatment of GBD

The most important in treatment of gas bubble disease is to determine the cause of the microbubbles for their quicker elimination.Once the source of the supersaturation of gas in the water is determined and corrected, the microbubbles will naturally disperse with time.Source of microbubbles could be aeration, water pumps, water flow back into the aquarium, filters. Spraying water to supersaturated pond help to release trapped gasses and equilibrate with air transfering the fish as soon as possible to a tank with the optimal oxygen concentration helps to relieve them from the suffering. Addition of activated carbon(AC) can effectively adsorp gas and accelerate supersaturated TDG dissipation.Gentle stirring of the water will help degas the water. Lowering the water temperature slightly will also help, as cooler water holds more dissolved gas naturally.

2. Body Deformities in Fish

Abberations and deformties in fish have been observed in nature. Fish with different types of abnormalities are frequently captured from natural water bodies all over the world. Skeletal anomalies in farmed fish is currently a major problem in aquaculture with economical, biological and ethical issues. Skeletal deformities can affect the normal physiological functioning of fish by disrupting buoyancy, ability to feed, reproduce,avoiding predators and overall its survival and growth. Skeletal anomalies in aquaculture are highly variable, in different species and under different rearing conditions. Skeletal deformities is an important factor that downgrade hatcheries' production. Furthermore, fishes with skeletal deformities live for shorter duration than the normal ones and are not preferred by the consumers. In general, in hatcheries skeletal deformities can affect 7–20% on an average of the produced juveniles, however occasionally this incidence increases to 45–100%. The real economic losses are difficult to estimate

Types of Deformities

A complex mixture of several types of bone disorders leads to deformities. Deformities include both vertebral and spinal malformation such as lordosis, skoliosis, vertbral fusion and platyspondyly. Minor deformities such as neck bend, compressed snout, short operculum, reduced or assymmetric fins, reduced or protruded jaws are also seen occasionally.

Diagnosis

It is normal to occasionally find deformed fishes in any population. Physical deformities are easily detected by morphological changes. Scoliosis of bone and spinal deformities are easily diagnosed because the fish has a distinct curve in its spine.Radiography and sonography are the two most rapid techniques available for detecting skeletal deformities. Other diagnostic techniques include histopathology, special staining and genetic analysis.

Causes of Body Deformities

Several factors may be involved in sketletal deformities, which in fact are not understood well. An insight into proximate composition of bone may throw some light on deformities. Fish bone mainly consits of calcium-phosphor hydroxyapatite salts.(65 %) embedded in a matrix of type 1 collagen fibres. Collagen represents more that 90 % of bone matrix which give resistance to the structure, and its relationship with hydroxyapatite is essential for bone strength. Diameter of collagen fibers determines biochemical strength of tissues. Bone strength is determined by its mineral contituents which include calcium, phosphate, carbonate with small amounts of magnesium, sodium strontium, lead, citrate, flouride and hydroxide.

Skeletal deformities can occur due to nutrition, environmental and genetic factors. Deformities in general can occur during embryonic and post embryonic phase while several structural abnormalities may actually occur throughout entire life cycle. Overall, nutrition, pollution, heredity and pathogens can cause body deformities infish.

Environment induces skeletal deformities in two ways. 1. by neuromuscular effects causing deformities of vertebral column without changing composition 2. by altering biological processes which maintains biochemical integrity of bones. Factors like temperature and velocity of water also known to influence bone deformities.

Nutrition has a major role in body deformities in fish. Among the nutrients calcium and phosphorus deficiency cause bone deformities as observed in carps, salmon and trouts. Bone formation by mineralisationis known to be regulated by vit K through osteocalcin synthesis by osteoblasts. Abnormalities such as khyposis, scoliosis and lordosis are linked to defiecny of vit C. However, overdose of some vitamins such as vit A is also known to induce vertebral deformities.

Pollutants in water can cause deformities in fish. Heavy metals are known to cause neuromuscular damage and skeletal deformities.Lead (Pb), cadmium (Cd), zinc(Zn) and selenium(Se) are common industrial pollutants impairing developmental processes and bone formation. Microplastic pollution in water is rampant today and is known to cause growth defects in fish including deformities.

Some rare genetic anamolies in a fish population are normal and inheritable also known to cause body deformities. Body deformities due to certain parasites such as myxosporidiens which affect cartilege of fish is well known. Association of infection with a bacterium *Flavobacterpsychophilum* with deformities of vertebral column including craniel, medial, and caudal part of vertebral columis well documeted.

3. Nutritional Fish Diseases

Nutritional diseases in fishes are attributed to deficiency, excess or improper balance of components and/or toxic comonents present in a diet. Broadly three types of nutrional diseases can be categorised. i,Under nutrition or dietary deficiencies in major components of food (protein,fat,carbohydrate and micronutrients), ii. Over nutrition and imbalnces due to biochemical changes in nutrients (amino acid toxicity, oxidation of unsaturated fatty acids, fatty acid toxicity, dietary vitamin toxicity and mineral toxicity), iii. Toxic components in the diet such as mycotoxins, algal toxins, anti-nutritional factors and anthropogenic chemicals.

i. Diseases due to deficiency

The disease usually develops gradually because animals have body reserves that make up for nutritional deficiency up to a certain extent. Disease signs develop only when supply of any diet component falls below critical level. Deficiencies of many amino acids suach as methionine, threonin, argininn, tryptopan,isoleucine, leucine, histidine have reflected in vertebral deformities, cataract, fin errosion and mortality in extreme cases. Deficiency of carbohydrate depres the digestion leading to enlarged liver, and hepatic degenerative changes with glycogen deposition. Deficiency of fat causes Lipoid hepatic degenerative disease with liver turning into yellow brown colour followed by anemia and pale gills. Micronutrients also have critical role in nutritional disesaes. Iodine deficiency causes thyroid tumor,with abnormal thyroid follicles.Pantothenic acid deficiency cause nutrional gill disease with respiratory difficulties. Vitamin A deficiency leads to under growth, blindness and hemorrhage at fin base. Folic acid deficiency causes anemia characterized by low values of erythrocytes and haemoglobin.

ii. Diseases due to over nutrition and imbalances due to biochemical changes in nutrients

Amino acid excesses have been known to produce toxic effects. Dietary excesses of leucine in rainbow trout (at 13.4 % of the diet) resulted in vertebral deformities and scale loss. Toxicity of fat soluble vitamins is contrastingly more pronounced in fish than that of water soluble vitamins. Accumulation of vitamins/fat in excess of requirements, leads to hyper-vitaminosis leading to reduced growths, drop in haematocrit values, fin necrosis, dark colorations, lethargy and reduced RBC count.

Oxidation of unsaturated fatty acids in diets produce products such as peroxides, hydroperoxides, aldehydes and ketones. These free radicals react with other dietary ingredients such as vitamins, proteins and lipids resulting in disorders associated with vitamin, protein (or) lipid deficiencies.

Several pathological effects such as hemorrhage on skin, base of fins, exophthalmia, loss of appetite, change in body colour, fatty liver, anemia and muscular damage have been recorded due to mere rancidity of oil in the feed stuff.

Cyclopropeonic acid in cotton seed oils has been shown to reduce growth and act as a synergist to carcinogenic effects of aflatoxins.

iii. Diseases due to Toxins in feed

Toxins in the feed include afflatoxins,algal toxins and chemicals. Afflatoxins in feed is common through feed ingredients and prepared feed stored under warm and humid conditions which encourage the growth of several moulds. Fungal contaminants like *Aspergillus* produce aflatoxins in the infested feed. Fish fed on aflatoxin contaminated feed often develop tumors in the liver known as hepatomas. Aflatoxins can also bring about significant pathological conditions in many of the organs like liver and kidney and condition is referred to as `aflatoxicosis. Aflatoxins in rancid feeds cause Red Disease in *Penaeus monodon* with gross signs of sudden drop in feed consumption, body weakness, poor growth, reddish colour of faecal matter.

Toxic minerals cause a host of nutritional diseases in fishes. Unconventional sources of feed ingredients often leads to dietary excess of minerals such as arsenic by using poultry waste, lead by pulp waste, and mercury, cadmium through fish meal is well known. Clinical signs include reduced growth, kidney stones, vertebral deforming, hyper activity and black tail. Contaminants such as Cd, Cu, Zn, K, known to cause Black gill disease in *Penaeus monodon* with gross signs of gill becoming reddish brown to black color, loss of appetite and mortality.

Incorporation of significant levels of silkworm pupae in the diet cause Sekoke disease with signs of cataract, degenerative changes in extrinsic eye muscle, retina. Anti-nutritional effect of cotton seeds with gossypol causing sudden anorexia and deposition of sudanophilic globules within liver and kidney has been recorded.

Occasionally residues of chemicals such as organochlorines or organophosphates-accidentally contaminatefeed can result in acute toxic effects. Acute level of pollutants (pesticides, heavy metals, hydrocarbons) in feed ingredients can directly bring about abnormalities and mortalities in fish. Pollutants even at low levels, can induce chronic stress and lead to elevated levels of plasma corticosteroids. Elevated levels of stress hormone can induce immunosuppression in these fishes pre-disposing them to several infections.

Common Nutritional Diseases Observed in Marine Fish Culture of Asia

Lipodosis

Lipodosis is a common nutritional disease in cage-cultured groupers in Indonesia, Thailand and the Philippines. Feeding with rancid formulated feeds or with fatty or poorly stored trash fish can cause lipodosis. Fish in the grow out stage are susceptible to lipodosis with clinical signs such as poor growth, lethargy, opaque eyes, slight distention of the abdomen and low mortality. Pale liver showing large fat droplets in histopathology is main feature of diagnosis

Fish Scurvy

Spinal deformity associated with ascorbic acid deficiency has been reported in grouper farms in Indonesia and Thailand. Affected fish exhibit gross signs such as anorexia, short snout, erosion of opercula and fins, hemorrhaging of eyes and fins, exophthalmia, swollen abdomen, abnormal skull, falling pharyngobranchials, severe emaciation and spinal column abnormality such as scoliosis and lordosis. Affected fish exhibit poor growth or severe emaciation in prolonged deficiency. Histologically, hyperplasia and detachment of gill epithelium can be observed and the hepatocytes may contain large fat droplets.

Essential Faty Acid (EFA) Deficiency

EFA are very esssential for normal growth and development of marine fish larvae. Fatty acids are important component of biomemembrane and as precursorof some physiological modulation are essential components of diets. Marine fish in general neeed unsaturated fatty acids as they can not desaturate saturated fats to unsaturated fats. EFAs particularly docosahexanoic acid (DHA,22:5(n-3) and ecosapentnoic acid (EPA-20:59 n-3) are found in live food and plankton. Adeffeciency of fatty acids in diet leads to larval mortality due to " shock syndrome". The syndrome due to unusual sensitivity to stress has been recorded in larvae of groupers in Thailand and Singapore.

Nutritional Myopathy Accompanying Ceroidosis

Diets with rancid fat or polyunsaturated fatty acids with low content of vit E has been linked to myopathy. Certain myopathies have been recorded due to unusual fatty acid peroxidation in the cell membrane compromising cellular integrity. Emaciation , darkening body colour, spinal cord deformities,petichia at the base of opercularwith low to high mortality in fingerling and brood are due to myopathy

Thiamine Deficiency

For a normal nerve function, thiamine which is a coenzyme catalysing carbohydrate metabolism is esential. Some species of trash fish such as sardine andanchovies contain thiaminase which degrade the thiamin contained in these fish. When such trash fish are fed for extended periodsin culture systems cause deficiency of vitamin B1 (thiamin). Groupers exhibited white body color, erratic swimming, hemorrhage on the body due to thiamine deficiency. In histology leisons found in brain with hemorrhage and degeneration of nerve cells.

4. Diseases Due to Harmful Algal Bloom(HAB)

Harmful algal blooms have been reported from natural and aquacultural systems of fresh, brakish and marine waters. And with increasing eutrophication due to higly active agriculture, aquaculture and industry new types of toxic algae are being reported.Algal blooms are harmful in several ways. Mainly toxins produced by algal blooms are harmful to fish and fish food organisms.Oxygen depletion during night time by algal blooms and also death and decomposition of bloom leading to fish kill is another area of concern. Harmful algal blooms often suppress useful algal production and even zooplankton in waters. algae produce compounds imparting off flavour to fish hurting marketting and consumption. Some algae are harmful even at low low density while some species are harmful at higher density. Some HAB impart red or blue green colour and hence more visible while some are undetectable with casual observations. Overal algal toxins have a variety of harmful effect on fish- fish kill and reduced growth due to gill damage and disrution of neurological functions. Sublethal effects of algal bloom also favour chronic effct on fish encouraging seconday infection by microorganisms.

Factors Favouring HAB

Factors that determine the growth of algae are nutrients, salinity or ionic strength, light conditions, turbulence and mixing, temperature and herbivory. Eutophication of natural water bodies with excess nutrients such as nitogen and phosporus have invariably caused algal blooms. Nutrient added through agriculture wash off and industrial and domestic waste are the main source for eutrophication. In aquaculture systems particularly in Asia , where most of the fish and shellfish depend on useful algae and zooplankton as food organism, pond fertilisation is a common practise. Besides good amount of artificial feed added goes unutilised and the waste generating good amount of nurients often in ecess encourage HAB.

Although there are several algicides advocated, their application is limited in natural watrs and even in aquaculture. Often algicides are also toxic to fish ahd fish food orgnaisms. Non-chemical treatments such as a) physical mixing and aeration, b) increasing flow rate c) decreasing or altering nutrient content and composition although available may not be viable at all sites and in all situations.

Algal Toxins, Their Nature, Mode of Action and Clinical Signs in Fishes

The most common toxin-producing algae are cyanobacteria, prymnesiophytes, and euglenoids and dinoflagellates.

Cyanobacterial toxins classified according to their chemical structures as cyclic peptides (microcystin and nodularin), alkaloids (anatoxin-a, anatoxin-a(s), saxitoxin, cylindrospermopsin, aplysiatoxins, lyngbyatoxin-a) and lipopolysaccharides. Most cyanotoxins are either neurotoxins or hepatotoxins.

Several genera of cyanobacteria produce neurotoxins which are known to affect nervous system of vertebrates and invertebrates. Three primary types of neurotoxins are : i) anatoxin-a, an alkaloid, inhibits transmissions at the neuromuscular junction by molecular mimicry of the neurotransmitter acetylcholine ii) anatoxin-a(s) blocks acetylcholinesterase similar to organophosphate pesticides; iii) saxitoxin a carbamate alkaloid that acts like carbamate pesticides by blocking sodium channels. In vertebrates neurotoxins cause rapid paralysis of peripheral skeletal and respiratory muscles. In fish loss of coordination, twitching, irregular gill movements, tremors, altered swimming, death by respiratory arrest are common clinical signs.

Blue green algae are known to produce cyanotoxins which have been linked to killing of fish, wild animals, livestocks and human worldwide. Microcystin or cyclic heptapeptides inhibits protein phosporylases type 1 and type 2 A, causing excess phosporylation of cytoskeletal elements, leading to liver failure. Microcystins are largest group of cyanotoxins which are produced by *Microcystis, Anabaena and Planktothrix*. Signs of poisoning by *Microcystis* include flared gills with difficulty in breathing, weakness to swim. In general fish are ore tolerant to algal toxinscompared to mamals. However, fish can accumulate algal toxins over time.

Prymnesiophytes, (the golden-brown algae).Prymnesium is comprised mainly of toxin-producing species that form harmful blooms usually in brackish water. Blooms of *P. parvum* have been responsible for fish kills and significant economic loss. The *Prymnesium parvum* produces at least two toxins, a cytotoxin and a hemolysin (a protein that lyses red blood cells). These toxins are collectively known as prymnesins and all can alter cell membrane permeability. The *P. parvum* ichthyotoxin affects gill-breathing aquatic animals such as fish, brachiated tadpoles and mollusks

Among the Euglenoids, *Euglena* species produce an ichthyotoxin in freshwater. The toxin kill many fishes, as a neurotoxin. The toxin is water soluble, non-protein, heat stable (30 °C for 10 min), freezing stable (-80 °C for 60 days).

Dinoflgellates, Marine environment. produce neurotoxins, causing paralytic shellfish poisoning having transient effcts on fish such as effect on feeding and poor refelx to predation

References

Firew Admasu and Mulugeta Wakjira 2021, Non -infectious disesaes and biosccurity managemnrt practices of fishes health in aquaculture, Journal of Fisheries Science vol 15 No 5

Franciele Fernanda Kerniske, Jonathan Pena Castro,Luz Elena De la Ossa-Guerra, Bruna Angelina Mayer, Vinícius Abilhoa, Igor de Paiva Affonso, and Roberto Ferreira Artoni, 2021Spinal malformations in a naturally isolated Neotropical fish.Published online 2021 Oct 20. doi: 10.7717/peerj. 12239

G. M. Hallegraeff. 1993, A review of harmful algal blooms and their apparent global increase Vol 32, Issue 2 Pages 79-99

Non infectious Diseases of warm water fish 2014, In book: Disesaes and disorders of finfish in cage culture, Editors: Woo P.T.K, Bruno D W, Publishers CAB international.

10

Toxicology in Aquaculture

K.M. Shankar

Introduction

Fish is the cleanest animal, gets bath 24 hrs a day provided water is clean without any toxic material. Perhaps this was the scenario world over 5 decades ago. However, in the last few decades due to industrialisation, modernisation of agriculture and animal husbandry, human health care industry and city developments all have contributed to toxic material to water endangering the fish and the consumer, leave alone clean environment. In recent years, even intensive aquaculture also add several chemicals, pesticides, antibiotics contributing to pollution. Natural water bodies such as rivers, lakes, reservoirs, scoastal/marine waters are ultimate receivers of pollutants which also serve as the source of water for aquaculture. Although pollution control by treatment measures and awareness have improved over the years still there is a long way to achieve the clean water. Effect of toxins in aquaculture has multiple angles, differing from that of toxins on land animal. Fish depends heavily on water for respiration, physiology, reproduction, food and feeding and its very sustenance- as a supporting medium.

Aquaculture is expanding world over growing at 6-8% gradually involving more areas and species contributing a major share to total fish production with gradual decline in capture fisheries from the natural water bodies. Invariably, most of the natural water bodies are polluted with one pollutant or the other in the country. In addition, several pollutants are added to the system by various aquacultural practices. Although large quantity of information is available on aquatic toxicology, toxicology in aquaculture from health point view of fish and the consumer is meager and not focused. Against this background, it is very important to study, understand various toxicants, their mode of action, accumulation and effect on fish and the consumer. Toxicology in aquaculture is incomplete unless it includes or addresses safety of the consumers. Further, poisoning of fish in farms and wild deliberately or otherwise needs forensic studies/approach to address the issues.

Kind and Source of Toxicants to Aquaculture Systems

Industry, agriculture including animal husbandry and aquaculture, other domestic human activities account for major source of toxicants to aquatic and aquaculture systems. It is important to remember that all compounds can be toxic if an organism gets too much of it. Important points to be considered for pollution are amount of compounds released and their water solubility for bioavailability, biomagnification, transformation and complex formation in the aquatic animals.

Industries

Metal and metalloids pollutants include essentials metals such as copper, zinc, iron and non essentials such as cadmium, lead, mercury and silver, and metalloids such as arsenic. Oil and its components from oil refineries, ship wreck, spillage from oil tankers accounts for major source in seas, bays and estuaries. Pharmaceuticals and personal care products such as antibiotics, soaps and detergents are other important pollutants in significant quantity. Paper mill effluents are a common these days containing phenolic and chlorinated compounds.

Domestic Activities

Domestic sewage with high organic load, detergents, hospital wastes is a significant polluter of natural water bodies. Although sewage treatment infrastructures are improving, they are inadequate. India according to Central Pollution Control Board produces 72,368 million litres of domestic sewage, could treat only 31,841 million litres/day leaving rest 40,527 million litres /day untreated to natural waters.(TOI, Dec 152021)

Agricultural Activities

Large quantity of fertlisers include nitrates, phosphates and potassium washed from agricultural land cause eutrophication, algal blooms, leading DO depletion and fish kill. Host of chemicals and antibiotics, hormones used in animal husbandry are also a concern. In addition, pesticides which include insecticides, herbicides and fungicides from agriculture are a major source of pollutants to aquaculture system.

Aquaculture Activities

Several chemicals are used in modern aquaculture including weedicides pesticides, antibiotics, hormones and other chemicals such as bleaches iodophores, KMnO4, malachite green Chlorinated hydrocarbons, organo-phosphorous compounds, bleaching powder etc. are used as piscicides, insecticides in aquaculture. Chlorinated hydrocarbons -Aldrin, Endrin are the most poisonous chemical for eradication of fishes in culture ponds, Organophosphates- three organophosphates, viz., thiometon, DDVP and phosphamidon have been found effective in killing fish. Gamaxene and Sumithion are effective in controlling aquatic insects. Bleaching powder with 30% chlorine content, kills all varieties of fishes, including catfishes, murrels, weed fishes and carps. Butox is used for the control of predatory aquatic insects. Antiseptics and antibiotics are used in the transport medium or short-term bath prior to transport. Acriflavin, Methylene blue, Formalin, $KMnO_4$ are used for foot dips (hatcheries, processing hall, etc.), treating plant substrates for egg collection, site of injection, spent brooders, etc. Malachite green is used at 2 ppm for treating trout eggs to control fungi. Iodophore having 100 ppm available Iodine is used as an anti-viral agent. Growth and sex converting hormones are employed in both table and aquarium fish industry. Most aquaproducts in Andrapradesh have significantly high amounts of antibiotics residues and heavy metals, a study conducted by Federation of Indian Animal Protection Organisation (FIAPO) for animal welfare, public health and environmental hazards. In addition, aquaculture itself generates pollutants/ toxins to aquaculture system and natural water bodies through the application of feed and fertilisers. Source and nature of these unusual toxins in aquaculture needs elaboration.

Aflatoxins

Aflatoxins are a family of toxins produced by fungi *Aspergillus flavus* and *Aspergillus parasiticus*, which are abundant in warm and humid regions of the world in foods such as maize (corn), peanuts, cottonseed, and tree nuts. These food ingredients are used in fish feed and serve as the source of aflatoxins in fish farm. Several types of aflatoxins are produced naturally. Aflatoxins are a family of toxins which are powerful carcinogens and mutagens. Consumption of the toxin at 1 mg/kg cause aflatoxicosis. The toxin is also health threat to livestock-liver damage and cancer in animals. In fish aflatoxin causes liver damage, tumor, slow growth and immunosuppression. Depending on the chemical characteristics of an individual, mycotoxin and the biotransformation abilities of the different fish species, certain mycotoxins could be found in the edible parts of a fish. In fish fed with the toxin, main target is liver , but the toxin also found in fish muscle concern to human health. Thus, the consumption of fish products increases the potential risk of mycotoxin exposure for humans due to their high prevalence, incidence, toxicity, and/or stability as they pass into the food chain.

Toxins from Algal Blooms

Fertilisers either from aquaculture or agriculture run off cause eutrophication encouraging algal blooms. Fish kills are common in farms due to algal blooms causing dissolved oxygen depletion and toxin production. Algal toxins can cause problems in aquaculture such as off-flavor, indirect toxicity through changes in water quality or direct toxicity to both vertebrates (fish) and invertebrates (shellfish) and fish food organisms (zooplankton). The most common toxin-producing algae are cyanobacteria (blue green algae), golden algae (*Prymnesium parvum*) and euglenoids inhabiting fresh, brackish, marine and hypersaline waters. Microcystin are neurotxins, dermatotoxins and hepatotoxins. Microcystin detected in liver of fish cause liver and spleen damage. Euglenophycin-neurotoxins, and hepatotoxins are Ichthyotoxic. Algal toxins are organic molecules produced by a variety of algae in marine, brackish and fresh waters. They are problem in aquaculture when they are produced in sufficient quantities, with sufficient potency, to kill cultured organisms, decrease feeding and growth rate, cause food safety issues, or adversely affect the quality. The term harmful algal bloom (HAB) is used to describe a proliferation of algae, or phytoplankton effects of which vary widely. Some algae are toxic only at very high densities, while others can be toxic at very low densities (a few cells per liter). Some blooms discolor the water (thus the terms "red tide" and "brown tide"), while others are almost undetectable with casual observation. HABs can cause serious economic losses in aquaculture if they kill cultured organisms or cause food safety problems in consumers.

Toxicokinetics of Toxins

Bioavailabilty, uptake, metabolism, storage and excretion of chemicals constitute toxicokinetics.

a) Bioavailabilty is the fraction that can be taken up by an organism compared to the total amount available. Factors affecting bioavailabilty of a chemical compound depend on route of uptake, whether chemical is in the water or sediment or in an organism. Bioavailabilty decrease with increased lipophilicity, amount of dissolved organic carbon or colloids in aquatic phase. In sediment grain size and organic matter determines the bioavailabilty. Main abiotic factors determining bioavailabilty are oxygen and pH.

b) Chemical uptake in fish as a generalisation is through gills and intestine. Uptake can also takes place through general body surface. Hydrophilic compounds reach uptake easily but rapid uptake requires a carrier.

Lipophilic compounds are poorly transported in aqueous media, but after reaching the vicinity of uptake sites, are effectively taken up and their primary source are sediment and food organisms. The lipophilic contaminants bioaccumulate in the organism and biomagnify along the trophic chain.

c) Chemical distribution in the organisms: Once in the organisms, chemicals are either metabolised or stored. Metabolism includes both toxic effects and detoxification. In addition to metabolism storing in inert material can be considered detoxification mechanism. Release of chemical in inert material amounts to delayed toxicity. The route of entry may affect distribution and effect of contaminants. Lipophilic compounds (PCBs, dioxins) stored in fat tissues and released slowly when animals starve. Storage sites of contaminants are the skeleton both exo and endoskeleton and fat tissues. The route of toxicant uptake affects metabolism, distribution and storage of chemicals. Toxicants taken up by the gills leads to faster absorption compared to those in food from intestine and then released to blood for further distribution to detoxifying centres, liver in fish and hepatopancreas in shellfish. Usually toxicant absorbed in intestine decreases as a result of first pass- metabolism which do not occur through gills.

Detoxification/ Biotransformation

Toxicants can be rendered less harmful by biotransformation, making non harmful complexes and then stored in inert state. Finally the toxicants are excreted. Thus detoxification has events before excretion and excretion closely linked to detoxification. Organic compounds are detoxified by enzymatic biotransformations, and metals via formation of complexes with short proteins - sulphur containing metallothineins. Detoxification by compartmentalisation also occurs in fish where compounds are transferred to inert fat tissues, microbes and algae stored in vacuoles.

Biotransformation of Xenobiotics in Fish and Its Consequences on the Residue

Exposure of aquatic organisms to xenobiotics is unavoidable and may occur deliberately (drugs, feed additives) or unintentionally (pesticides, feed contaminants, industrial pollutants). In fish, the most important routes in which chemicals can enter the body is through the gills and the gastrointestinal tract, although the skin can play a role in the uptake of some lipophilic compounds such as polychlorinated biphenyls (PCBs). After entry into the blood circulation, xenobiotics are distributed into the organs upon which they exert their biological actions and which are responsible for their metabolism and elimination from the body.

Absorption and distribution from the circulation to the tissues generally involve passive diffusion across the membranes resulting in a trend for highly lipophilic compounds for bioaccumulation in body fat. Once a xenobiotic has entered the organism, it may be: (a) eliminated unchanged (b) be retained unchanged or (c) undergo chemical or biochemical transformation. Of these, enzymatic transformation predominates and is the main defense mechanism for fish to avoid concentration of foreign compounds in tissues. It consists in the metabolism of lipophilic chemicals into more polar, readily excretable molecules and is characteristically a biphasic process. In this process the xenobiotic first undergoes a functionalization reaction of oxidation, hydrolysis, or reduction (functional metabolism) which serves to introduce or reveal within the compound a functional group able to participate in the second reaction of conjugation. This involves linkage of the metabolite to an endogenous moiety, the conjugating agent. Some compounds will undergo only functionalization while others already possess a functional group which is suitable for conjugation.

Many of the biotransformation reactions which have been described for xenobiotic substances in mammals have been demonstrated in fish in both *in vitro* and *in vivo* experiments. Several of these biotransformation reactions have been shown to occur in fish at rates which are sufficient to have significant effects on the toxicity and residue dynamics of selected chemicals. Inhibition of these reactions can lead to increased toxicity and bioaccumulation factors for certain chemicals. Several classes of compounds, including some polychlorinated biphenyls, are metabolized slowly, and their disposition in fish may not be influenced to any great extent by biotransformation. Metabolites of compounds which are biotransformed rapidly may appear in certain fish tissues, and in many instances these are not accounted for by conventional residue analysis methods. Microsomal mixed-function oxidases in several species of fish have been demonstrated to be induced by specific polycyclic aromatic hydrocarbons and by exposure of fish to crude oil. Induction of these enzymes in fish can result in both qualitative and quantitative differences in the metabolic disposition of xenobiotics to which fish are exposed.

Xenobiotic metabolism takes place in many parts of the fish. Biotransformation can occur in several extra-hepatic tissues, however, in fish as in mammals the liver is the most important organ for biotransformation and the rate of metabolism is substantially higher in hepatocytes than in any other cells.

Following biotransformation, metabolites of xenobiotics are eliminated in the urine, via the bile into the feces, or by diffusion through the gills into the aquatic environment. Although metabolism usually enhances excretion of foreign compounds by converting them into more water- soluble forms, in some cases the metabolites are more persistent than the parent chemical, for example benzo(a)acridine.

Factors Associated with the Metabolism of Xenobiotics in Fish

Several factors are associated with the metabolism of xenobiotics in fish. A number of chemical, physiological and environmental factors influence the metabolism of xenobiotics in fish. Chemical factors such as exposure and dose, molecular weight, water solubility, lipid solubility, ionisation and protein binding determines the fate of toxicants in fish. Environmental factors such as temperature, salinity, other contaminants play a key role in metabolism of toxicants in fish. Water temperature influences the biotransformation of chemicals by fish-a lower uptake and elimination at lower temperature and higher uptake and elimination at higher temperature. A number of contaminants can increase the activity of xenobiotic metabolizing enzymes. These contaminants include polycyclic aromatic hydrocarbons, dioxins, PCBs, steroid-like compounds, etc. These compounds may also stimulate their own metabolism. For example, exposure to the carcinogenic benzo(a)pyrene can lead to induction of the oxidation to the active metabolites, resulting in a marked increase in the amount of residues bound to DNA may influence the metabolism and should be kept in mind when assessing the fate of exogenous compounds in fish.

Nature of fish such as age, sex, lipid content, species and stress determines the metabolism of toxicants. Profound qualitative and quantitative species differences in metabolism may exist from one species to another. Therefore, there is inherent dangers in extrapolating xenobiotic metabolism and residues data from one species to another unless the relative behavior of the chemical in the different species has been established.

The rate of metabolic transformation is a dominant factor in determining the extent to which xenobiotics concentrate in fish. Biotransformation pathways may produce toxic residues and/or persistent residues. Physicochemical, physiological and environmental factors influence the metabolism of xenobiotics in fish. Consequently, the identification of metabolic capabilities is a crucial point in evaluating the persistence of residues in fish and could be helpful in exploration of data from species to species. Analytical schemes for monitoring residues of xenobiotics in fish must be designed to detect metabolites in addition to the parent compound. Bile is a good source of biotransformation products (metabolites). Bile often contains high concentrations of metabolites. Bile could be useful both as an aid in residue monitoring studies and to determine metabolic profile of a chemical/drug used in aquaculture. The excretion of toxicants from fish is often called phase 3 of biotransformation/ detoxification. Excretion includes both cellular excretion and excretion from the organisms by the gills, kidney and intestine mainly in bile. Fish excrete hydrophilic compounds via gills.

Interaction Between Environmental Factors and Toxicity

Temperature, oxygen, salinity, pH modify toxicological responses. Often these environmental parameters change reaching stressful levels and consequently the environmental stresses are conpounded with toxicant stress.

Temperature affects the rate of chemical reactions within the organism, where the detoxification and overall metabolism are temperature dependent. Temperature also influences the influx and efflux of contaminants to and from organisms. Interaction between oxygen and contaminants depend partially on the fact that hypoxic conditions are reducing and hyperoxic oxidising. Changes in the redox balance influence both the stability and effects of contaminants. Contaminants and oxygen transport also interact at the level of gills, as many toxicants affect gill structure and function. In addition to oxygen transfer, gills are the major site of ion and pH regulation whereby contaminants effect on them result in interaction between contaminants and ion and pH regulation. Salinity particularly affects metal toxicity partly differences in uptake and efflux in marine and fresh water environments.The effect of PH on toxicity depends partly on the toxicities of acid and base forms of weak acid are different, with the dissociated acid form being more toxic.

Residues of Toxins

Persistent residues correspond to compounds which cannot be extracted from the fish tissues by extensive treatment with organic solvents. Usually such residues are due to the metabolic activation of a chemical to reactive intermediates which can bind covalently to cellular macromolecules. The task of characterizing bound residues is bedeviled by the low concentrations and low solubility of these materials which can be detected through the use of radiolabeled xenobiotics. However, toxicologists remain suspicious that a portion of the bound residues could possibly retain the activity of the parent compound or may cause adverse effects to the consuming organism. A typical example is furazolidone, a nitrofuran drug used for disease control in aquaculture. In general, the determination of xenobiotic residues is performed in the edible part of fish, which mostly is muscle and skin, and less frequently liver and eggs. However, it may be attractive to analyze other specific tissues or body fluids (e.g. bile) in which the parent compound and/or metabolite concentrations are elevated.

Effects of Toxicants on the Fish

Both acute and chronic toxicity scenario occur depending on concentration and exposure duration of toxicants. Higher concentration of toxins directly cause mass mortality. Lower concentration of pollutants cause chronic toxicity which affects growth, reproduction and general well being of fish. Genotrophic effects range from point mutations to DNA strand breaks. DNA damage due to formation of DNA adducts by many toxicants. Reproductive and endocrine disruption caused by disturbances of reproductive hormones cycles as a major cause. Others involve in disturbances of sperm function and development of egg. Nerve functions are specially affected by insecticides. Toxicants also hinder energy production. Some toxicants affect the fluidity of cell membrane affecting membrane permeability and lateral movement of proteins. Several toxicants cause apoptotic cell death. Toxicants known to affect a range of immune function. Terratogenesis (formation of abnormal structures) and carcinogenesis can have several causes. Toxicant also affect on the behaviour of fish

Investigation of Fish Kills Due to Toxins and Pollutants in Farms and Natural Water Bodies

Fish in aquaculture and natural water bodies often are killed due to toxins from outside or those generated in the system. Occasionally fish are also poisoned by people for theft or enmity. Besides local poisonous plants, often pesticides are employed for killing. Under these circumstances the fish kills need to be investigated and analysed for further prevention. Certain standard procedure needs to be followed for investigation.

1. First step, visiting the sites and collect information on case history. 2. Noting the species involved and their biology, habitat, requirements 3. Death status body condition and collecting preferably freshly dead or moribund species in sufficient numbers.4. Analysing the water *in situ* particularly dissolved oxygen, pH, ammonia and collecting water samples for further detail analysis in the laboratory. 5 Transporting all the samples(water, fish) to the laboratory in ice.6. Initiating detail laboratory analysis of water for DO, pH, CO2, ammonia nitrite.7. Subjecting the fish samples for microbiological examination - TPC specific isolation for pathogens.8. Noting down external damage to surface

gills and subjecting samples to histopathology.9. Subjecting the samples for chemical analysis for suspected poison/ toxin- pesticides. 10. Recommending remedial measures such as dilution of water, aeration/oxygenation of water detoxification using antidotes, oxidising chemicals such as KMnO4, Ozone application of algicides for bloom control.

Consumer Safety Regulations of Drugs and Chemical Used in Fisheries/ Aquaculture

The use of drugs in food animals is fundamental to their health and well-being and to the economics of husbandry. A food contaminant means any substance not intentionally added to food, which is present in such food as a result of the production, processing, preparation, treatment, packing, transport or holding of such food or as a result of environmental contamination. Furthermore supply of foods of animal origin from 'farm-to-table' also requires use of one or the other xenobiotic viz., antimicrobials, growth promoters, hormones ectoparaciticides, food processing chemicals, preservatives etc. Thus, it would result in entry of residues of active ingredients and or metabolite(s) in human food chain. In addition to these xenobiotics, pesticides, herbicides, industrial and agriculture and other environmental contaminants also present risk to the food supply which pose public health concern because of their potential risk to carcinogenesis, mutagenesis, teratogenesis, altered or contributing to bacterial resistance to antimicrobial agents. Therefore, there is necessity to establish Maximum Residue Limits (MRL) to define safe residue concentration in food /fish that will protect the consumer from ingesting harmful residues.

Withdrawal Periods

Several national and international regulations have been implemented to approve drugs and secure safe products. In the existing regulations a main element is the establishment of scientifically based 'Withdrawal Periods'. These withdrawal periods are defined as the minimum time to elapse between terminated treatment and harvest of the organism in question, and are specific for each drug. For each substance, the period to reach acceptable residue level is prolonged with an extra safety margin in case of individual fish to fish variations in the elimination of the drugs. Withdrawal periods (time) should be established for each antimicrobial drug, which make it possible to produce food in compliance with the MRLs. Withdrawal periods have to be established by taking into account: i)MRLs established for the considered antimicrobial drug, ii)pharmaceutical form, iii)target animal species, iv)dosage regimen and the duration of treatment and route of administration. The establishment of an MRL allows the setting of a withdrawal period for the product. The withdrawal period the time between last medication and earliest date of harvest must be sufficient to deplete residue levels below the MRL. The withdrawal period refers to the specific formulation and presentation of a complete product. In most food animal species, withdrawal periods are given in elapsed days, but this differs with fish as these are poikilotherms metabolic rate is determined by temperature. Therefore, for fish with drawl period should be based on trials conducted with at least two water temperatures relevant to the proposed conditions of use. Furthermore, very important withdrawal period should be mentioned in degree days which depends on temperature and time.

Depuration of Aquatic Animals for Human Consumption

Fish raised in toxic or polluted waters are often risky to fish and the consumer. Either water source is polluted or certain toxins are produced in farm due to aquacultural practises such as addition of drugs, chemicals or eutrophication. Under these circumstances fish have to be depurated in fresh clean water for considerable time before marketing. This process will purge toxic materials including harmful microorganisms from the fish. Depuration depends on type of fish(lean or fat). Normally it is essential to depurate common pollutants-compounds causing off flavour/bad taste, heavy metals and banned chemicals including antibiotics. Exact depuration has to best standardised for each toxicant and species of fish against a toxicant.

Terpinoids, geosmin, 2 methylisoborneol -produced in pond/RAS by microbiota cause off flavour/taste in several fish, salmon, trout, baramudi, tilapia. These off flavour concentration is high in fatty tissues and is reversible. Generally off flavor can be reduced with 16 days depuration. Heavy metals(Cu, Fe, Zn, Mg and Pb) in cultured fish often from industrial pollution can be removed by 2-3 months of depuration. Banned antibiotic such as Chloromphenicol (CMP) can be reduced in shrimp to 90 % by depuration in 5 days. Microbial contaminants normally are removed in 48 hrs of depuration

Codex Committee on Residues of Drugs

The *Codex Alimentarius* is a global institution initiated by the UN and affiliated under the World Health Organization (WHO) and the Food and Agriculture Organization (FAO). The main aim of the *Codex Alimentarius* is to protect the health of the consumer and facilitate fair trade by the elaboration of common international food standards. At the international level, the Codex Committee on Residues of Veterinary Drugs in Foods has the responsibility of setting acceptable limits to concentrations of drug residues in foods for human consumption, including farmed fish. These 'acceptable concentrations' are expressed as Maximum Residue Limits (MRL) and are based on a careful toxicological evaluation of the substance, combined with information on food consumption patterns. Consumption of food with drug residues below the MRL should not, with large safety margins, give any negative health effects for the consumer. The MRLs are given for each substance either for all food producing species or for groups of related species such as salmonid fish. Products containing illegal drugs or legal therapeutics in concentrations above the MRL cannot be sold for human consumption.

Maximum Residue Limits (MRL)

A Maximum residue limit (MRL) is 'a level of residue that could safely remain in the tissue or food product derived from a food-producing animal that has been treated with a veterinary drug. This residue is considered to pose no adverse health effects if ingested daily by humans over a lifetime. *The* Codex Committee on Residues of Veterinary Drugs in Foods (CCRVDF) determines priorities for the consideration of residues of veterinary drugs in foods and recommends MRLs for veterinary drugs. A Codex Maximum Limit for Residues of Veterinary Drugs (MRL) is the maximum concentration or residue that results from the use of a veterinary drug (expressed in mg/kg or g/kg on a fresh weight basis) recommended by the *Codex Alimentarius Commission* to be legally permitted or recognized as acceptable in or on a food.

An MRL is based on the type and amount of residue considered to be without any toxicological hazard for human health as expressed by the Acceptable Daily Intake (ADI), or on the basis of a temporary ADI that utilizes an additional safety factor. An MRL also considers public health risks as well as food technology issues. When establishing a MRL, residues that occur in food of plant origin and/or the environment are also considered. Furthermore, a MRL may be reduced to be consistent with good practices in the use of veterinary drugs to the extent that practical and analytical methods are available.

The Official MRL is defined for each country viz., USA, Canada, and Australia, Japan or group of countries in European Union (EU). MRLs for a given drug/ chemical residue frequently change, although international MRL database is updated frequently, the information in it may not be completely up-to-date or error free. Additionally, commodity nomenclature and residue definitions vary between countries.

Establishing MRLs for fish presents several problems, including the identification of what are edible tissues and the complex pharmacokinetic properties and metabolism of veterinary drugs in fish. Currently, the only full CAC MRLs for aquaculture species are for the administration of oxytetracycline at 100 µg/kg to 'fish' and 'giant prawn', but several additional MRL proposals from JECFA are now within the CAC system. Only two aquaculture MRLs are approved for fish and shellfish in Japan: 0.2 ppm for oxytetracycline and 0.2 ppm for spiramycin.

MRL Fixing Relies on Three Essential Concepts

When establishing a MRL, residues that occur in food of plant origin and/or the environment are also considered. Furthermore, a MRL may be reduced to be consistent with good practices in the use of veterinary drugs to the extent that practical and analytical methods are available. MRL for a given residue in target species is fixed in order to ensure drugs consumed in food (through milk, meat, fish, egg etc.) does not exceed acceptable daily intake (ADI). Fixing of MRL relies on three things viz., search for the No-Observed-Effect Level (NOEL) with different biological tests. b. NOEL and safety factors (of 100 and 1000) calculation of an ADI is made and c) knowledge of the average food ingestion per inhabitant and the repartition in the different tissues and organs. The final fixing of official MRL will be the lowest obtained between toxicological MRL and bacteriological MRL. Experience shows that, for antibiotics, in most cases the bacteriological MRL is lower than the toxicological MRL.

Government of India Food Safety and Standards: MRL for Aquaculture

India is signatory to World trade organization (WTO) and to meet a country's sanitary and phytosanitary requirements, food must comply with the local laws and regulations to gain market access. The Government of India vide extraordinary Gazette Notification has amended the earlier Order(s) pertaining to Ministry of Commerce (GOI) relating to maximum residual limits (MRLs) for antibiotics, pesticides and heavy metals for trade of fresh, frozen and processed fish and fishery products (Tables 1 to 4). In the said Order vide Gazette Notification (dated 10th July 2002) under Schedule – I, for clause (e), the following has been inserted namely: Maximum Residual Limits (MRLs) for pesticides, heavy metals and antibiotics and other pharmacologically active substances in fish and fishery products shall meet the requirements as given below. However, if the MRLs fixed by the importing countries are more stringent than these prescribed limits, the standards specified by those countries will be complied with.

Table 1: Pesticides

Pesticides	Max. permissible residue (ppm)
BHC	0.3
Aldrin	0.3
Dieldrin	0.3
Endrin	0.3
DDT	5.0

Table 2. Heavy Metals

Heavy Metals	Max. permissible residue (ppm)
Mercury	1.0
Cadmium	3.0
Arsenic	75
Lead	1.5
Tin	250
Nickel	80
Chromium	12

Table 3: Antibiotics & other pharmacologically active substances

Antibiotics	Max. permissible residue (ppm)
Tetracycline	0.1
Oxytetracycline	0.1
Trimethoprim	0.05
Oxolinic acid	0.3

Further, the use of any of the following antibiotics and other pharmacologically active substances shall be prohibited in the culture of; or in any hatchery for producing the juveniles or larvae or nauplii of; or in any unit manufacturing feed for or in any unit pre-processing or processing shrimp, prawns or any other variety of fish and fishery products: All nitrofurans (including furaltadone, furazolidone, furylfuramide, nifuratel, nifuroxime, nifurprazine nitrofurantoin, nitrofurazone), chloramphenicol, neomycin, nalidixic acid and sulphamethoxazole. *Aristolochia* spp and preparations thereof: chloroform chlorpromazine, colchicines, dapsone, dimetridazole, metronidazole ronidazole ipronidazole, other nitroimidazoles, clenbuterol, diethylstilbestrol (DES) sulfonamide drugs (except approved sulfadimethoxine, sulfabromomethazine and sulfaethoxypyridazine), fluoroquinolones and glycopeptides.

The Indian Parliament has passed the Food Safety and Standards Act, 2006 that overrides all other food related laws. The new regulatory requirements with respect to all food commodities are under consideration.

Zero Residue and MRL Concepts

The transition from the concept of **'zero residue'** to that of MRL became unavoidable by the development of analysis methods. Indeed, further to a treatment, the elimination of the molecule by the organism progressively tends towards 'zero' but over very long periods. The detection threshold was progressively divided by one hundred or a thousand with the development of the analysis methods.

With the available methods, the concept of 'zero residue' *de facto* tended to make the use of many medicines impossible for economic reasons. The transition to the MRL concept became possible with the development in toxicology knowledge: a. Demonstration of the absence of effect at low doses/ exposure and b. progressive acquisition of methodological knowledge allowing to perform assays guarantee the **'No-observed-effect-level'[NOAEL]**.

Table 4: MRL for few commonly found residues of veterinary drugs in food producing species (FPS) including fish

Drug	Foods	Tissue	MRL μg.kg^{-1}
Amoxicillin	All FPS	Muscle	50
Ampicillin	All FPS	Muscle	50
Benzyl-penicillin	All FPS	Muscle	50
Chlortetracycline	All FPS	Muscle	100
Oxytretracycline	All FPS	Muscle	100
Enrofloxacin	All FPS	Muscle	100
Florphenicol	Fish	Muscle + skin	100
Thaimphenicol	All FPS	Muscle	50
Flumequine	Fish	Muscle +skin	600
Oxacillin	All FPS	Muscle	300
Oxolinic acid	Fish	Muscle	100
Sarafloxacin	Poultry/Salmonids	Muscle +skin	30
Trimethoprim	All FPS	Muscle	
Tilmicosin	All FPS	Muscle	50
Tylosin	All FPS	Muscle	50
Azamethiphos	Salmonids	Muscle	
Cypermethrin	Salmonids	Muscle+skin	50
Deltamethrin	Fish	Muscle+skin	100

Toxicological MRL vs Bacteriological MRL

Toxicological MRL is set to ensure the consumer's safety. This concept includes all elements related to the molecule's toxicity in the short or long term and whatever the nature of the effects observed on the individual or his descendants. Bacteriological MRL aims at guaranteeing the absence of antibiotic residue effects on human digestive flora; whether this flora modification has an effect on the human being or not, the limit is taken into account independently.

Acceptable Daily Intake (ADI)

The FAO/WHO Expert Committee on Food Additives (JECFA) defines the amount of a veterinary drug, expressed on a body weight basis that can be ingested daily over a lifetime without appreciable health risk (standard man–60kg). In other words, it is the dose of drug or residue that during the entire lifetime of a person appears to be without appreciable risk to health on the basis of all the facts known at that time. Normally a two year chronic study is required to be conducted to determine ADI. Additionally, a safety coefficient of 100 or 1000 (if it is a teratogen) or more is applied to determine the ADI. The safety coefficient takes into account the transfer from animal to human and the population heterogeneity (age, sex, weight).

When setting ADIs and MRLs for veterinary antimicrobial drugs, the safety evaluation is carried out in accordance with international guidelines and should include the determination of microbiological effects (e.g., the potential biological effects on the human intestinal flora) as well as toxicological and pharmacological effects. An acceptable daily intake (ADI) and a maximum residue limit (MRL) for appropriate food stuffs (i.e. meat, milk, eggs, fish and honey) should be established for each antimicrobial agent.

ADI = NOEL mg.g-1 body weight / 100

Once the ADI is determined, then safe concentration (SC) or tolerance is constructed. Tolerance is essentially lent to maximal residual limit (MRL).

Safe concentration (SC) = ADI x Human weight/ Daily tissue intake

Extra-label Use of New Animal Drugs in Food-Producing Animals

'Extra-label use' is defined as 'actual use or intended use of a drug in an animal in a manner that is not in accordance with the purpose approved on the label'. Under the provisions of the Animal Medicinal Drug Use Clarification Act of 1994 **(AMDUCA)**, these drugs can be used for purposes beyond those approved in a situation if the health of an animal is threatened, or suffering or death may result from failure to treat. Extra-label use of medicated feed in aquaculture is limited to medicated feed products approved for use in aquatic species. Since no tolerance is set for off label use, adequate withdrawal times must be given to avoid any illegal drug residues occur in food-producing animal subjected to extra-label treatment.

However, under AMDUCA provisions there also exists to specifically prohibit the extra-label use of specific drugs in food-producing animals. This prohibition particularly when lack of adequate analytical methods for the drug or if the extra-label use of the drug resents a risk to the public health. Currently Chloramphenicol, Dimetridazole, Furazolidone and Nitrofurazone, and, Fluoroquinolones are prohibited for extra-label animal and human drug uses in food-producing animals.

Minimum Performance Limit (MRPL)

Minimum performance limit (MRPL) is defined as 'minimum content of an analyte in a sample, which at least has to be detected and confirmed' and is the reference point for action in relation to the evaluation of consignments of food. For several substances which have been expressly prohibited from use in food producing animals in the EU (*e.g.* chloramphenicol, nitrofurans), or not authorized (*e.g.* malachite green), the concept of the minimum required performance limit (MRPL) has been established in Commission Decision 2002/657/EC (Table 5). With regard to each of these EU limits/levels, Member States are required to ensure that they have validated laboratory analytical methods in place which are capable of meeting these thresholds.

Table 5: Some of the established Minimum Required Performance Limits (MRPLs)

Substances and/or Metabolite	Animal Product	MRPL
Chloramphenicol	Meat, Eggs, Milk, Urine, Honey, Aquaculture products	0.3 $\mu g.kg^{-1}$
Medroxyprogesterone acetate	Pig kidney fat	1 $\mu g.kg^{-1}$
Nitrofuran metabolites:Furazolidone Furaltadone Nitrofurantoin Nitrofurazone	Poultry meat All aquaculture products	1 $\mu g.kg^{-1}$
Sum of the malachite green and leucomalachite green	Meat of aquaculture products	2 $\mu g.kg^{-1}$

References

Cravedi, J. P. 2002, Role of biotransformation in the fate and toxicity of chemicals: consequences for the assessment of residues in fish. Revue Méd. Vét., 153:419-424.

Jimenez, B.D., Cirmo, C.P and Mc Carthy, J.F. 1987, Effects of feeding and temperature on uptake, elimination and metabolism of benzo(a) pyrene in the bluegill sunfish (Lepomis macrochirus). Aquatic Toxicol., 10:41-57.

Mikko Nikinmaa, 2010, An introduction to Aquatic toxicology, Academic Press, Elsevier,ISBN:978-0-12-411574-3, PP 233

Prakash Nadoor, 2016, Biotransformation of xenobiotics in fish and its consequences on residue analysis. In Aquatic medicine-a training manual, Special publication No 01/2016, Editor Prakash Patil, Aquatic Animal Health Management Laboratory, College of Fisheries, Karnataka Veterinary, Animal and Fisheries Sciences University, Mangalore,575002

Prakash Nadoor, 2016, Drug - chemical residues in foods of animal origin including aquaculture and their regulatory implications, In Aquatic medicine-a training manual, Special publication No 01/2016, Editor Prakash Patil, Aquatic Animal Health Management Laboratory, College of Fisheries, Karnataka Veterinary, Animal and Fisheries Sciences University, Mangalore, 575002

Times of India,15.12.2021. Domestic waste discharge to natural water bodies in India-A study by Central Pollution Control Board.

Times of India 2021- Aquaproducts from AP- A study by Federation of Indian Animal Protection Organisation (FIAPO)

11

Epidemiology and Biosecurity in Aquaculture Health Management

K.M. Shankar

Epidemiology is the study of diseases in a population in its natural setting. In epidemiology, the population is the patient and hence epidemiology is population medicine. All epidemiological studies are field based. In aquatic epidemiology, aquaculture ponds serve as the laboratory. This can be a great advantage over laboratory studies, which cannot be easily extrapolated to field conditions.

Diseases are among the greatest deterrents to the sustained production in aquaculture. Diseases such as Epizootic Ulcerative Syndrome (EUS) in fishes and White Spot Syndrome (WSS) in shrimps have devastated aquaculture with bad impact on productivity and profitability. Various diagnostic and health management tools are being used to devise disease control strategies. It is becoming increasingly difficult to manage pathogens that have become endemic. Epidemiological approaches which have been very successful in human and veterinary medicine are now being increasingly recognised as important weapons that can be used in aquaculture to formulate disease control programmes.

Concept of Cause for a Disease

Cause from an epidemiological perspective is interpreted in quite a wide sense. This is somewhat different to the more traditional view of the role of cause being restricted to etiological agents. An epidemiological definition of a cause of a disease is "an event, condition or characteristic that plays an essential role in producing an occurrence of the disease". Epidemiologists avoid defining the word cause for any disease outbreak, but prefer to use words such as determinants, exposures and risk factors. Alternatively, they will categorise causes as direct or indirect; necessary or sufficient; and single or multiple rather than defining primary etiology.

For most diseases, there is strong evidence that disease outbreaks occur only when a number of causal factors combine. The multifactorial nature of disease causation can be represented using the concepts of necessary cause component cause and sufficient cause. Each combination of various causal factors (component causes) which together cause a disease collectively as a "sufficient cause" for that disease. It is important to recognise that, under different circumstances, different combinations of "component causes" may constitute sufficient cause for a disease. Disease outbreaks will occur only if there is a sufficient cause for that disease. Moreover, all sufficient causes for a particular disease have in common at least one component cause, known as "necessary cause". This necessary cause must always be present for that disease to occur. Mere presence of necessary cause alone will not always cause the disease.

Under such a definition, the presence of say WSSV (necessary cause) alone in a pond of shrimp may not of itself be a sufficient cause for WSS to occur. It may require a stress trigger to cause an outbreak. Under this concept, the WSSV is a necessary cause (no disease would occur if WSSV was not present) but not a sufficient cause of the particular syndrome, whereas the stress is neither necessary nor sufficient but can be a component of a sufficient cause. In fact, for any particular expression of a particular disease, there may be a range of possible sufficient cause complexes.

Similarly, presence of *Aphanomyces invadans* (necessary cause) alone is not sufficient for EUS to occur. For EUS to occur, combinations of causal factors must ultimately lead to exposure of dermis, attachment to it by *Aphanomyces invadans* propagules (the only currently recognised necessary cause) and subsequent invasion by the fungus of dermis and muscle. The resulting mycotic granulomatous dermatitis and myostitis are, by case definition, EUS. Interventions aimed at eliminating any one of the component causes can prevent occurrence of EUS despite the presence of the necessary cause.

On the other hand, the etiology of a disease outbreak may be well defined, however, the cause of the same outbreak may be confusing. For example, two ponds may test PCR + for WSSV but only one pond may experience a disease outbreak. In this instance, WSSV (necessary cause) must be present for the outbreak to occur but the presence of WSSV in the pond doesn't necessarily lead to an outbreak. Likewise in two PCR+ ponds, one may experience high mortalities while the other low mortalities. Therefore, there are other risk factors (component causes) which would determine whether the disease is expressed and if it is expressed, the severity of the outbreak.

Risk Factors Associated with a Disease

Risk factors are those characteristics on the basis of population based evidence which are associated with increased risk of disease. Protective factors are those which are associated with decreased risk of disease based on a population evidence. Risk factors may be either causal or non-causal. Epidemiological studies have the objective of identifying these risk factors, quantifying their effect on outcome, and formulating intervention strategies in a pond, farm region or country level. The challenge for the epidemiologist is to help identify some of the more important components of sufficient causes for a particular disease with the view to devise cost-effective intervention strategies at critical points to either prevent disease expression or reduce the production effects.

Epidemiologists use three rules to reject a hypothesised cause (risk factor). The first is association. If a risk factor does not occur more frequently in diseased animals, then it is rejected as a cause. The second is time order. If the risk factor does not precede the outcome, it is rejected. The third criteria used is common sense or coherence. Risk factors are often rejected if they do not make any biological sense.

Outcomes and Associations

Epidemiological studies define certain outcomes and use them for separating the population into groups for comparison. Frequently used outcomes in epidemiology are mortality and disease outbreaks. For example, shrimp ponds may be divided into successful ponds and failure ponds based on disease outbreaks, yield, weight at harvest, length of production cycle, etc. Using epidemiological and statistical tools, it is possible to identify variables (factors) associated with defined outcomes. Variables statistically associated with successful outcomes are regarded as protective factors. Variables statistically associated with failure outcomes are regarded as risk factors. The risk/protective factors identified based on a population are mere associations with the outcome and should not be regarded as either cause of the outcome or solution to the outcome.

Disease as Outcome

In aquatic epidemiological studies it is very common to use disease as outcome. Several aquatic epidemiological studies are being conducted to identify factors associated with outbreaks of WSS and EUS. Whenever, disease is used as an outcome in epidemiological study, it is very important to have a case definition for the disease outcome to maintain consistency. A case definition is a set of standard criteria for deciding whether an individual unit of interest (fish/pond) has a particular disease. If different criteria are used, errors are likely to be introduced. For example for WSS, case definition can include clinical white spots, pink coloration, increasing mortalities, emergency harvest, PCR positivity and intranuclear inclusions of WSSV. Similarly for EUS, surface ulcers, mycotic granulomas in histology, isolation of *Aphanomyces invadans* seasonality and species range can be included in the case definition.

Quantitative Epidemiological Studies

Quantitative epidemiological studies can be broadly grouped into observational intervention and theoretical studies.

Observational Studies

In observational studies, nature is allowed to take its course while the differences or changes in characteristics of the population are studied without intervention from the investigator. Observational studies can be descriptive longitudinal, cross-sectional or case-control. Descriptive studies describe the distribution and frequency of a disease in a population in terms of animal place and time. The described patterns may lead to an hypothesis (for example EUS occurs normally during cooler months of the year). Longitudinal or cohort studies follow study units through time, observe and record the course of natural events, record frequency of disease of interest, look for differences (factors) between groups with and without the disease. Cross-sectional study gives a snapshot in time, prevalence of disease is measured and compared among those with and without the risk factor of interest (ex: PCR positivity of PL). Case-control study selects units with the disease of interest as cases and units without the disease as controls, frequencies of suspected risk factors are then measured for the two groups and compared.

Intervention and Theoretical Studies

Intervention studies are epidemiological experiments imposed at the population level. The purpose is to evaluate some preventive or treatment or management strategy. This population based approach can be effectively used to screen and evaluate some of the commonly used preventive and therapeutic chemicals and formulations (for example Vitamin C, immunostimulants, etc. in shrimp culture). Theoretical approaches use mathematical modelling using computers to answer "what-if" type of questions.

Steps in Designing and Execution of an Aquatic Epidemiological Study

The first thing is to have clear questions or objectives for the study (for example identification of risk factors for WSS outbreaks). This is followed by identification of the target population (shrimp farmers) and the study unit (pond).

At this stage, it is very important to involve and understand the population. Most epidemiological studies of this type will be observational studies. The next step is deciding on the study design. This will depend on the physical and financial resources. Once the study is designed, data collection needs to be planned through questionnaires, measurements, sampling, laboratory tests etc. The data gathered has to be analysed using epidemiological models to identify risk factors associated with defined outcomes.

Sampling is a crucial issue in epidemiological studies. The sample should be random and should represent the target population (Table 1). The results obtained from the sample should have to be generalised to the whole population. In order to achieve this, the sample size should be such that the results could be generalised with 95% confidence to the whole population and should have the power to show the statistical differences between groups. Problems of bias and how to avoid it is of central importance to the validity of all epidemiological studies. Bias occurs when observations do not reflect the true situation. This is because of systematic error such as sample selection bias, sampling bias and measurement bias.

The aim of data analysis is to identify risk factors for disease outbreaks. Risk factor is any variable having statistical association with the disease outbreak of concern. Positive association suggests increased risk of disease (risk factor) while negative association suggests decreased risk of disease (protective factor). Such identified associations are only statistical associations and not the same as the cause. The strength of the association is measured by relative risk (RR)/odds ratio (OR). The RR/OR of 1 indicates no association, > 1 indicates increased risk, <1 indicates decreased risk.

Univariate and multivariate analysis are the methods used for data analysis. In univariate analysis, all the variables are examined individually and looked for associations with the defined outcomes. ANOVA, Chi square, Fishers exact and contingency table are some of the univariate tests. In multivariate analysis, effect of each variable when adjusted for the effects of all others is looked into. Logistic and linear regression models are some of the multivariate tests.

Sampling Size for Screening an Apparently Healthy Fish and Shellfish Population

For diagnostic/epidemiological studies, sampling of moribund/clinically diseased fish, at least 10 specimens that exhibit clinical signs typical of the disease affecting the fish in each lot to be selected for testing. In epidemiological study to screen or estimate the infected appropriate sample size depends on the suspected rate of infection say 5,10 or 15 % and population size in 1000, 1lkh, crores. (Table 1).

Table 1: Sample size based on assumed pathogen prevalence in the lot for epidemiology (source AFS-FHS Blue Book-2014)

Lot size	Sample size at 2 % prevalence	Sample size at 5% prevalence	Sample size at 10% prevalence
50	50	35	20
100	75	45	23
250	110	50	25
500	130	55	26
1000	140	55	27
1500	140	55	27
2000	145	60	27
4000	145	60	27
10000	145	60	27
10000 or more	150	60	30

Relevance of Epidemiology in Aquaculture

Several pathogens have become endemic and established in natural populations. WSSV in shrimp is a good example. It is almost impossible to eliminate some of these well established endemic pathogens like WSSV which is the necessary cause of WSS. Eliminating/avoiding/ minimising any one of the component causes (risk factors) can prevent the occurrence of WSS despite the presence of WSSV. The challenge for aquatic epidemiologists is to conduct population based studies, identify component causes (risk factors) for disease outbreaks, devise cost effective intervention strategies at critical points. These could include strategies to prevent disease expression, to minimise the severity of disease or to minimise production losses. Recent findings from epidemiological studies on EUS have found out several risk factors associated with EUS outbreaks in culture ponds. By managing some of these risk factors, it has been possible to prevent EUS outbreaks despite the presence of the necessary cause. Similarly, epidemiological evidences coming from studies on WSS have successfully identified several risk factors. Managing some of these risk factors it has been possible to minimise the impact of WSS outbreaks despite the presence of the necessary cause (WSSV). Epidemiology holds great potential for aquatic animal disease management.

Biosecurity in Aquaculture

Biosecurity in aquaculture has become a buzz word today, mainly due to rapid expansion and growth in aquaculture activity round the world with an increase in incidences of diseases. Therefore, future dependence and growth projection of aquaculture depends on biosecurity measures which are also included in World Trade Organisation (WTO).

Need and awareness for biosecurity should be viewed from the growth development in disease loss and aquaculture. Aquaculture has witnessed a 6-7% growth annually, contributing close to 55% of total fish production in the country. Further aquaculture in non conventional water bodies such as saline soils, or culture after acclamatising fish to salinity is steadily increasing. Furthermore, in the last decade or so there is a 3 fold increase in number of species adopted for aquaculture in India and elsewhere in the world. Fish crustaceans, molluscs and several fish food organisms constitute, a total of 300 cultivable sp in the world, about 210 sp in the Asia pacific region and about 40 in India.

Fish dominates the culture followed by crustaceans and molluscs. In general fish immunity is not well developed compared to that of higher vertebrates. Furthermore, within fishes there exists difference in immunity development- carps, tilapia catfishes having well developed immunity compared to trouts and salmon. Immunity in crustaceans lacking specificity and memory can not be compared even with that of fish. Aquatic animals depend more on non specific immunity compared specific immunity. Overall, we are playing with these very delicate animals in aquaculture.

Modern aquaculture in the last 30 years has witnessed several pathogens in fish and shellfishes. Pathogens represented by parasites are most predominant (78 %) followed by bacteria (12 %), fungi (3.5 %) Viruses (2.7 %). World Organisations such as NACA, OIE and FAO have already listed 8-10 pathogens seriously affecting fishes and shrimp. Among these control of parasites and viral pathogens has been a difficult task

Biosecurity (BS)

Biosecurity is protection of the aquaculture industry through establishment of safeguards to fish health, the facilities in which fish are reared and the environment. It includes prevention of new pests and diseases and eradication or control of those already present. In simple it is protection of fish and shellfish or aquatic plants from infectious (viral, bacterial, fungal or parasites) and non infectious (toxins) agents. In general, aquatic animals have less developed immune system. Specific immunity with memory exists in fish but not well developed as in higher vertebrates. Specificity and memory yet to be demonstrated in shellfish and hence developing vaccine for shellfish is difficult. Furthermore, for viruses by virtue of their unique biology, developing chemotherapy is very difficult task. Therefore, in the absence of drugs and vaccines against viruses in shrimp, avoidance of the viruses is best strategy. Avoiding the virus through brood, seed, water and other biological materials such as fish food organisms by proper screening before use is suggested. Aquatic environment has great influence on coldblooded fish and shellfish particularly on the immune system. Lack of information on health and pathogens and poor developments in preventive and curative measures further a concern for pathogen management. Therefore there is a need for additional biosecurity for aquaculture compared to that of higher warm blooded animals

Levels of Biosecurity

i.) Farm level

Biosecurity is site specific. It depends on operation type, business needs, farm layout, species, life stages, disease profile of the region and geographic location. Biosecurity at farm level depends on good husbandry, a good understanding of host, pathogen and environment for preventing, control and eradication of the pathogen. Furthermore biosecurity should consider pathogens of both native host and non native host.

There are several internal barriers within a farm such as vaccination for prophylaxis to activate immune system, helping to reduce load of pathogen and reducing infection pressure on host. Treatment options with drugs and chemicals employing flow through or recirculatory methods are also internal barriers for spread of pathogen in a farm.

Eradication of pathogen is a costly process and involves eradication in water, equipment and host. Timely removal of dead and decaying animals along with contingency measures help eradication.

Above all farm managers has an important role- his sincerity, basic knowledge, disease awareness, husbandry awareness, up to date knowledge goes a long way in biosecurity at farm level.

ii. National and Regional level

a) Crop holiday

Whenever disease epidemics are reported due to a pathogen in an area, normally it is advised to go for culling, and practising crop holiday for one or two seasons. With the crop holiday, due to non availability of host, pathogen can not replicate and spread. As a result pathogen load in the environment, in water, other carriers goes down. Once this is ensured, fresh crop begin with good preparation such as disinfection of hatchery/farm, disinfection of water or obtaining new source of water, PCR screening of mother and seed. There are good number of examples of successful crops following crop holiday in trout hatcheries affected with IPN virus and shrimp hatcheries/farm with from White spot virus.

b) Specific pathogen free (SPF) and specific pathogen resistant(SPR) brood stocks

Fish or shellfish SPF stock can be developed against a specific pathogen or pathogens. This exercise involves first selecting pathogen stocks from nature or farms followed by screening for pathogen with PCR. Following building up virus free stock, they are bred and F1 progeny are screened for the pathogen by PCR. The F1 generation are grown and bred to get F2. Again progeny are screened for the virus at several stage of development and bred to F3. The F3 generation are studied for growth and other desirable traits, again screened vigorously for the pathogen. The F3 pathogen free generation serves as nucleus stock which can be distributed to government / private organisations of a country for breeding under standard conditions and distribution of stock to farms.

Similarly, Specific pathogen resistance (SPR) fish/shrimp can be identified and selected from selective regions/farms known to be resistant to a pathogen. Stocks are raised from this original stock to F1, F2 and F3 generations with adequate screening for pathogens. Raising SPR stocks also involves lot of work, patience and time, but effort worth undertaking.

Employing SPF and, SPR stocks crop can be raised successfully with less incidences of diseases provided farms take up other biosecurity measures. Successful SPF stocks of *Penaus vannamei* and *P monodon* for WSSV developed and used worldwide.

iii) International level

Biosecurity is best achieved at farm level, regional level and national and international level. International organisations such as NACA, OIE, FAO and WTO help the member countries in achieving biosecurity. Biosecurity at farm level can be effective by adopting farm measures coupled with policies and regulation of national governments and international level organisations. Effective compliance only with regional, national and international regulation

At regional and national level government institutions conduct regular surveillance of diseases and report to their central organisations. Information from these organisations flows back to institutions at regional/state level for creating awareness about status of various diseases and necessary remedial biosecurity and other management measures. In turn central government of a country feed the data to international organisations and also interact with them for biosecurity measures.

International Organisations Support Biosecurity Measures for Member Countries

a) Network of Aquaculture Centers in Asia (NACA)

Diseases are impediment and a concern in the expanding aquaculture activities world over particularly in Asia. Information on the existing and emerging diseases, their spreading, nature and seriousness are collected from member countries and documented. In Asia, Network of Aquaculture Centers in Asia also collects information from the member countries for providing help and guidelines. The NACA/FAO/OIE Quarterly Aquatic Animal Disease (QAAD - Asia-Pacific) reporting system lists all diseases listed by the OIE plus diseases of concern to the region. The information generated through the regional reporting system, participated by 21 countries, provides information on important diseases in the Asia-Pacific region and also serves as an early warning system for emerging pathogens (e.g. KHV, TSV). Emphasizing biosecurity approaches by regional level organisations is integrated within and among nations. NACA, established during 1990 for Asia, Africa and Latin America has been a good information centre for aquaculture development including health. NACA has aquatic animal health as an important component. Mandate of NACA includes a). assisting members countries to reduce risk of disease affecting its trade b) economics c) and environment through providing technical advice, service, and information. In addition to providing service through Asia specific quarterly aquatic animal disease reporting and information, NACA also helps capacity building, import risk analysis, forming Asia regional animal health advisory group and response to disease emergencies,.

b) The World Organisation for Animal Health (OIE), France. World Organisation for Animal Health(OIE) an intergovernmental organisation established 1924, has 167 member countries. Its objective is to safeguard world trade by providing a) scientific information i) providing manual of diagnosis tests for aquatic animal ii) Aquatic animal health code(aquatic code) iii) zoonosis information b) Technical support to member countries, c) supply of experts to poor countries. OIE has biosecurity guide lines for import-import licence, import based on risk analysis- nature of agent, source of agent, disease profile of the country, transport requirement, laboratory to handle transport of animals and containment facility. Each member countries has provincial machinery to collect data on fish health. Each year the OIE updates the diseases, pathogens, diagnostics, remedial measures and communicate to member countries

c) World Trade Organisation (WTO)

Sanitary and Phytosanitary measures(SPS) of WTO ensures basic rights to a member country to protect against diseases, protection based on a) risk assessment(international standard, b) Codex Alimentaris Commission(codex) c) International plant protection convention(IPPC). WTO also involves in biosecurity measures at international level.

Overall Biosecurity

Overall biosecurity measures include - surveillance, quarantine and restricted access, appropriate husbandry, disinfection, disease treatment, vaccination eradication, contingency measures, and more important a good awareness among the stake holders. A farm should build external barriers by surveillance monitor pathogens in water in culture and wild population including secondary hosts. Level of surveillance depends on species cultured, diagnostics available-use of simple farm level diagnostics v/s sophisticated PCR.

Quarantine includes on site quarantine procedure for importing tested and certified fish. Separate quarantine measures required for fish, personnel, water supply and feeds. It should be noted always prevention of spreading possibly through fish, staff, visitor, equipment, vehicles, birds and animals, water and air should be kept an eye. Disinfection of water, men and equipment is very essential.

Surveillance, monitoring and diseases reporting: Surveillance is defined as a systematic collection, analysis and dissemination of health information of a given population of aquatic animals and is an ongoing process involving handling of health information from different sources, including surveys.

Surveillance is not same as surveys. Passive (general) surveillance is the collection, analysis and dissemination of existing disease information. It includes all the routine disease investigation activities that may be undertaken in a country/ state such as field investigations of disease incidents and results of laboratory testing. It is important that passive surveillance is undertaken on a continuous basis throughout a country/state and that the disease information produced is effectively captured, analysed and used for mounting an early response. Active surveillance (targeted surveillance) refers to active collection of disease data following a structured surveillance design, often targeting specific diseases. Active surveillance collects specific information about a defined disease or condition so that its level in a defined population can be measured or its absence reliably substantiated. Disease surveillance should be an integral and key component of all national aquatic animal health strategies. This is important for early warning of diseases, planning and monitoring of disease control programs, provision of sound aquatic animal health advice to farmers, certification of exports, international reporting and verification of freedom from diseases. It is particularly vital for animal disease emergency preparedness. Information generated from surveillance systems must be housed in a national database, from where the Central Administration will be able to make use of the surveillance data for the purpose of implementing national disease control programs or for meeting regional and international disease reporting obligations. Implementation of surveillance systems will directly and indirectly contribute to improved disease diagnosis, better research collaborations, reliable advice to primary producers, capacity building at the level of extension workers and primary producers, development of an early warning and emergency preparedness system. Disease reporting and information sharing can go a long way in minimizing the impact of serious aquatic animal health emergencies. By international agreement, diseases listed by the OIE should be reported by member countries and are subject to specified health measures that are intended to limit disease spread and assure sanitary safety of international trade in aquatic animals and their products.

New diseases of unknown etiologies emerge. Existing diseases often exhibit outbreak on a large scale. Diseases are single largest impediment in aquaculture due to very delicate nature of aquatic animals and the altered environment. Therefore, there should be continues monitoring and surveillance of the disease with a good reporting system in all over the country. Reporting system starts from farms and ends with highest authorities in states and the centre and therefore all reporters particularly at farm level should be trained in data collection, basic information on clinical signs, sample collection and detection.

India has started surveillance and monitoring of fish and shellfish diseases during 2014 on par with that of veterinary diseases with funding support from the Ministry of Agriculture through the nodal institution National Bureau of Fish Genetics Research(NBFGR), Lucknow. To begin with, the programme initiated during 2014 involving ICAR institutes and State agriculture universities, in all 24 institutions with an idea to eventually transfer the responsibility to Sate Fisheries Departments. The states conduct surveillance similar to that of system existing in Department of animal husbandry. Along with surveillance, awareness and training of state government personnel and farmers in sample collection have been initiated in the programme.

Quarantine, Health Certification and Movement of Stock

Quarantine is defined as maintaining a group of live aquatic animals in isolation with no direct contact with other aquatic animals, in order to undertake observation for a specified length of time and, if appropriate, conducting tests and treatment, including proper treatment of the waste waters. Quarantine process involves pre-border, border and post-border activities, including pre-movement certification, movement, confinement on arrival, checking during confinement, releases, and subsequent monitoring as appropriate. Quarantine measures are to facilitate trans-boundary trade in living aquatic animals, while minimizing the risk of spreading infectious diseases. An effective system of quarantine measures also increases protection of surrounding resources such as harvest fisheries, non- exploited species and other components of the environment.

Need of a good database required for quarantine, certification and stock movement. A good database of native and exotic species, their pathogens, list of diseases of concern including national and OIE list and list of prohibited species (eg Pirana, African catfish). SPS agreement of WTO implies that introduction of a species not restricted unless import risk proved through a good database scientific evidence.

In recent years due to rapid growth in aquaculture and intense associated activities spread of existing diseases and emergence of new disease is increasing. The Government of India has introduced a number of measures similar to that in animal husbandry to contain spread of diseases. Now the country has begun a quarantine of exotic species before their introduction in the country. CIBA, Chennai and NBFGR, Lucknow of ICAR have been identified as nodal institutions who have quarantine facilities with ponds/tanks in isolated region with facilities for eradication of water. The new species are watched after receiving from port of entry for known time before introduction in the country.

Diagnosis for Quarantine

There should be national level laboratories, network of laboratories and national experts for developing diagnostics for quarantine. Diagnostics should be sensitive, specific, rapid and adequate enough for sensitive detection of asymptomatic carriers. The diagnostics should meet both OIE listed and national important pathogens. Diagnosis for quarantine should meet at 3 levels. Level

1. Observation of animals, environment and gross clinical examination. Level II For internal pathogens, not identified by level 1, including parasitological,

bacterial, mycological and histopathological tests. Level III to detect subclinical pathogens using virology, EM, molecular biology and immunological techniques

Guidelines for Quarantine Facilities

Quarantine centers should have personnel with good awareness of legislation. Location of quarantine sites should be safe from flood and not located near of any farms. Quarantine centers should have effective disinfection facilities. Quarantine facilities required to watch animals and not let out animals before set quarantine period

Awareness and Training of Farmers and Capacity Building for Quarantine and Certification

It is very essential and important to create awareness amongst farmers, aquarium breeders who are in large numbers and unorganised, not registered with government, businessmen, common man about illegal introduction, through web page, training of officials, brochures, pamphlets, mass media and newspaper. Quarantine and Certification system should have Diagnosis of level 1 to 3, Establishment of national referal laboratories, trained persons to carry out import risk analysis, trained persons for operating quarantine units, Accreditation system private labs for certification.

Institutional Mechanism and Linkages for Quarantine and Certification

Need of essential "National Administrative frame work" for implementation of Q and C by 1. Linking various institutions already working to cohesive units 2.Assigning responsibility to each institution, 3 Coordination efforts of all organisations by a nodal agency 4. Including all stake holders. In India seven national organisations identified 1. Dept of Animal Husbandry, Dairying, and Fisheries, GOI, 2. ICAR, 3. State Fisheries Departments 4. Fish Health Laboratories of Universities, 5 MPEDA, 6, Industry and farmers 7. NGOs involved - Aqua Foundations, Swaminathan Research Foundation,

Legislation for Q and C

For effective implementation of quarantine , with clear responsibility, a good legislation support for enforcing the act. Legislation should meet international standard in the light of WTO and OIE. Such legislation should be enforced by state governments particularly legislation for overseeing introduction of Exotics into the country and also movement of indigenous varieties within the country. India has "Livestock Importation Act 2000"

Health Certification

These certificates are documents issued by the Central administration of the exporting country attesting to the health status of a consignment of live aquatic animals. A health certificate is a legal document which is used especially for the purpose of applying quarantine measures in trans-boundary trade of live aquatic animals and their products, for minimizing the risk of spread of infectious diseases. Health certification is also one of the strategies aimed to protect the natural environment and native fauna from the deleterious impacts of exotic species and/or diseases. Because of the diversity of species, the purposes for which the aquatic animals are being traded (import-export, local market, and other variable factors), health certificates should be comprehensive and be able to accommodate all the required information. Model health certificates are provided in the OIE Code. Certification is an essential requirement under WTO which includes certification essential for OIE listed diseases and other concerned diseases for both export and import. Certification is from experts from identified laboratories in the country.

Import risk analysis and movement of stock: The importation of live aquatic animals always involves a degree of disease risk to the importing country. Import risk analysis (IRA) is the process by which hazards (e.g. pathogens) associated with the introduction of a particular animal are identified, the paths and likelihood of introduction and establishment are described, consequences are defined and management options are assessed. The results of these analyses are communicated to the Central administration and stakeholders (importer/ exporter). Typical risk analysis process involves four components: hazard identification, risk assessment, risk management and risk communication. Import decisions based on scientific risk analysis will minimize the risk of introducing exotic pathogens to a country. Import risk analysis is carried out for 1). Ecological risk, and 2. Disease risk. Risk may be low, medium or high. The analysis also develop steps for risk management. There should be elaborate standards for transporting aquatic animals with leak proof bags, no contamination with other animals/plant clear information , document, health certificate.

Through movement of stocks over the last 50 years large number of fishes, shellfishes, ornamental fishes have entered the country legally or illegally. Furthermore, within the country there is a large movement of these species. These species harbour parasites and pathogens of their own and often behave badly in foreign environment. It is always safe to certify a fish that it is free from pathogens of that country before introducing. Species are selected for introduction in a country if benefit outweighs the risk. Regulating entry or movement of infected animals is implemented by the Sanitary and Phyto sanitary (SPS) Act under the World Trade Organisation (WTO).

Zoning for Movement of Aquatic Animals

Zoning is a program for delineating areas within countries on the basis of aquatic animal disease status. The advantage of zoning is that it allows for part of a nation's territory to be identified as free of a particular disease, rather than having to demonstrate that the entire country is free. In the past, outbreaks of disease could impact on trade from the entire country, but by zoning restrictions may only apply to animals and products from the infected area. Zoning is particularly helpful for diseases where eradication is not a feasible option.

Zoning is clearly delineated geographical area having animals with a particular health status- infected or free of specific disease. Delineation of zones may vary for different diseases as each disease has different mode of spreading. Usually a zone is never free of all diseases. Zoning is to facilitate trade while minimizing introduction and spread of disease. Disease free zone continues to trade while others restriction on movement of stock. One of SPS under WTO is "Regionalisation" ie, identification of region of pest/disease free region. It means if a disease is absent in a country, which can be used to refuse import from countries having that disease. Zones should be notified and should have legislation for enforcement.

Contingency Biosecurity Measures During Disease Outbreak

Always, emergency preparedness and contingency planning(similar to fire fighting preparedness) similar to that in animal husbandry should be in place for aquaculture. A disease emergency exists when a population of aquatic animals is recognized as undergoing severe mortality events, or there is otherwise an emerging disease threat where urgent action is required. Infectious disease emergencies may arise in a number of ways, including: introductions of known exotic diseases, sudden changes in the pattern of existing endemic diseases, or the appearance of previously unrecognized diseases. A contingency plan is an agreed management strategy and set of operational procedures that would be adopted in the event of an aquatic animal disease emergency. This should be developed during "peace time" (i.e. not at time of emergency). When there is an emergency, the response should proceed according to the plans that have been developed. For effectively dealing with aquatic animal health emergencies, governments should have the capability to develop contingency plans and build the required operational capacity to effectively implement the plan. Through a well-documented contingency action plan agreed upon by all major stakeholders, it would be possible to minimize the impact of an aquatic animal disease emergency.

Mere establishment of contingency plan without appropriate skills and capacity development would be of little value. The aim of early warning is to allow the recognition of a potential threat and a rapid detection of a disease emergency. For establishing an effective early warning program, a strong technical capability is fundamental in the areas of disease diagnostics, disease surveillance, epidemiological analysis, aquatic animal health information systems, national and international disease reporting and information communication and sharing.

Early response is identified as all actions that would be targeted at rapid and effective eradication/containment/mitigation of an emergency disease outbreak. The responses may be of different types depending on the disease agent and the likely impact. Operational capabilities at different levels (farm/ village/ province/national) are vital to mount an effective early response. Contingency measures(firefighting measures) should be in place whenever a disease outbreak occurs. Culling of animals, destruction to prevent further spreading is essential. To perform this a well placed system with trained personnel, information, infrastructure money for managing the show be available all time. The system is in good state in human and other land animals health management. A very good example in India is epidemic of birds flue in poultry and contingency measures taken to control the outbreak. Such a system is being planned for fisheries with trained people, farmers and infrastructure.

Optimum Contingency Plan-Requirement

Contingency planning is required for both OIE and national diseases of importance. An optimum contingency plan requires i) Effective surveillance system, ii). Adequate reporting structure iii). Adequate experienced and trained personnel, iv) An emergency work plan, v) Adequate funding for various activities, vi) Legal support to execute an eradication programme/campaign, viii) Public support and cooperation ix) Adequate diagnostic sources xi) Culling/slaughtering of diseased stock and compensation. Importantly, a national level task force comprising state and central officials with expert members is established well in advance to implement contingency measures.

Biosecurity with Best Management Practices (BMP)- A New Trend in Health Management

Altogether a new concept in aquaculture in India adopted from population medicine of human and animal health management. This strategy studies elaborate details of risk factors associated with disease in a scenario verses population of fish and its health. From the risk factor data, best management practices (BMP) are developed. Overall BMP includes pond preparation such as drying, bleaching, seed screening for pathogens, monitoring water and feed quality.

Best Management Practices for Biosecurity for Shrimp Health Management

White spot virus has caused devastation of culture both *Penaeus monodon* and *P vannamei* world over. The virus is endemic now and hence living with the virus in shrimp culture with biosecurity is the only way ahead. In India College of Fisheries, Mangalore with Overseas Development Agency (ODA),UK conducted several detailed epidemiological studies on risk factors associated with WSSV disease and loss in shrimp culture. Based on the studies, best management practices(BMP) recommended for biosecurity against WSSV and a host of other pathogens. Biosecurity measures should be part of every shrimp farm's health management program. Shrimp farm biosecurity involves applying sets of targeted, science based procedures to eliminate or reduce the risk of a particular pathogen that is, a disease-causing infectious agent such as a virus — (a) entering the farm, and(b) spreading within a pond, between ponds, to other farms, or to the wider environment. BMP has been followed in culture of shrimp with prevalence of white spot virus has achieved 60-100 % successful crop.

Because of the costs (in time and money), many farmers generally implement biosecurity programs to reduce the risks associated only with dangerous pathogens. In an ideal world, they might aim for zero risk, but in reality they will need to balance the costs of any biosecurity program against the uncertain costs of future disease outbreaks. Development and implementation of a biosecurity program therefore requires a clear appreciation of the technical issues and involves compromises in which costs and benefits must be carefully considered. Effective implementation requires long-term commitment from the farmer as well as discipline from the farm workers. Effective biosecurity relies on a secure farm design, hygiene and quarantine, regular health testing, recordkeeping, and control of disease vectors. Using specific pathogen free (SPF)seed is the first important step in reducing risk.

Implementation of BMP for Shrimp Health Management in India

National Centre for Sustainable Aquaculture(NACSA), under the Marine Product Export Development Authority(MPEDA), a Government of India organisation introduced BMP in shrimp culture in India. Following are the detailed key instructions.

1. Pond preparation is the first step in biosecurity. liming, ploughing, drying of ponds help in reducing pathogens load to a great extent.

2. Stocking only post larvae that have acceptable test results in terms of pathogen absence at optimal stocking densities. Screening of seed in hatcheries is very vital. PCR screening of adequate number of PL based on statistics protocol is very important to avoid the pathogen at 95% level of confidence.

3. Eliminating or reducing risk from potential vectors (infection carrying agents) on the farm, using water management practices that prevent or reduce contamination by the pathogens, Reducing the risk of spreading infection between ponds by restricting movements of people, equipment and other possible agents and finally implement a health management program that aims to minimise stress to prawns by optimising the pond environment. Installing physical barriers to prevent crabs with crab fence (gill nets with mesh size of less than 1 cm of more than 0.5 meter height), birds and other animals with bird nets provided at a height of 2 m from the dyke. with gap between bird lines should not be more than 10 cm have helped reducing WSSV risk to a large extent.

4. Following a good farm sanitation and hygiene is centre point to successful farming. Area surrounding the farm should be kept clean without accumulated garbage, with good toilet and good sanitary condition. Avoiding contamination of domestic sewage into grow out pond, reservoir and canal is advised. Disinfectants such as bleach or KMnO4 (one table spoon for10 L water) and plain water should be arranged at the entrance of each pond for workers and visitors to sanitise their hands and legs. Using separate, marked equipment for each pond (nets, feed buckets, water sampling jars, feeding float etc.) helps to eliminate the risk of contamination between ponds. In case separate equipment are not available disinfect them with bleach or KMnO4 solution before using in another pond. Maintaining separate water sampling container. for each pond is advised. Regular checking the pond bottom on weekly basis and removal of any black soil or benthic algae accumulated at pond corners is recommended.

5. Health of shrimp should be assessed on daily basis by observing feed trays- poor feed consumption for consecutive three to four days indicates health problem of shrimp. Additionally general health and growth of

shrimp also monitored by sampling using cast net. Healthy shrimp have normal colour, with full gut(>80%).

6. Farmers association which facilitate cooperation and communication with shrimp farmers and farm workers with regular meetings is the strength for successful implementation of BMP. Discussion and exchange of information on disease outbreaks, emergency plans, releasing water from infected ponds, disinfection of ponds, disposal of infected shrimp movement of equipment and host of other well established preventive measures help in containing disease and loss in an area to a great extent.

References

AFS-FHS (American Fisheries Society-Fish Health Section). 2014. FHS blue book: suggested procedures for the detection and identification of certain finfish and shellfish pathogens, 2020 edition. Accessible at: https://units.fisheries.org/fhs/fish-health-section-blue-book-2020.

Arthur, J.R. and Bondad-Reantaso M.G. 2012. Introductory training course on risk analysis for movements of live aquatic animals. FAO SAP, Samoa. 167p.

Better Management for Sustainable L.Vannamei Culture, 2017, National centre for sustainable Aquaculture (NACSA), MPEDA, Ministry of Commerce and Industry, Government of India, Kakinada 533003, 48 pp

J.Richard Arthur, F Christian Baldock, Rohana P Subasinghe and Sharon E Mc Gladdery, 2005 Preparedness and response to aquatic animal health emergencies in Asia: Guidelines FAO Fisheries Technical paper ,486,Rome 40 p

J.Turnbull, 2011, Principals of health control- cause ,risks and control, In Course manual Training Programme on Biosecurity in Aquaculture-Aquatic medicine, Editors,

K. M. Shankar, K.S. Ramesh, Naveen Kumar and Abhiman Ballaya, Department of Aquaculture College of Fisheries, Karnataka veterinary, Animal and Fisheries Sciences University Mangalore,575002

K. M. Shankar, 2016 , Importance and Scope for Aquatic medicine in India ,In Aquatic medicine-a training manual, Special publication No 01/2016, Editor Prakash Patil, Aquatic Animal Health Management Laboratory, College of Fisheries, Karnataka Veterinary, Animal and Fisheries Sciences University, Mangalore - 575002

K. M. Shankar and C.V Mohan,2002, Fish and Shellfish Health Management , ISBN 81-7525-328-2, Fish Pathology and Biotechnology Laboratory, Dept of Aquaculture, UAS, College of Fisheries, Mangalore - 575002

Mary Nickum and John G Nickum,2008, Finfish aquaculture Biosecurity; Part II The Farm, Aquaculture Magzine, April 2008 p 32-36, National aquaculture Association Washington, USA

Mary Nickum and John G Nickum, 2008, Finfish aquaculture Biosecurity; Part I National and International overview, Aquaculture Magzine, Feb 2008 p 28-36, National Aquaculture Association, Washington, USA

S Ayyapan, 2020, Drivers of Aquaculture Development in India, In Indian Aquaculture. Editors, V.V Sugunan, V.R. Suresh and C. K. Murthy, The Society for Indian Fisheries and Aquaculture, Hyderabad

Subasinghe, R.P and Bondad Reantaso M G 2008. The FAO/NACA Asia regional technical guidelines on health management for responsible movement of live aquatic animals: lessons learned from their development and implementation. Rev Sci. off. Int. Epiz, 27 (1) 55-63

12

Zoonotic Diseases Originating from Fish and Shellfish

K.M. Shankar and Prakash Patil

Introduction

Aquaculture and Zoonotic Diseases

Diseases that can transfer from wild or domestic animals to humans are referred to zoonosis (Fig 1). Zoonosis also includes biotoxin producing agents from fish to human beings which is well documented. There are a number of pathogens which spread from fish and shellfish to human and other domestic animals causing diseases.There are also many other infectious organisms of fish origin that have not been reported but have potential to infect and harm man. Zoonotic diseases originated from fish and shellfish have been documented in the past mostly from wild caught fish and shellfish. However, in the last 30 years as the capture fisheries is declining and aquaculture is growing steadily worldwide, zoonotic incidences from the latter are increasing.

Aquaculture growing at 6-7% annually world over including India is contributing to 50-60% of total fish production today. Additionally, aquarium industry is booming worldwide. Therefore,incidences and scale of zoonosis due to aquaculture are quite different from that from capture or wild fisheries. Disease outbreaks in cultured fish are often by pathogens indigenous to aquatic environment. In aquaculture, disease outbreaks are associated with quality and quantity of nutrients and high stocking density which increase pathogen load leading to easy transmission to human. Apparently healthy fishes may also harbor bacterial pathogens, especially in their kidneys and intestines. World Health Organisation (WHO) considers many diseases in aquatic animals as emerging type. Emerging diseases appear in a population for the first time or it may have been existed before, but is now increasing rapidly. Usually information on zoonotic potential of emerging diseases is limited which is essential to disseminate to human health managers. Potential biological contamination of aquaculture products can occur from bacteria, fungi, viruses, parasites and biotoxins. The location of the farm, the species being farmed, water temperature, husbandry systems, postharvest processing, and habits in food preparation and consumption are among the main factors influencing the zoonotic risk associated with aquatic animals and their products.

With increasing aquaculture production, the consumption of aqua products also increases which in turn increases possibility of contracting zoonotic diseases. Zoonotic diseases spread either by consumption of aqua products ingestion or contact with infected waters. Growing world aquaculture is associated with large movement and trade of live aquatic animals and their products across national and international borders. Therefore, zoonosis from growing aquaculture along with increasing human population in the world is a major concern and needs special attention.

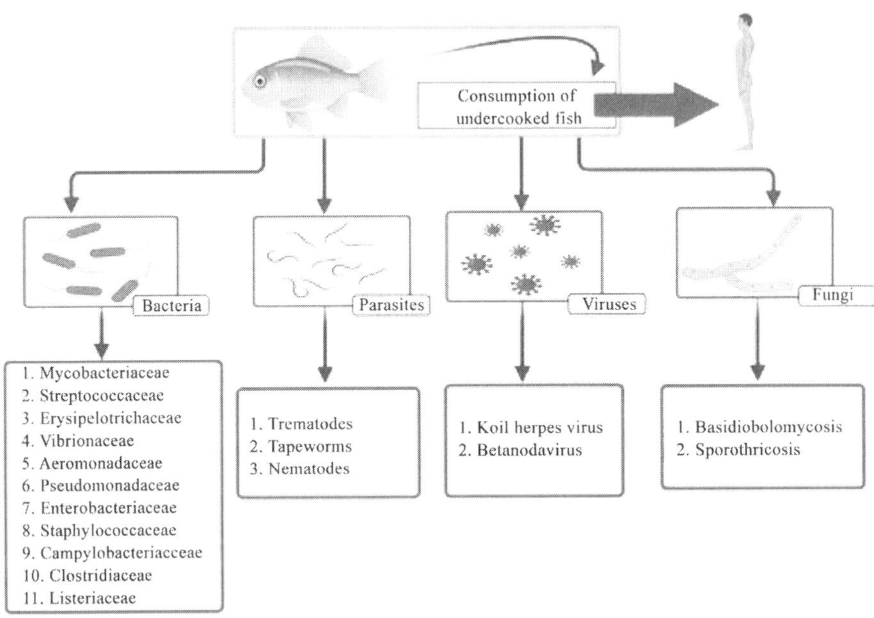

Fig. 1: A broad classification of fish zoonotic agents including bacteria, parasites, viruses, and fungi. (*Source*: Mina Zirati *et al* 2022)

Exposure Routes to Zoonotic Agents of Fish and Shellfish

Several organisms and toxins and exposure routes have been reported and have the potential to infect humans. The major exposure routes include ingestion and introduction of organisms through open wounds or abrasions. More specifically, ingestion includes consumption of raw or undercooked infected fish tissue, ingestion of fish tissue contaminated with feces from infected fish, and ingestion of water harboring infectious organisms. Dermal exposure includes introduction of infectious agents into open wounds or abrasion through handling infected fish or infected water.

Reported cases in humans through ingestion of fish tissue (undercooked or feces contaminated), bacterial- *Clostridium*, *Vibrio*, *Pleisomonas shigelloides*, *Escherichia*, *Salmonella*, *Edwardsiella*, parasites-*Anasakiasis*, *Eeustrogyloides*, Cestodes, teamatodes, and toxins- ciguatera poisoning. Reported cases in humans through ingestion of infected aquaria water include *Plesiomonas shegelloides*, *Aeromonas*, *Edwardsiella*. Although there are no cases in humans but potential risks exists include *Staphylococcus*, *Pseudomonas* and protozoa. Reported cases in humans through dermal contact of infected fish include *Erysipelthrix*, *klebisiella*, *Edwardsiella*. There are no known cases in humans but the potential risk exists through dermal contact infected aquaria fresh/ sea water include *Mycobacterium*, *Vibrio*, *Aeromonas*, *Edwardsiella*. An important feature of many bacterial and protozoan organisms is their opportunistic nature., where the development of disease in human host often requires a preexisting state that compromises the immune system. Persons with immunocompromised medical condition or taking medicines that impair immune system-steroids, immunosuppressive drugs are at risk of contracting fish borne diseases. The following is a list of known and potential fish borne zoonosis. Despite an increased awareness of zoonotic disease agents, their diagnosis in humans by clinicians and medical practitioners is often hampered by a poor knowledge of the zoonotic potential of disease agents in aquatic species and the associated clinical signs.

Food Borne Zoonotic Diseases (FBZDs)

The risk of FBZDS has been rising due to several factors including pathogen behaviour, rapid population growth, and increased global trade in foods and farm animals from the countries where appropriate microbiological safety procedures are not followed. The improved transport logistics and conditions facilitate some microorganisms to survive in food products for longer periods and reach consumers in viable form. The change in consumer eating habits, for eg, consumption of lightly cooked food and a shift from low to high protein based diets(meat and fish products) have also increased FBZDs. Further the increased movements of human also facilitates easy spread of pathogens. Therefore preventing FBZDs and deaths a major public health challenge. FBZDS pose constant threat to public health, socioeconomic development and international trade. More than 200 known diseases are transmitted to human beings through food contaminated microbes and their toxins and parasites. Estimated global burden by 31 FBZDs is over 600 million and 420000 deaths (WHO,2015). FBDZs burden is heaviest in Africas (43 %) followed by south

East Asia (24%) and Eastern Mediterranean sub regions (16%). Around 40 % FBZDs were reported under 5 years of age with India recording death of 1,00,000 during 2015. FBZDs costs India about Rs1,78,000 crores or around 0.5 % of GDP every year. Current FBZD represents about 100 million cases per year, corresponds to one in 12 people falling ill, which is an underestimate as most of FBZDs are not reported.

Bacterial Zoonotic Diseases

Bacteria are the main zoonotic agents of fish. Among them families of Gram positive bacteria include *Mycobacteriaceae*, *Streptococcaceae* and *Erysipelothricaceae* and Gram negative bacteria include A*eromonadaceae*, *Vibrionaceae*, *Pseudomondaceae*, *Enterobacteriaceae*,and *Hafniaceae*.

Mycobacterium spp

Mycobacterium spp has a large number of pathogens infecting human, mammals, reptiles and fish. Two species *M fortuitum and M marinum* are the primary pathogens of fish. Important species of mycobacterium causing fish zoonosis are *M. avescencs*, *M. chelonae*, *M. fortuitum*, *M. gordonae*, *M. marinum*,

M. ulcerans, *M. septicum*, *M. peregrinum* and *M. avium*. *Mycotuberculosis* causing mortality in fresh, brackish and marine waters in wild and farmed fish is common.In addition, there are several non tuberculosis Mycobacterial infection (NTM), recorded in 150 species of fish worldwide. Zoonotic potential of these infections is a concern and more often human get infected when in contact with fish or water. Common example is *Mycobacterium sp* infection in human from handling aquarium fishes. Cut wounds or skin lacerations favour development of granulomatous nodule in 6-8 months at the site of infection usually on appendages. Infection may spread to lymph nodes and disease spread is quicker in immunocompromised individuals.

Nocardia sp has been isolated at very low frequencies from tuberculoid lesions of both fish and man. Infections with these organisms maybe misdiagnosed as mycobacterial disease because of similarity of clinical signs and the positive reactions to acid fast staining. These mycobacteria can occasionally cause transitory superficial infections of cuts on the hands of aquarists. Lesions can later found in lungs and bronchial lymph nodes.

Streptococcus inae

The organism causing streptococcosis leading to meningoencephalitis in fish worldwide besides economic damage is a public health concern. The pathogen infects both fresh and marine wild and cultured fish causing morbidity and mortality. Nearly 10 % of the wild fish harbour this pathogen. *S inaea* is an emerging zoonotic agent in human, due to direct contact with infected fish or indirectly through contaminated water. Fish handlers and processors infected with the organism show symptoms of fever, endocarditis, meningitis, septicemia and rarely death.

Erysipelotrix sp

The organism is not a serious pathogen of fish although can infect edible and aquarium fishes. It is a saprophyte in soil, which also grows well in the mucus and surface of fish surviving for long periods. Fish handlers including veterinarians getting in contact with fresh, dead fish or products, particularly during hot season get erisepeloid disease. These personnel are advised to use protective gloves and wash hands frequently. In animals and human the organism cause septicemia, endocarditis and skin infection. In mild cases rash appear on the skin which may become severe with painful purple spots.

Aeromonas spp

Aeromonads are facultative anaerobic, gram negative rods which cause septicemia and dropsy in infected fish. *Aeromonas* spp are ubiquitous aquatic bacteria, also found in intestine of normal healthy fish. Fish play a key role in transmission of *Aeromonas* sp to human. Although *A hydrophila* is a common zoonotic agent, other species *A. caviae*, *A. jandaei*, *A. sorbia*, *A. salmonidae* and *A. veroni* have zoonotic potential.The reservoir of the speciesinclude wide range of fish, environment and water. Human infected with *A hydrophila* show a variety of clinical signs, but the most common syndromes are gastroenteritis and localised wound infections. *Aeromonas*have great potential for causing infection in wounds. Immunocompromised persons are more susceptible for wound infection. Symptoms include pain, swelling or redness at the site and possible fever and chills. *Aeromonas* sp enter through cuts, ulcers or ingestion for infection. Infections are rare in human usually in the range of 5-6 %. However, multi antibiotic resistance in *Aeromonas* is increasingly emerging in human infections which is a public health concern. Usually infections are mild but serious in immunocompromised individuals. A wide range of symptoms have been reported such as muscle necrosis, cellulitis, septicemia, respiratory infections, gastroenteritis, urinary tract infection and edima to swelling at the site of infection.

Vibrio sp

Vibrio sp causing vibriosis in human and animal are potential zoonotic agents with risk to aquaculture professionals and fish consumers. The species are present in fresh, brackish and marine waters fish and shellfishes which include *V. alginolyticus*, *V. anguillarum*, *V. campbellii*, *V. harvey*, *V. vulnificus*, and *V. parahaemolyticus*. The premier species of *Vibrio* in marine fish are *V. vulnificus* and *V. parahaemolyticus*,. Most common human infections are caused by *V. cholerae*, *V. vulnificus* and *V. parahaemolyticus*. Among more than 100 species of *Vibrio*, *V cholerae*, *V parahemolyticus*, and *V vulnificus* cause food borne diseases. The reservoirs include normal brackish water flora(micro and macrophytes) and fauna(zooplankton and crustaceans), humans and animals. Raw oysters, crustaceans, crabs and undercooked mussels are common source of *Vibrio* infection in human and animals. Under refrigeration Vibrio can survive for weeks. Complications of infection include severe dehydration cardiac and circulatory failure,(that occur due to loss of potassium)septicemia wound infection, ear infection, cellulitis, peritonities, necrotising fascitis cholecystis, endophthalitis and meningitis. *Vibrio vulnificus* can infect wounds sustained in coastal or estuarine waters. Lesions varies from mild , self limited lesions to rapidly progressive cellulitis and myositis that can spread rapidly and destructively. Raw fish and shellfish consumed may serve as source of *Vibrio* infection. Most common zoonotic infections are due to *V. parahaemolyticus* and *V. vulnificus* which cause primary or secondary septicemia when wounds are exposed to sea water.

Edwardsiella tarda

The organism is an established fish pathogen, which has zoonotic potential causing human infection. The organism has a wide host range including aquatic mammals, reptiles, fishes and occasionally human. It spreads through consumption of raw fish, ingestion or contact with water. The pathogen can cause gastroenteritis associated with meningitis, septicemia and wound infection.

Enterobacteria

Escherichia coli, *Klebsiella* and *Salmonella* are the common zoonotic agents in fishes. These organisms commonly found in digestive tract and aquatic environment infect human through open wounds, contact at scratches, or through systemic infection. Ingestion of fish products such as dried fish also can be source of infection eg *S. typhimurium Escherichia coli* is transmitted through fish or aquatic products. Most common spread of *E coli* is through ingestion of food or water contaminated by human feces. *E coli* although not a natural microbiota of fish, is often isolated from digestive tract and even from tissues of gills, kidney, muscle and blood through infiltration from contaminated water. However, infection depends on season and immune status of individuals. This organism becomes pathogen infiltrating peritoneum or urinary tract some time causing diarrhea, food poisoning by producing toxins. Fish has been identified as a new vector of *E coli* transferred to other animals.

Intestinal diseases in human due to *Salmonella* is due to consumption of wild or processed aquacultural products(fresh, frozen or smoked) and contaminated water. Although the pathogen is not a common bacterium of fish, it is carried on the surface or gut if water is contaminated. Fish also carry and persistently shed *Salmonella* to aquatic environment. *S. typhimurium* and *S. enteritidis* are the most common zoonotic agents causing salmonellosis in human through consumption of fish. Consumption of *Salmonella* infected fish cause gastroenteritis, abdominal fever, cramps, fever and bacteremia. Infection also cause complex clinical signs such as sepsis, abdominal pain diarrhea and vomiting. Occasionally infection leads to meningitis, urinary tract infection particularly in immunocompromised individuals.

Klebsiella pneumoniae and K. oxytoca

The species particularly the multi drug resistant *Klebisiella sp* (*Kpnemonia* complex) is zoonotic in human and is of public health concern. *Klebsiella* has been isolated from untreated water from dams and sea, sediment, intestinal content of fish and shellfish. The organism also has been isolated from wounds of aquarium fish and edible fishes.

Other Bacteria Associated with Fish-derived Zoonoses

There are several zoonotic bacteria such as *Staphylococcus*, *Listeria*, *Clostridium*, and *Campylobacter* associated with fish consumption. *Staphylococus* spp are not usually fish pathogens, however incidences *of S. xylorum*as primary pathogen causing death is emerging. The organism is source to fish in two ways. *Staphycoccus* particularly methicilin resistant ones are increasingly isolated from contaminated waters. Fish processing personnel also contribute the organism from their wounds, skin and mucous membrane. Enterotoxin from *Staphylococus is* heat resistant and cause gastroenteritis with contaminated fish and fish products.

Listeria monocytogenes is another economically important bacteriumwhich spreads through fish and fish products. The organism as such is indigenous flora of water and can be found on or inside fish.The organism is very versatile capable of growing in a wide range of temperature and in a variety of fresh and salt water. The pathogen is a public health concern, causing septicemia, meningitis, pnemonia and abortion. Elderly persons with chronic diseases and immunocompromised conditions are considered high risk groups of human listeriosis.

The spore forming *Clostridium* sp are ubiquitous in soil, aquatic sediments and natural anaerobic environments are fish origin pathogens. They have been isolated from fish surface, fresh and canned fish. *Closteridium perfringens* enterotoxin cause gastroenteritis in human. The other *Clostridium botulism* identified in intestine of healthy fish, fresh water and sea sediments produce botulin toxins (type A to H) cause flaccid paralysis. Botulism toxin A to F areconsidered to be more toxic to human. Botulinum is a neurotoxin which is heat resistant has been isolated from fish intestine and therefore improper cooking or processing pose risk of botulin, manifested early as diarrhea, vomiting, dizziness, bloating and constipation.

Bacteria such as *Campylobacter*, *Plesiomonas shigelloides*, *Legionella pneumophila* can be considered as zoonotic associated with water or fish consumption. However, infections with these organisms associated with fish are very rare.

Zoonotic Viral Diseases

Compared to large number of bacteria, few viruses Norovirus and Hepatitis A have been identified to cause zoonosis in human. Norewalk virus, a non enveloped RNA norovirus known to infect human causes gastroenteritis, muscle pain, head ache, low grade fever and loss of taste. The symptoms appearing 12-48 hrs post consumption of contaminated fish are self limiting except in individuals with poor immune status. Consuming ready to eat fish and shellfish products with this virus from fecal contaminated waters are increasingly becoming public health concern.

Hepatitis A virus (HAV) can spread through consuming fresh and frozen fish and shellfish is a potent zoonotic agent. The virus cause hepatitis, an inflammation of liver manifested as fatigue, nausea, jaundice, dark urine, fever and abdominal pain lasting for several weeks. Rarely liver damage and failure especially in elderly and in people with chronic liver disease due to the virus also recorded

Zoonotic Fungal Diseases

There are few zoonotic fungi of public health concern which are naturally transmitted between animals and human. Basidiobolomycosis and Sporotrichosis are the two well known zoonotic fungal diseases of fish origin.

Basidiobolomycosis caused by *Basidiobolus ranarum* is an ubiquitous saprophyte belong to the class zygomycetes. The fungi also been recorded in GI tracts of amphibians, reptiles, fishes and mammals including human. The fungus enters damaged skin or through insect bite, producing an enlarged node beneath the skin leading to subcutaneous and gastrointestinal infection. In histopathology dermal granulomatous inflammation, infiltration with broad septate fungal hyphae and yeast like structures are seen. Infection induces IgG and IgM specific antibodies which could be detected as diagnostic test. The infection is common in tropical regions of Asia, Africa, Europe, South and North America.

Sporothrix schenckii causing sporotrichosis is a dimorphic saprophytic fungus quite prevalent in tropical and subtropical regions. The fungus is also isolated from fish and dolphins. Transmission of the fungus to human beings is associated with nature related activities such as hunting, fishing, farming, and gardening. The fungus usually enters cut wounds leading to local infection in most of the cases and rarely ulcerative with edema and pain. In several regions of the world the fungus infection is endemic in farmers, forest dwellers and fishers. Basically three clinical types of sporotrichosis identified, 1) lymphocutaneous sporotrichosis, 2) fixed-cutaneous sporotrichosis, and 3) multifocal/disseminated-cutaneous sporotrichosis. Systemic sporotrichosis also recorded , where fungus spreads from local infection to lymph nodes.

Zoonotic Diseases Caused by Parasites

A large number of fish parasites, some found rarely are capable of zoonotic diseases in human. Nearly 40 taxa of fish parasites are posing public health risk affecting millions of fish consumers around the world. Some of these parasites are mild, often their infection gounnoticed and hard to diagnose. Others cause severe allergic or gastrointestinal diseases such as indigestion, abdominal pain, diarrhea, including brain hemorrhage and cancer. Anisakiodosis and gnathostomiasis are two serious fungal diseases of public health concern. Helminth parasites are of special concern to fish eaters because of their diversity, abundance and easy transmission to human. A study in Vietnam indicates that 268 helminth species occur in 213 fish species. This abundance and easy transmission is mainly because of the orientation of parasite life cycle to trophic system food web. Secondly, some helminths have life cycle where fishes are intermediate hosts.

Main transmission of fish derived parasites -tapeworm, round worm and flukes is through consumption of raw fish or improperly cooked fish and fish products. More and more emerging fish zoonotic parasites are reported due to increase in aquaculture production, growing appetite for fish and fisheries products and also their transport in frozen condition across the world. Although freezing inactivates many parasites, some might persists and hence need of sufficient care through surveillance and screening to avoid spread of zoonotic parasites.

Trematodes (flukes)

An estimated liver fluke infection in the world is around 45 million people. And at least 600-700 million people are at risk with fluke infection. Fish zoonotic trematodes are found in fresh, brackish and marine fish. Consumption of raw fish and shellfish is the main source of infection in human. Considering the growing aquaculture production and consumption of fish, WHO has listed fish trematodes as emerging zoonotic infectious agents. Important fish zoonotic trematodes are *Clonorchis sinensis*, *Centrocestus formosanus*, *Haplorchispumilio*, and *Haplorchisyokokawi.*

Two trematode families *Opisthorchiidae* and *Heterophyidae* have large number of genera noted as fish zoonotic agents. Common examples of fish/crab liver flukes are *Clonorchis sinensis*, *Opisthorchis viverrini*, and *Opisthorchis felineus* and lung flukes *Paragonimus westermani* and *P. heterotremus.* Infection from flukes range from low level to intensive infection.In the latter, liver damage lead to inflammation and damage to the epithelial bile duct, leading to gastrointestinal problems and liver damage and may lead to major clinical problems such as cholangitis, choledocholithiasis, pancreatitis, and cholangiocarcinoma (CCA). *Paragonimiasis* is lung fluke infection acquired through fish/crab harboring metacircaria of the fluke. In general zoonotic fish trematodes prevalence is high in Asia region. As visual diagnosis is inadequate to detect life stages such as circaria and metacircaria, novel PCR have been developed for sensitive detection of different stages of the parasite in fish and fish products.

Cestodes (tapeworms)

Cestodes are common group of zoonotic parasites of fish worldwide infecting more than 20 million people. Diphyllobothriosis caused by *Diphyllobothrium* spp with nearly 14 species are human pathogens. Most pathogenic specis are *D. dendriticum*, *D. nihonkaiense*, *D. latum*, *Adenocephalus pacificus* and *Diplogonoporus balaenopterae*. The tapeworms cause mild disease but quietly rob nutrients from the human. Usually infected people are asymptomatic but in some cases diarrhea, abdomen pain, anemia, weight loss and vit B12 deficiency are common symptoms.

Nematodes (round worms)

Zoonotic nematodes infecting human worldwide is common, particularly in region where people consume raw fish or improperly cooked fish and shellfish. Some of the fish nematode infections are life threatening. These nematodes are common in fresh and marine waters with their larval stages having wide host range. They are capable of migrating from gastrointestinal tract to mucosal layer, viscera and even muscle posing risk for human.

Anasakidosis caused by *Anisakis* spp, *Pseudoterranova* spp and *Contracaecum* and gnathostomiasis by members of *Gnathostomatidae* are the most common fish nematode infection of public concern from marine fish worldwide. Anasakidosis is caused by third stage larvae(L3). Human are considered accidental host in the parasite life cycle. The parasite fails to develop in the human gastrointestinal tract. Usually live larvae entering intestinal mucosa cause anisakidosis. Even dead larvae also known to cause the disease. Gastrointestinal symptoms of anisakidosis mimics food poisoning lasting for days to months. Hypersensitivity associated with anisakidosis is an important concern. Anisakidosis although recorded worldwide is most common in Japan and Europe. Raw, frozen and smoked fish are source of the parasite. Exposing raw fish to temperature more than 60 °C for 2-3 min known to significantly reduce load of the nematode.

Gnathostomiasis is another important zoonotic fish nematode of high significance infecting large number of Asians consuming raw fresh and brackish water fishes. L3 larvae of the parasite family Gnothostomatidae having several species *Gnathostoma spinigerum*, *G. doloresi*, *G. hispidum*, *G. binucleatum*, *G. nipponicum* and *G. malaysiae* as well as *Echinocephalus* sp. cause the disease. Symptoms are almost similar to anisakidosis , but more severe developing 24-48 hrs post infection with nausea, abdominal pain, vomiting. The larvae of the parasite can pass through gastrointestinal tract migrating to several vital organs including brain, some time even causing death.

Besides these worldwide fish zoonotic parasites, several endemic infections have been recorded in Philippines and Thailand. Capillariasis caused by a zoonotic fish nematode *Capillaria philippinensis* by consuming fresh water fish. Fish is intermediate host while fish eating birds are definitive host. The larvae of the nematode multiplies in the gastrointestinal tract causing diarrhea, abdominal pain and edema. Chronic infection robs protein and electrolyte from the patient, often causing death in serious cases.

Ciguatera Fish Poisoning

Biotoxin accumulated in some varieties of fish when consumed cause serious disorders in human. These toxins called ciguatoxin affect digestion, muscular and neurological systems of the human. More than 400 species of fishes living in the low lying shore waters or coral reef in tropical and sub tropical regions are known to cause ciguatera poisoning. Few edible fishes such as seabass snappers and perches are also included under this category. The poisoning has a seasonality occurring at certain time of the year. The origin of the biotoxin is a dinoflagellate *Gamabierdiscus toxicus* living in coral reef and low lying shore areas on which fish feeds and the biotoxin get biomagnified in fish muscle. The toxin is heat stable may present in different forms.

Usually symptoms of acute ciguatera fish poisoning begins 30 min post consumption of contaminated fish. A number of initial symptoms have been recorded such as itching, tingling, numbness of lips, tongue, appendages. Several symptoms are also recorded within 6-7 hrs such as abdominal cramps nausea, vomiting, diarrhea, red skin rashes, blurred vision, photophobia and myalgia. Symptoms are related to quantity of toxin consumed. Ingestion of large quantity of cigutera fish leading to sever cases within 24 hrs with breathing difficulties, muscular paralysis, slow heart beat, respiratory arrest and convulsion. In patients after initial sufferings, neurological symptoms may continue for several months. Ciguatera poisoning recorded in all age groups but elders are more susceptible. The transmission of ciguatoxin between individuals is interesting and a concern too. Ciguatoxin has been detected in semen of infected persons which was later traced in females after intercourse. Furthermore, mothers carrying the toxin also known to pass it to children through breast milk.

Other Toxins of Fish Origin

Tetradon poisoning and scombroid fish poisoning are the other poisoning associated with consumption of fish. Tetradon poisoning is due to eating puffer fish containing tetradon toxin. Scombroid toxin is produced due to bacterial decay of fish under poor refrigeration. Because of similarity of symptoms often these disorders are confused with that of ciguatoxin. Therefore there is urgent need for differential diagnosis of the toxins for deciding appropriate treatment

Prevention and Control of Fish and Shellfish Origin Zoonoses

Following measures are known to reduce zoonoses of fish origin to a large extent (Fig 2).

i. **Good management Practices (GMP):** With increasing activities in aquaculture and ornamental fish industry, zoonotic diseases of fish origin are increasing worldwide. Fig. 2 illustrates methods of prevention and control of zoonotic pathogens. With Good Management Practices(GMP) in farms a good number of zoonotic diseases can be prevented. Chemical disinfecting of aquaculture systems, cleaning, drying of ponds and facilities disrupt cycles of pathogens and parasites. Furthermore, proper surveillance of zoonotic diseases in the wild and farms in the region, in different seasons helps in preventing the diseases to a large extent.

ii. **Fish meat quality control:** Monitoring consumption of fish in the population and information on quality of fish marketed provide adequate information for prevention of zoonotic diseases. Proper practices during harvesting, handling, processing, storage, are helpful to reduce risk of zoonotic disesaes to a great extent. In this direction good manufacturing practices(GMPs) and HACCP system can reduce the risk of zoonosis. Freezing fish and cooking at more than 60 °C for 15 Sec is adequate for inactivation of a number of parasites/pathogens.

iii. **Education and awareness:** Educating the public about zoonotic diseases is the first step in prevention. Consumers should be educated / informed about ill effects of eating raw fish or improperly cooked fish. People working in aquafarms, aquarium farms should be educated about zoonotic diseases existing around. Adequate safety measures such as wearing gloves, washing hands frequently, avoiding eating while handling fish should be followed. Very important infected people should provide clear history of disease to physicians for appropriate diagnosis and treatment.

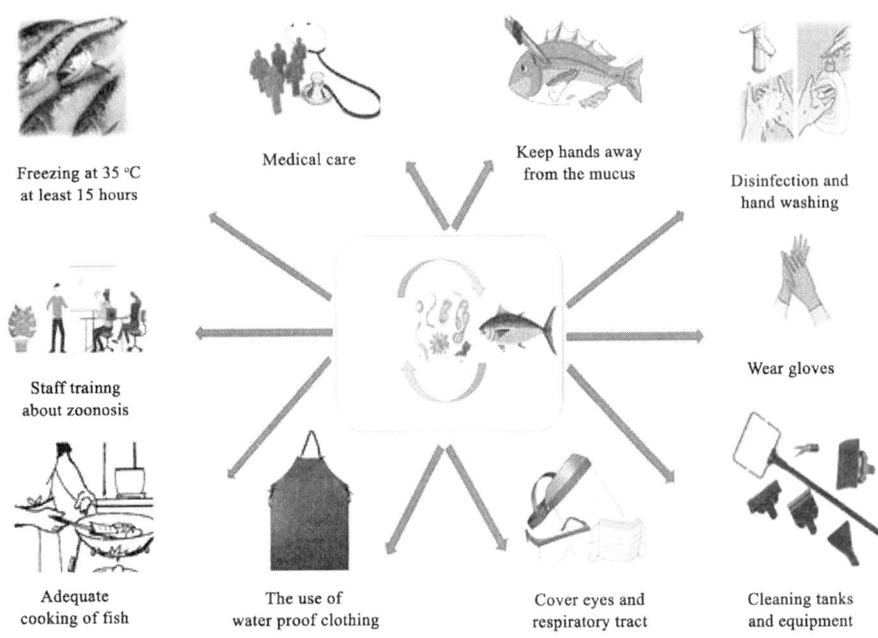

Fig. 2: A schematic representation of possible methods and ways for control and prevention of zoonotic diseases (*Source*: Mina Zirati *et al* 2022)

One Health (OH) Approach

With increasing incidences of fish origin zoonotic diseases in general world over, world is focusing on one health (OH) programme covering animals and human beings (Fig. 3). In this direction OH programme should strengthen and incorporate fish zoonotic diseases as well.Maintaining healthy animals including fish with better welfare in good environment and healthy human population with strong health care can easily prevent potential zoonotic outbreaks. Fig 3 illustrates strengthening of OH system to mange zoonotic diseases. Towards this goal zoonotic diseases are managed with organisational control, interruption of transmission, diagnosis and detection. Accordingly,guidelines developed for effective cooperation between human and animal health officials (WHO 2021).With recent Corona virus pandemic suspected to be of animal origin, OH programme is gaining worldwide importance. Training university students, researchers, international agencies is an important component of OH programme.

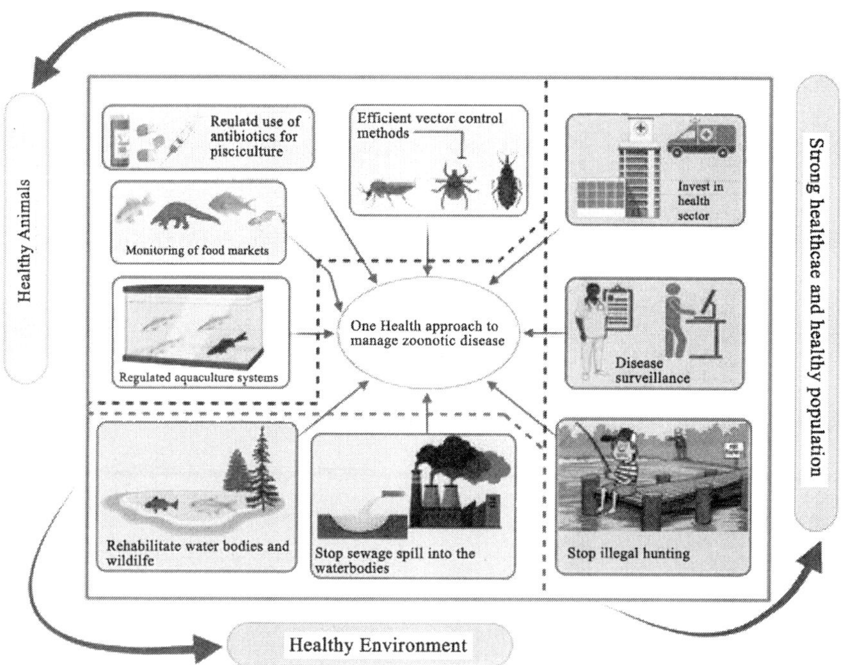

Fig. 3: One Health (OH) approach in managing fish zoonotic disease
(*Source*: Mina Zirati *et al* 2022)

References

Food borne Zoonotic diseases. Policy paper 95, National Academy of Agricultural Sciences, New Delhi, July 2020.

Jong Yil Chai , K Darwin Murrel and Alan J Lymbrey 2005 Fish borne parasitic zoonoses: status and issues, International Journal of Parasitology, 1233-54

Mina Ziarati, Mohammad Jalil Zorriehzahra, Fatemeh Hassantabar, Zibandeh Mehrabi, Manish Dhawan, Khan Sharun, Talha Bin Emran, Kuldeep Dhama, Wanpen Chaicumpa, Shokoofeh Shamsi, 2022. Zoonotic disease of fish and their prevention and control - A review Vet. Quaterly Vol 42,2022, issue 1:95-118

O.L.M. Haenen, J.J. Evans & F. Berthe, 2013. Bacterial infections from aquatic species: Potential for and prevention of contact zoonoses. Rev. Sci. Tech. Off. Int. Epiz., 32 (2), 497-507

Shokoofeh Shami 2019, Sea food borne parasitic diseases : A "one health " approach is needed, Fishes 2019 4:(91).

Wondeema Tessema 2020. Review of parasites of fish and their public health importance, ARC Journal of Animal and Veterinary Science, vol 6 (2) P23-27

13

Conventional Methods of Disease Diagnosis

K.M. Shankar

Disease diagnosis refers to the various procedures and techniques used to identify the nature of disease and to precisely pinpoint the primary and secondary pathogens involved. Disease diagnosis is an integral part of health management. Proper diagnosis helps to adopt accurate therapy and avoids indiscriminate use of chemotherapeutics. Information on case history and clinical signs should be carefully used while examining samples for diagnosis. By following rational and scientific sampling protocols, one can hope to arrive at accurate diagnosis. Sampling methodology is key to the success of proper disease diagnosis. Dead fish/shrimp should never be used for any diagnostic purpose. Autolytic post-mortem changes, saprophytic invaders, spoilage organisms, etc. mask the actual etiological agents and can lead to false diagnosis.

The three commonly used conventional diagnostic approaches are fresh examination using microscope, microbiological methods and histopathology. With little facilities and infrastructure, fresh microscopical diagnostic approach can be done at the farm/hatchery site itself. Gross examinations along with simple skin and gill preparations examined under microscope can many a times give very good information about the disease and the etiology. All fresh preparations should be made from live or moribund fish and the preparations should be wet and moist. The fish meant for examination should be kept wet and not allowed to dry. Many of the commonly occurring external disease problems can be easily diagnosed by this method.

Systemic bacterial diseases need to be positively diagnosed by adopting the microbiological approach. Samples for bacterial isolations should be taken from kidney, blood, body cavity, etc. from live or moribund fish following strict aseptic conditions. In shrimp, it is ideal to use hemolymph for bacterial isolations. Dominant bacterial isolates are identified following standard procedures and their antibacterial sensitivity tested before recommending any antibacterial therapy. This method requires a minimum of 4-5 days.

Histopathological diagnosis by light microscopy is an ideal strategy for routine health monitoring and diagnosis. The changes at the cellular and tissue level due to the pathogen is interpreted to arrive at diagnosis. Many a times, the pathogen may also be seen in tissue sections. This method involves fixing tissue samples from live fish in suitable fixative, processing the tissue, section cutting, staining, and studying the sections under microscope. This method also requires a minimum of 3-4 days. In addition electron microscopy(EM) can provide ultrastructure information on causative agent and pathology. However, EM is expensive, time taking process and can not be advocated for routine diagnosis.

Importance of Case History and Clinical Signs in Disease Diagnosis

Case history information is very vital for proper disease diagnosis and for taking proper remedial steps. It is very essential to regularly record information on water quality, feeding percentage, feed intake, fertilization schedule, time of last algal bloom, liming details, treatment details, source of seed, stocking details, size at stocking, new fish introductions, etc. During a disease outbreak, careful scrutiny of the case history information will help to precisely pinpoint the circumstances under which the disease has developed.

Clinical signs refer to the external features of a disease. Aquaculturists should be well versed to recognise the clinical signs at the earliest stages. Clinical signs are largely non-specific and similar clinical signs can be expected from two unrelated pathogens. In majority of the cases clinical signs are the result of pathological processes that are developing in some organ or tissue. For example, gasping is a well recognised non-specific clinical sign and can be because of low dissolved oxygen, gill ectoparasites, blockage of branchial circulation, necrosis of kidney, spleen hemopoietic tissue and haemolysis of RBC. There is always a very good, intricate relationship between the pathogen, the target tissue, mechanisms of pathogenesis, pathological changes and the resulting clinical signs. Rational interpretations of clinical signs can tell a lot about the nature of disease and will certainly augment the process of diagnosis.

1. Techniques in Gross and Fresh Microscopical Disease Diagnosis

By making a gross examination of moribund fish using a magnifying glass, one can easily recognise external conditions like ulcers, deformities, hemorrhages, fin and tail rot, large parasites like *Argulus*, leech and *Lernaea*.

Fish should be killed quickly by cutting through the spinal cord with a sharp scalpel in the region immediately posterior to the gills. Blood can be collected at this stage from the heart or major vessels using a Pasteur pipette.

Procedure for Examination of Fish Skin

i) Entire fish is examined under low power using a stereo microscope. Fins as well as other areas should be examined carefully, as large metazoan parasites and *Argulus* can be seen in this way.

ii) Fish is scraped with a sharp scalpel in an anterior to posterior direction and place mucus and epithelial cells on a slide in a drop of water. Scraping scale is avoided as these reduce the visibility of small protozoa. Thin preparations are essential. "Scrapings" spread thinly and covered with a cover slip are examined under high power. Ectoparasitic protozoan and metazoan agents can be easily diagnosed by this method.

Procedure for Examination of Fish Gills

i) Operculum is removed and examined inside.

ii) A whole gill is removed and placed on a slide or in a petri dish (water added if necessary) and examined under low power stereo microscope. Primary lamellae are separated with needles to observe large monogenea and curstacea and any lesions in detail.

iii) After cutting off lamellae and gill arch is removed. Lamellae is placed on a slide and spreaded thinly, chopped if necessary, covered with cover slip and examined under high power. Ectoparasitic protozoan and metazoans can be easily diagnosed by this method.

Procedure for Examination of Other Organs in Fish

i) Incision is made along the ventral side from vent to head and abdominal wall removed to expose viscera. Visceral surfaces, abdominal cavity and pericardial cavity are carefully observed under low power stereo microscope. Any abnormalities or cysts, spots etc. can also be examined in detail under high power.

ii) Alimentary canal and associated organs are removed and surface and scrape contents are examined on a slide under high power. Compressed sections of alimentary canal between slides are also examined under high power.

iii) Squash preparations from heart, liver, gall-bladder, spleen, kidney gonads, urinary bladder and swim bladder are examined carefully.

iv) Eyes and nares are dissected for examination under low and high power for helminths.

v) Skin is removed and muscle sliced, squashed and placed between slides or glass plates, for examination for helminth larvae and protozoan cysts.

Procedure for Examination of Shrimp Larvae and Post Larvae

i) Samples are examined in a petri dish with a dissecting microscope. Activity, gut contents, fecal strands, surface fouling, deformities, broken melanized or missing appendages are examined carefully.

ii) Abnormal larvae are selected, transferred to a slide, coverslip and observed under low and high magnification. Surface fouling, bacterial lesions, larval mycosis, cuticular deformities, melanised appendage tips can be identified by following this approach.

iii) Squashed preparation of hepatopancreas stained with 0.01% malachite green can be used for MBV and BP diagnosis.

Procedure for Examination of Shrimp Juveniles and Adults

i) Shell and appendage deformities, melanisation, discoloration of gills midgut and hepatopancreas , muscle and gonads are examined.

ii) Preparations from gill and appendage (for fouling, necrosis, melanisation) hepatopancreas (for BP, HPV, MBV, melanised HP tubules, swarms of bacterial rods), mid gut (for BP, MBV, melanised masses of haemocytes indicating vibriosis), muscle or gonad (for Microsporidians) and haemolymph (for bacteria) are examined by microscope.

2. Techniques in Histological Disease Diagnosis

A. Light Microscopy Tissue Fixation

The aim of fixation is: i)To prevent autolysis, bacterial decomposition and putrefaction. ii) To coagulate the tissue so as to prevent loss of easily diffusible substances such as glycogen.iii)To safeguard the tissue against the damaging effects of tissue processing. iv) To leave tissues in a condition which facilitates differential staining with dyes and other reagents.

Tissue for histological examination should be placed in fixative for at least 24 hours prior to cassetting. Neutral buffered formalin (10%) is normally used. One volume of tissue per 9 volumes of fixative is recommended. Individual tissues must be of a suitable size to allow permeability of fixative. It is very important to label the samples.

Cassetting

Before cassetting, it is essential to allocate case number for all the samples. The allocated case number is entered on the cassette using a pencil (ink will be removed by solvents during processing). Tissue samples should be trimmed to a suitable size and must not be overcrowded in cassettes as this will lead to ineffective dehydration and ultimately difficulty in sectioning. Small samples are wrapped in tissue paper before placing in the cassette. Soft and hard tissues should be kept separate. Cassetted samples should not be allowed to dry out and must be left in a bowl of water or fixative until loading onto the processor.

Processing

The aim of processing is to impregnate the tissue with an embedding medium which will give support to the tissue during section cutting. Paraffin wax embedding is most commonly used. First, water is removed from the tissues by immersing them in a graded series of alcohol, ending in absolute alcohol. This is followed by immersion in a clearing agent (Chloroform) which is miscible with both alcohol and wax. The clearing agent is therefore easily removed by the molten wax in the final stage. This procedure is carried out by placing the cassettes into a basket which is moved round automatically by a tissue processor at the appropriate time intervals.

Blocking

Cassettes are removed from the tissue processor and placed in molten wax until ready to block out. The metal lid is removed and the appropriate size of base mould selected to give an adequate margin of wax around the tissue. The base is filled with wax and the tissue sample pressed into the wax. The empty cassette is placed on top of the mould and topped up with wax.

By placing the mould on the cold plate the wax gets solidified and the sample is held in position. Orientation of the tissue depends primarily on the type of section required. Tissues such as skin should be embedded so that the skin edge is uppermost in the block. This makes sectioning easier as the soft tissues underneath is cut through before the hard skin surface. The wax is allowed to solidify on a cold plate before removing to cut sections.

Microtomy

The surface layer of wax has to be first removed to expose the complete surface of the specimen. The rate of advancement of the block towards the knife is determined manually at this stage. Before sections can be obtained from blocks of hard tissue it is often necessary to surface decalcify them. This is carried out by placing blocks face down in a vessel containing a layer of decalcifying solution for approximately 1 hour. Blocks of soft tissues which have been hardened excessively during processing can be soaked in water.

Blocks are cooled using a cold plate prior to sectioning. Specimens are clamped into the block holder which is automatically advanced every rotation of the operating wheel (normally 5 µm). When a 'ribbon' is obtained, this is removed and floated on a water bath. The best section is selected and picked up on a clean glass slide. The case number is marked on the slide using a diamond pencil and the slide is placed face down on a hotplate. Racked slides are then dried in an oven at 60°C for at least 1 hour before staining.

Staining

In order to examine sections effectively under the microscope they require to be stained. Many different staining techniques can be used, the most common being Hematoxylin and Eosin.

Upon completion of staining and cover slipping, sections are labeled with the case number, and mounting fluid allowed to dry before examination by microscope.

B. Electron microscopy

In biological applications there is always need for ultrastructure information, particularly in virology, cell biology and pathology. High resolution and magnification provide a way for ultrastruture information. Electron Microscopes were developed during 1940s due to the limitations of light microscopes which magnify 500x or 1000x with resolution of 0.2 micrometers. An electron microscope (EM) uses a beam of accelerated electrons as a source of illumination. The wavelength of an electron can be up to 100,000 times shorter than that of visible light photons, can achieve higher resolving power than light microscopes to reveal the structure of smaller objects. Electron microscopes are used to investigate the ultrastructure of a wide range of biological and inorganic specimens including microorganisms, cells, large molecules, biopsy samples, metals, and crystals. Besides ultrastructure information, EM are quite useful for understanding pathological processes at ultrastruture level. However, electron microscopy has limitations for routine diagnosis as the procedure is time consuming and expensive.

Types of Electron Microscopes

There are two main types of electron microscopes – the transmission EM (TEM) and the scanning EM (SEM). TEM is used for study of ultrastruture of biological samples- image of the interior of cells, the structure of protein molecules, the organization of molecules in viruses and cytoskeletal filaments and the arrangement of protein molecules in cell membranes. SEM is normally used to study surface structures of metals and biological surfaces such as compound eye of insects, pollen grains, leaf surface etc.

Transmission Electron Microscope (TEM)

The TEM uses a high voltage electron beam produced by an electron gun to illuminate the specimen and create an image. The electron beam focused by electrostatic and electromagnetic lenses, is partly transmitted through the specimen which is transparent to electrons and partly get scattered out of the beam. Emerging from the specimen, the electron beam carries information about the structure of the specimen that is magnified by the objective lens system of the microscope. The spatial variation in this information (the "image") may be viewed by projecting the magnified electron image onto a fluorescent viewing screen. Alternatively, the image can be photographically recorded on a photographic film sensor of a digital camera which may be displayed on a monitor or computer. EM is used in conjunction with a variety of ancillary techniques (e.g. thin sectioning, immuno-labeling, negative staining) to answer specific questions.

A TEM has achieved better than 50 picometer resolution in annular dark-field imaging mode. Resolution below 0.5 angstrom (50 picometres), enabling magnifications above 50 million times has been achieved using a high-resolution transmission electron microscope (HRTEM). Electron microscopes use shaped magnetic fields to form electron optical lens systems that are analogous to the glass lenses of an optical light microscope.

Limitations of EM

EM is a powerful tool for ravelling the hidden ultrastruture of a cell or microbes. Furthermore, EM are very powerful tool to explain pathology processes damage to a cell at ultrastructure level. However EM has its own following limitations i. Electron microscopes are expensive to construct and maintain, need stable housing. ii. The samples are ideally viewed in a vacuum, as the molecules in the air would scatter the electrons. iii. Extremely thin sections of the specimens about 100 nanometers are required which is technically challenging. iv. All biological specimens have to be processed -stabilising, ultrathin sectioning contrast staining which may result in artifacts. However, analysis of cryofixed vitrified specimens can confirm the validity of the artifacts.

Preparation of Sample for Electron Microscopy

Samples for EM could be viral/bacterial purified preparations or infected tissues and accordingly sample preparations vary

1. Purified Viral/Bacterial Suspensions

Viruses grown in cell lines are harvested, purified by ultracentrifugation. If suitable cell lines are not available, viruses can also be harvested from infected tissues and purified by ultracentrifugation. Virus purified to high density with very low impurities, usually a particle density above 10^{10} ml required for studying size, shape etc.Suspensions containing nanoparticles or fine biological material (viruses and bacteria) are subjected to negative staining briefly by mixing with a dilute solution of an electron-opaque solution such as ammonium molybdate, uranyl acetate (or formate), or phosphotungstic acid. This mixture is applied to a suitably coated EM grid, blotted, dried and viewed immediately under TEM. Negative staining is important in microbiology for morphological identification and for high-resolution 3D reconstruction using EM tomography.

2. Tissues Infected with Pathogens

Tissues to be viewed under an electron microscope may require processing to produce a suitable sample requiring following steps. These processing steps are different from those followed for light microscopy:

i. A Chemical fixation of biological specimens stabilize mobile macromolecular structure by chemical cross linking of proteins with aldehydes such as formaldehyde and glutaraldehyde, and lipids with osmium tetroxide.

ii. Dehydration of tissues with organic solvents such as ethanol or acetone, followed by critical point drying with embedding resins or freeze drying.

iii. Embedding specimens after dehydration for ultrathin sectioning. Embedding involves immersion in a 'transition solvent' such as propylene oxide (epoxypropane) or acetone followed by infiltration with an epoxy resin such as Araldite, Epon, or Durcupan. Tissues can also be embedded directly in water-miscible acrylic resin which polymerises (hardens) the specimen which has to be ground and polished to a mirror-like finish using ultra-fine abrasives. Polishing is performed carefully to minimize scratches and artifacts that reduce image quality.

iv. Sectioning – produces ultra thin slices(60–90 nm thick) of the electron semitransparent specimen using an ultramicrotome with a glass or diamond knife. Disposable glass knives are made in the laboratory are cheaper for routine sectioning. Ultrathin sections usually acquire gold or silver gold colour when floated on water are used for staining.

v. Sections are stained using heavy metals such as lead, uranium or tungsten to scatter imaging electrons and thus give contrast between different structures. Biological materials are nearly "transparent" to electrons and hence stained "en bloc" before embedding and also later after sectioning. Typically thin sections are stained for several minutes with an aqueous or alcoholic solution of uranyl acetate followed by aqueous lead citrate for viewing by TEM on sample grids.

Cryoelectron Microscopy

Artifacts are common in electron microscopy due to series of steps in sample preparation such as chemical fixation, embedding, sectioning. Alternative to chemical fixing of tissue ,croyofixtions are used to minimise artifacts. In Cryofixation – a specimen is rapidly frozen, in liquid ethane so the water forms vitreous (non-crystalline) ice which preserves the specimen in a snapshot of its solution state. With the development of cryo-electron microscopy of vitreous sections (CEMOVIS), it is possible to observe any biological specimen close to its native state with least artifacts. Freeze fracture technique is used to expose ultrastructure for photography. Alternatively, freeze fractured sections are imunolabelled for specific detection.

3. Techniques in Microbiological Disease Diagnosis

Before examining a fish internally, the external body surface should be examined for the presence of any lesions. The gills, tail and fins should also be observed for any visible signs of infection. Samples from these sites can be taken by searing the surface with a hot scalpel blade followed by insertion of a sterile bacteriological loop. Material from the loop is then placed out onto suitable agar medium by the spread plate technique.

Once external examination and sampling has been carried out, the body surface is opened to expose the internal organs. Care must be taken not to puncture any part of the intestinal tract. In the absence of any visible internal lesions a sample of kidney is taken and inoculated onto suitable agar medium. Material from other organs can be taken if any abnormality is evident in these. Ideally the surface of any internal organs to be sampled should be seared with a hot scalpel blade before insertion of a sterile loop.

Agar plates containing the streaked out samples should be incubated and examined daily for any evidence of growth. Majority of fish pathogens will grow on Tryptone Soya Agar within 7 days.

Direct examination of Gram stained kidney smears may give an indication of bacterial septicemia. A small portion of kidney is emulsified in a loopful of distilled water on a microscope slide, allowed to air dry and then fixed. The fixed smear is then stained by Gram's method. Direct observation of the slide is carried out at x 400 and x 1000 magnification.

Samples for bacterial isolations from sick shrimp should be taken from hemolymph and suspected tissues following strict aseptic conditions. Marine Agar or TCBS medium is normally used for all isolation purpose. The isolated bacteria is characterised and identified by following standard morphological and biochemical tests some of which are detailed below.

Gram's Staining

The morphology of bacteria is difficult to observe in wet, unstained preparations and these are not permanent. It is usual to stain thin films of organisms in order to examine them. The Gram's stain is a differential stain which demonstrates bacteria of different types. This staining procedure is the most important and most widely used in bacteriology for it divides nearly all bacteria into one of two categories.

i.	Gram- positive	-	Which resist decolourisation by ethanol or acetone and stain blue/ purple (positive = blue/purple)
ii.	Gram– negative	-	Which are decolourised by ethanol or acetone and are stained red/ pink (negative= red/pink)

This difference in colour reaction is due to the different chemical composition of the cell wall and membrane. This procedure is used to show the general morphology of the bacteria as well as demonstrating their Gram reaction.

Motility Test

This test demonstrates whether a bacteria is capable of independent movement. Many species of bacteria are capable of motility by the movement of external appendages called flagella. This motility can be observed directly under the microscope using a suspension of living bacteria in a "hanging drop slide". A drop of bacterial suspension is hung from the underside of a cover slip and mounted using soft paraffin, on a microscope slide.

Direct observation of the slide under the x 40 objective lens of a microscope will then reveal whether the bacteria are motile. Motility tests should ideally be carried out on actively growing broth cultures although a suspension of growth prepared from an agar plate is generally suitable. It is important not to confuse true motility with vibration or Brownian movement. All bacteria in suspension exhibit Brownian movement which is quite random and non-directional whereas only some bacteria exhibit true motility which is non-random and directional.

Oxidase Test

This test demonstrates whether a bacteria possesses certain oxidase enzymes that are involved in electron transfer from electron donors. Growth from an agar culture of the organism under test is smeared over filter paper soaked in a 1 per cent aqueous solution of the hydrochloride of the dye (dye tetramethyl-p-phenylane diamine). The colour of the smear is noted after 1 minute. A solution of the dye is highly unstable in the presence of light and the test must be carried out immediately after preparation of the solution. Alternatively, growth of the organism can be smeared onto pre-prepared freeze-dried reagent strips and a similar result obtained. A deep blue or purple colour developing within 1 minute indicates oxidation of the reagent and a positive result.

O-F Test

This test demonstrates whether bacteria can break down glucose aerobically (by oxidation) or anaerobically (by fermentation). A culture of the organism is inoculated into freshly prepared tubes of O-F medium by a single stab with a straight wire. One tube is incubated in the presence of air (open, aerobic tube) the other is covered with a thick layer of liquid paraffin to exclude air (closed, anaerobic tube). The medium contains bromothymol blue pH indicator to indicate the formation of acid from the breakdown of glucose. After suitable incubation the results are interpreted as follows:

Open Tube	Closed Tube	Result
Green	Green	No reaction on glucose
Blue at top	Green	Alkaline reaction
Yellow	Green	Oxidative
Yellow	Yellow	Fermentative

Antibiotic Sensitivity

The susceptibility of bacteria to different antibiotics and other antibacterial agents can be useful in (1) identification of an organism and (2) suggesting a possible treatment for a bacterial disease. Sensitivity to a range of antibiotics is determined by placing filter paper discs impregnated with the antibiotics on an agar plate which has been spread with a suspension of the organisms under test. After incubation, the plate is examined for the presence of clear areas of no growth (Zones of inhibition) surrounding each disc. If present, this indicates that the organism under test is sensitive to that particular antibiotic.

The absence of a clear zone indicates that the organism under test is resistant to that particular antibiotic.

Although this is a very useful test, care must be taken when interpreting results. Sensitivity tests do not attempt to reproduce conditions present in fish undergoing antibiotic treatment but they do provide information to suggest a possible effective treatment for diseased fish.

References

Ayallew Assefa and Fufa Abenna 2018. Maintenance of fish health in aquaculture; Review of epidemiological approach for prevention and control of infectious diseases of fish Veterinary Medicine International article Id 5432497, 10 pages

Evans J , Klesiusp H, Shoemaker C A, 2006. An overview of Streptococus in warm water fish Aquaculture Health International , Issue 7 page 10-14

K. M. Shankar and C.V Mohan, 2002. Fish and Shellfish Health Management , ISBN 81-7525-328-2, Fish Pathology and Biotechnology Laboratory, Dept of Aquaculture, UAS, College of Fisheries, Mangalore-575002

Katja R. Richerd, Poggar et al., 2019. Electron microscopy methods for virus diagnosis and high resolution analysis of viruses. Mini review article, Fron. Microbiol 07.

14

Antibody Based Disease Diagnosis

Omkar Vijay Byadgi, Naveen Kumar B.T. and Raj Reddy

Introduction

Development and standardisation of diagnostics for aquaculture health management have undergone great improvements over the last three decades. Diagnosis in aquaculture health management involves several stages starting from farm-level observations to conventional laboratory identification and histopathology to sophisticated modern molecular tools. Accordingly, diagnosis can be broadly categorised under three levels depending on the level of scientific development.

1. level-I detection system, farm site observation and record-keeping;
2. level-II system, conventional laboratory identification employing microbiology, parasitology, and histopathology and
3. level-III system, advanced and specialized immunological and biomolecular techniques.

The interaction of an antibody with an antigen forms the basis of all immunochemical techniques. Antibody binds specifically to an antigen at epitope or antigenic determinant level and this property has been used for development of immunodiagnosis. Furthermore, as antibodies react with antigen at ambient conditions and do not require sophisticated equipment ideal field test kits can be developed. Biomolecular tools using nucleic acid probes although very sensitive are costly and time-consuming requiring sophisticated equipment and trained personnel. Against this background, there is always a need for developing simple, low-cost, rapid yet sensitive farm-level diagnostics. Such field-level tests facilitate screening a large number of samples by a large number of poor and medium-sized farmers who constitute a majority in developing countries. Large-scale screening besides empowering farmers is considered to be ideal for effective health management. Antibody probes provide an ideal format for developing simple field-level diagnostics as antigens ideally react with antibodies at ambient conditions without requiring sophisticated equipment. Available technology for raising large quantities of monoclonal antibodies (MAbs) against pathogens has increased the scope for developing field detection kits in human, animal and agricultural health management. Several farmer-friendly detection systems also referred to as point-of-care (POC) detection method based on antibodies have been innovated in the recent past. POCT is defined as 'medical diagnostic testing at the time and place of patient care.

Antibodies

Antibodies are proteins produced by vertebrates in response to the presence of foreign molecules in the body. They are synthesised primarily by plasma cell, a terminally differentiated cell of B-lymphocyte lineage, and circulate throughout the blood and lymph where they bind to foreign antigens. Antibodies are a large family of glycoproteins (immunoglobulins–Ig) that share key structural and functional features. Functionally, they can be characterised by their ability to bind both to antigens and to specialised cells or proteins of the immune system. Structurally, antibodies are composed one or more copies of a characteristic unit that can be visualised as forming a Y shape. Each Y contains four polypeptides – two identical copies of polypeptide known as the heavy chain and two identical copies of polypeptide called light chain.

Basic structure of an antibody can be explained easily with immunoglobulin G (IgG). IgG molecules have three protein domains. Two of the domains are identical and form the arms of the Y. Each arm contains a site that can bind to an antigen, making IgG molecules bivalent. The third domain forms the base of the Y, and this region is important in certain aspects of the immune response. The two domains that carry the antigen binding sites are known as Fab fragments (fragment having the antigen binding site), and the protein domain that is involved in immune regulation is termed the Fc fragment (fragment that crystallizes). The region between the Fab and Fc fragments is called the hinge. This segment allows lateral and rotational movement of the two antigen-binding domains. The two heavy chain polypeptides in the Y structure are identical and are approximately 55,000 daltons. The two light chains are also identical and are about 25,000 daltons. One light chain associates with the amino-terminal region of one heavy chain to form an antigen-binding domain. The carboxy-terminal regions of the two heavy chains fold together to make an FC domain. The four polypeptide chains are held together by disulphide bridges and non-covalent bonds.

In addition to the IgG, the serum contains other classes of antibody molecules. They are IgM, IgA, IgE, and IgD, distinguished by the number of Y-like units and the type of heavy chain found in the molecule. IgG molecules have heavy chains known as γ-chains, IgMs have μ-chains, IgAs have α-chains IgEs have ε-chins, and IgDs have δ-chains. The differences in the heavy chain polypeptides allow these proteins to function in different types of immune responses and at particular stages of the maturation of the immune response. The protein sequences responsible for these differences are found primarily in the Fc fragment. Different classes of antibodies may also vary in the number of Y-like units that join to form the complete protein. Fish antibody is unique in that it is an IgM like molecule which has either four or five arm (tetramer or pentamer). In diagnosis, IgG and IgM are employed. However, monovalent IgG is preferred to IgM which cross react due to its polyvalent nature.

Source of Antibodies

Conventionally, specific antibody probes are raised in rabbits, goats or guinea pigs. The target antigen/pathogen is grown and purified. The purified antigen is then injected to rabbit in several doses. Blood is collected from the immunized animal 30 days after the first dose and antiserum containing specific antibodies is collected. Since antiserum is polyclonal derived from several plasma cells against different epitopes, there is always background and cross reaction in immunological tests. Therefore, monoclonal antibody production through hybridoma technology is considered as the best strategy as source of antibodies.

Basis of Antibody Based Diagnostics

The interaction of an antibody with an antigen forms the basis of all immunochemical techniques. Antibody binds specifically to an antigen at epitope or antigenic determinant level and this property has been used for development of immunodiagnosis. The structure of the antigen-antibody complex has been studied by measuring the affinity of binding between an antibody and a series of related antigens, by using affinity labeling reagents by site-directed mutagenesis of the antibody combining site, by molecular modeling, and most compellingly, by x-ray diffraction studies of antibody-antigen co-crystals. The antigen binding site of an antibody is formed by the variable regions of the heavy and light chains. Affinity labeling and x-ray crystallography of immune complexes have established that the antigen binding site is formed by the heavy and light-chain variable regions. The two variable regions are closely associated and are bound to each other by noncovalent interactions. The remainder of the heavy and light chains forms other domains that are not involved in antigen binding and correspond to the amino acids of the hypervariable regions determined from protein sequencing. The hypervariable regions are known as the complementary determining regions (CDRs). Binding strength of antibody with the antigen determines the success of all immunodiagnosis. Binding force between antibody and antigen are non-covalent depending on hydrogen bonding, electrostatic force, Van der waals and hydrophobic interactions. These binding forces depend upon pH ionic strength and temperature. These parameters are given due importance during development of immunodiagnosis.

The region of an antigen that interacts with an antibody is defined as an epitope. An epitope is not an intrinsic property of any particular structure, as it is defined only by reference to the binding site of an antibody. The size of an epitope is governed by the size of the combining site. From x-ray studies of the structure of cocrystals between small antigens bound to antibodies, the size of combining site was thought to be relatively small (6-16 amino acids long). The site is visualised as a cleft or pocket into which the epitope docked. Relatively few of the amino acid side chains of the CDR are in close contact with the antigen. Because antibodies can recognise relatively small regions of antigens occasionally they can find similar epitopes on other molecules. This forms the molecular basis for cross reaction. Although similar epitopes can occur on related molecules, the presence of similar epitopes does not necessarily imply a functional relationship.

Antibody Based Tests

There are two kinds of antibody-based tests; solid-phase and liquid-phase tests. In recent years, solid-phase assay systems have become by far the most common (Plate 12). However, liquid-phase assays such as radioimmunoassay are rarely used in diagnosis. In addition, antibody based in situ tests are becoming popular in diagnosis (Plate 13).

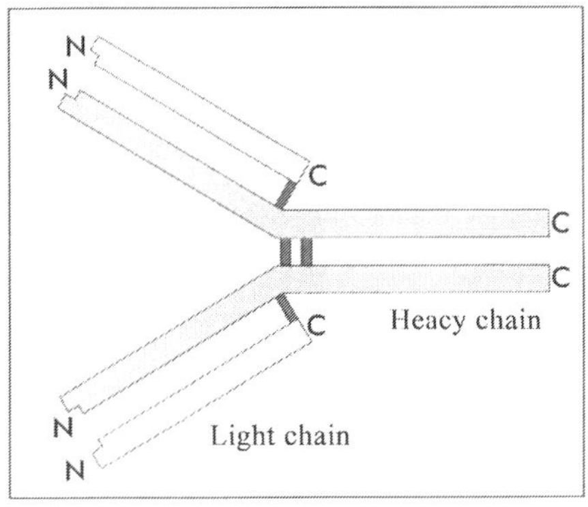

Basic structure of Immunoglobulin.

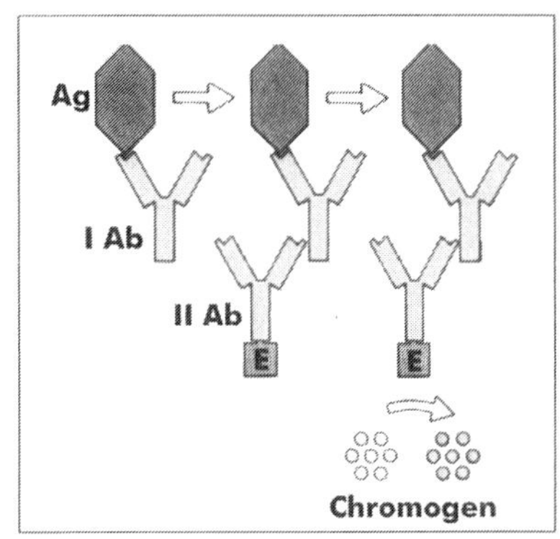

Principle of solid phase immunoassay.

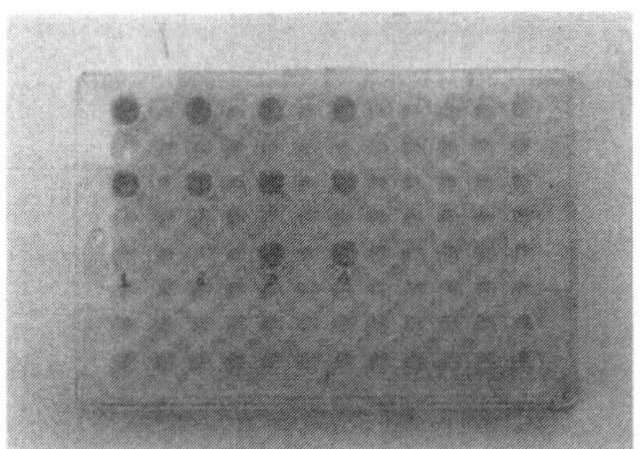

ELISA of WSSV.

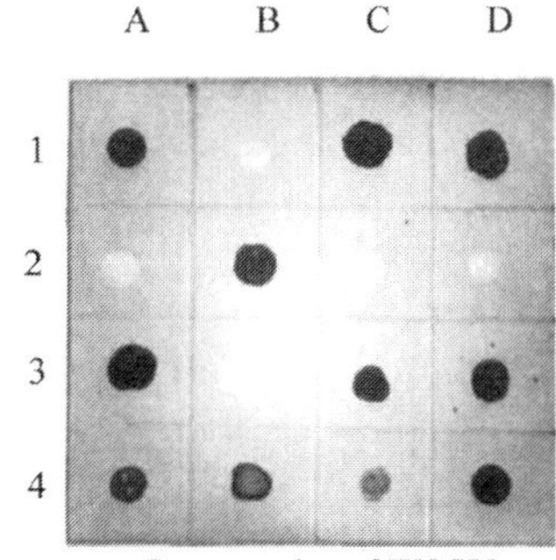

Immunodot of WSSV

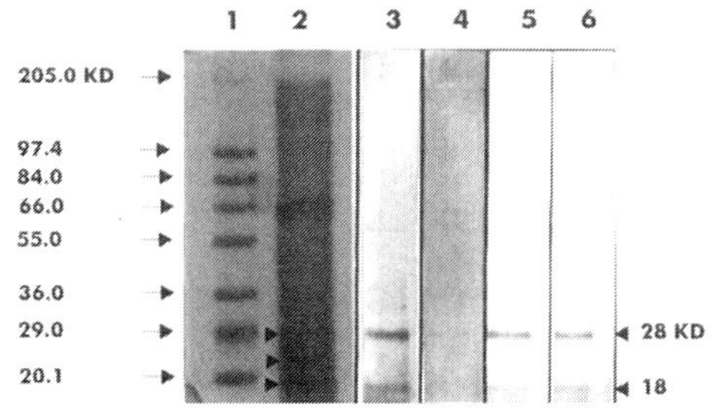

Western blot analysis of WSSV by Monoclonal antibodies

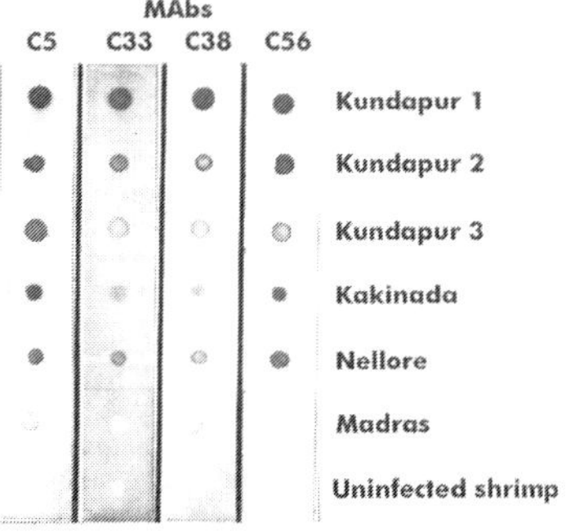

Serotyping of WSSV isolates by MAb based immunodot

Plate 12: Antibody based diagnostics (Solid Phase) (See colour version on page 377)

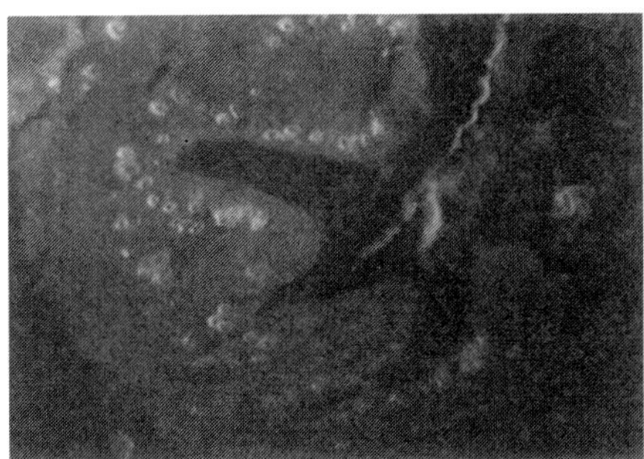

Monoclonal antibody (MAb) based immunofluorescence of *A. Hydrophila* in carp gut

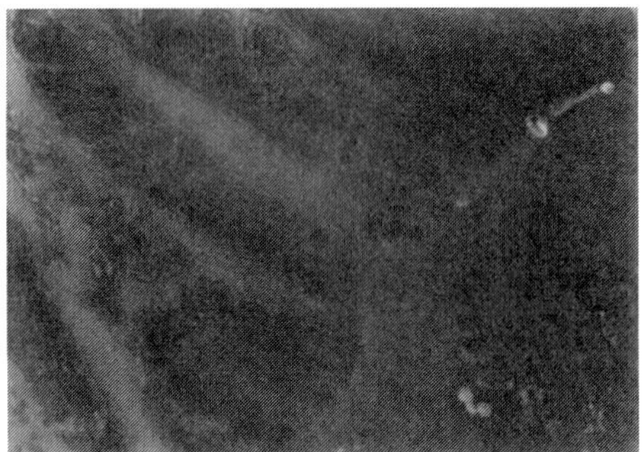

MAb. based immunoperxidase of *A. hydrophila* in carp gut

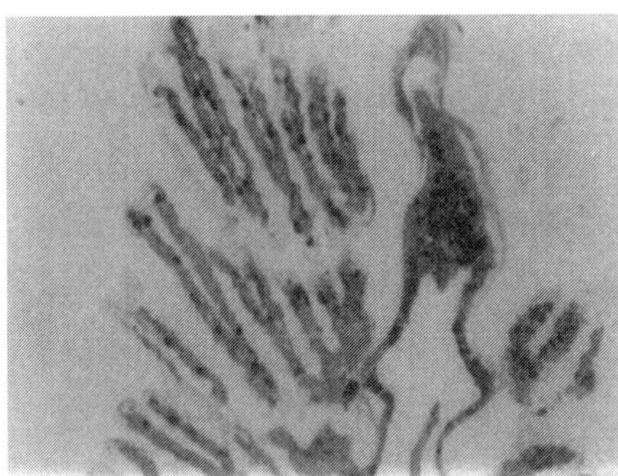

MAb based immunoperoxidase of WSSV in gills

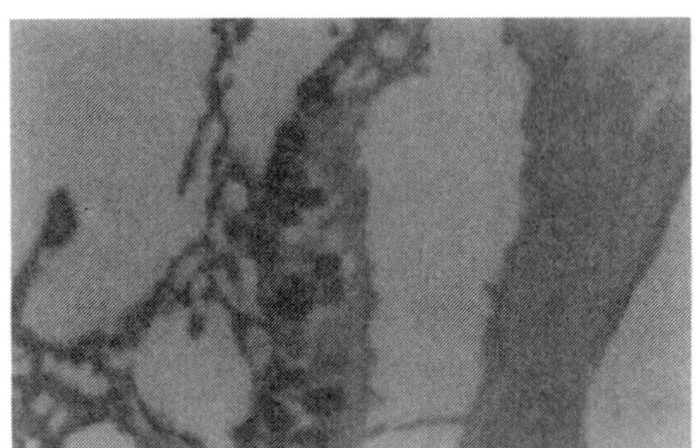

MAb based immunoperoxidase of WSSV in cuticular epithelium

Plate 13: Antibody based diagnostics (*in situ*) (See colour version on page 378)

Solid-Phase Assays

There are three classes of solid phase immunoassays, 1) antibody capture assay, 2) antigen capture assay, 3) two-antibody sandwich assay.

In an antibody capture assay, the antigen is attached to a solid support, and labelled antibody is allowed to bind. After washing, the assay is quantified by measuring the amount of antibody retained on the solid support. In an antigen capture assay, the antibody is attached to a solid support, and labelled antigen is allowed to bind. The unbound proteins are removed by washing, and the assay is quantified by measuring the amount of antigen that is bound. In a two-antibody sandwich assay, one antibody is bound to a solid support, and the antigen is allowed to bind to this first antibody. The assay is quantified by measuring the amount of labelled second antibody that can bind to the antigen. The choice of class depend upon whether one needs to detect antigen or antibody.

The solid-phase assay system is adaptable to almost any antigen with almost any antibody. In ELISA or radioactive binding assay, the antigen is bound to the plate and incubated with an antibody. A second antibody coupled to an enzyme or isotope is then used to detect the first. In the sandwich assay, one antibody is bound to the solid support, the antigen is then added, and a second labelled antibody directed towards different epitope on the antigen is used to detect the bound antigen. The first system is generally used for screening and quantifying antibody and the second one for quantifying antigen. However, if both antibodies are of different species, the second assay may be adapted for screening with the use of a third labelled antispecies antibody. A sand witch assay is required to detect fish antibody. In all solid phase assays, suitable negative and positive controls are required for validating the results. Negative control without antigen or antibody will yield negative result, compared to positive control with all the steps resulting in a positive.

The Nature of the Solid Support

The chemical nature of the solid support may be glass, nylon, sepharose, cellulose, cyanogen bromide, DBM activated paper, polystyrene or nitrocellulose paper. However, the most common are polyvinyl or polystyrene plates and nitrocellulose paper. The plates are generally coated with a material which encourages the binding of the antigen or antibody.

Solid Phase Assays

All solid phase assays (ELISA, Immunodot, Western blot) have common basic steps such as attachment of antigen to solid phase, blocking of remaining sites, incubation with first antibody, incubation with second antibody linked to an enzyme and finally reaction with a substrate. In all these steps due importance should be given to performance of assay at appropriate pH, temperature and ionic strength to ensure successful interaction of antigen antibody. Incubation of antigen/suspected sample and antibody in a particular buffer of known ionic strength at an appropriate temperature is crucial. Further, non-specific binding between antibody and solid support leading to false positive is common. Such non-specific bindings can be avoided by washing the solid support at each step with a buffer containing detergent such as Tween-20. It is in this context, important to maintain positive and negative controls (without antigen or antibody steps) in all the immunodiagnostic assays, to validate the tests. In *in situ* immunodiagnosis, the above essential steps are also largely followed.

Incubation of Antigen

Most soluble protein and nucleic acid antigens are passively adsorbed onto the solid surface. The capacity of the various supports varies widely. In addition, the amount adsorbed depend on the diffusion coefficient of the protein, the surface area to volume ratio and the time and temperature of the adsorption/incubation. It must be remembered that the binding is non covalent and in most systems, protein will leach off the plate during subsequent incubation with the antibody. This leaching often results in inconsistent readings in the final result, which is a drawback of the system.

Storage of Antigen Bound Solid Supports

Most antigen bound microtitre plates can be stored at 40°C for several weeks. A 0.02% sodium azide is included with the blocking buffer to avoid bacterial contamination and the plates are stored with blocking buffer to avoid antigen desorption, which is washed off immediately before use. Dried antigen bound to nitrocellulose paper may also be stored for 2 to 3 months.

Blocking of Remaining Sites on Solid Support

Most solid supports adsorb proteins non-specifically by hydrophobic interactions. It is evident that they will also adsorb the antibody if the plates are not fully saturated with antigen. Once the antigen has been bound, it is therefore usual to block any other sites on the support by the use of bovine serum albumin (1% w/v), gelatin (0.2-0.5% w/v), nonspecific serum (1% w/v) or bovine gammaglobulin (1% w/v), dissolved in phosphate buffer saline. The use of 0.01% Tween 20 or Triton X-100 in subsequent buffer also reduces the background as it discourages the formation of further hydrophobic interaction between the solid support and the first and second antibodies.

Incubation with Antibody

It is desirable to have an assay which allows fairly long periods of incubation with an antibody so that small amount of antigen may be detected. In a solid-phase assay, this has to balance against the tendency of the antigen to leach off the plate during the incubation. The extent of the incubation can be monitored by performing the checkerboard assay with duplicate plate incubated for a different periods of time (overnight at 4°C and 2 h at room temperature) with the antiserum from the immunised animal

Incubation with Second Antibody

The bound antibody is usually detected with the second antibody directed against it. Covalently linked to this second antibody can be an enzyme reacting with a chromogenic substrate. Sheep anti mouse or goat anti mouse antibodies linked to enzyme are thus used to detect murine antibodies. Alternatively, radiolabelled second antibody may be used. The enzyme substrate reaction facilitate visual observation of the result.

Enzymes Used in Immunoassays

There are three enzymes generally used with a second antibody for colour development. The earliest was alkaline phosphatase and the commonest now is horse raddish peroxidase. The third is beta-galactosidase from *E. coli* which is active at the higher pH values preferred for antigen adsorption.

Reactions with Substrate

The substrate generally employed for alkaline phosphatase and beta-galactosidase are p-nitro phenyl phosphate and o-nitro phenyl beta-D-galactopyranoside, respectively with colour development being detected at 405 and 420 nm. In the case of horse raddish peroxidase o-phenylenediamine (OPD) and tetramethyl benzedine (TMB) are most commonly used. More sensitive ELISA detection system may be obtained by the use of fluorogenic substrates for alkaline phosphatase or beta-galactosidase. Most ELISA are read in a spectrophotometer adapted for microtitre plates. This is an excellent method for obtaining printed results for storage and removes the subjective element.

Common Solid Phase Assays

1. Enzyme-Linked Immunosorbent Assay (ELISA)

ELISA is a solid-phase assay where an immunoreagent (Ag) is immobilised on the solid carrier system (polystyrene plate). Next, the reciprocal immunoreagent (Ab) linked to an enzyme is reacted with the antigen. The specific binding of Ag with the Ab is determined by enzyme-substrate reaction. There are two types of ELISA; Direct and Indirect ELISA. In a direct ELISA, antigen is coated on to ELISA plate followed by incubation with a specific antibody which is linked to an enzyme. The bound antibody is determined by using enzyme reaction with a chromogenic substrate. In indirect ELISA, a second univalent antibody to first antibody linked to an enzyme is used to increase specificity and sensitivity. ELISA can be used for quantitative or qualitative estimate of antigen/antibody. The test can be used for detection of both antigen and antibody. The assay at a time can analyse 70-80 samples. Overall, the assay is specific, sensitive and rapid which can be easily streamlined for use at field level. ELISA have been developed for several fish and shellfish pathogens employing rabbit antiserum or monoclonal antibodies.

2. Immunodot Assay

The immunodot assay in which the antigen is attached/bound to nitrocellulose paper by dotting is a system of choice for screening on a limited budget. It is claimed to be equally sensitive to or more sensitive than ELISA. The method is similar in principle to ELISA except the use of nitrocellulose paper as solid support. The method involves applying dots of the antigen to nitrocellulose paper and performing the test in a similar manner to ELISA. Alternatively, the dotted sheet can be cut into squares which can be transferred to microtitre wells for incubation with the antibody. The first antibody is reacted with a second antibody linked to an enzyme, usually horse raddish peroxidase. The second antibody is estimated by reacting the enzyme with a substrate such as 4 chloro 1 naphthol which gives a purple blue dot. The dot assay is sensitive at picogram level, giving consistent result ideal for low concentration of antigen. Results are easier to read, store and the test can be simplified further for field level use by farmer. Compared to ELISA, the test is ideal for field level use as it does not require any special equipment.

3. Western Blot

In an antigen-antibody reaction in ELISA or Immunodot, false positive result can occur due to several reasons (non-specific binding of antigen/antibody to solid phase or background). Therefore, it is difficult to visualise and verify the specific binding of antigen and antibody. Further, antibody especially rabbit antiserum is polyclonal where antibodies are directed to both antigen and background, giving false positive. Similarly, a pathogen is present with other proteins in the host tissue. In order to avoid this confusion and check false positive, presence of pathogen and its specific reaction with antibody can be demonstrated in a Western blot. In Western blot, tissue proteins mixed with pathogen are first separated by Sodium dodecyl sulphate polyacrylamide gel electrophoresis (SDS-PAGE) and later antigen separated are transblotted to a nitrocellulose paper. The separated antigens are reacted with a first antibody (rabbit antiserum or monoclonal), followed by second antibody and substrate similar to that in an immunodot test. Protein bands of the pathogen reacting specifically with the first antibody gives purple blue color with substrate such as 4 chloro 1 naphthol. This will precisely indicate the antigen recognised by antibody. This method though specific has its own disadvantages as it is time consuming and cumbersome. Further, antigen gets denatured during SDS-PAGE by SDS reaction and in the process discontinuous epitopes fail to react giving a negative result. The technique may not be ideal for routine diagnosis. However, this technique is very much required to resolve doubtful false positive result in ELISA or immunodot.

4 Flow Through Assay (FTA)

A flow-through immunoassay (FTA), an improved version of immunodot where nitrocellulose membrane (pore size 0.25 to 0.45 u) baked onto adsorbent pads enclosed in a plastic cassette is employed to detect pathogens (Fig 1). FTA in which the reactants are drawn through nitrocellulose membrane due to its contact with an absorbent bed is both simple to use and manufacture. Furthermore, since flow of reagents is with gravity, there is higher scope for reducing pore size of nitrocellulose membrane for increasing sensitivity of test. Overall in FTA configuration diffusing time and distance between Ag and Ab dramatically reduce facilitating a focused reaction with a sharp dot. Sharp purple dots developed with antigen of a pathogen against the white background of the nitrocellulose membrane provides sensitivity better than lateral flow assay.

In the FTA device, (Fig 2) the upper half of the circular portion of the membrane enclosed in the cassette is demarcated for doting a small volume (2-3 ul) of positive and negative controls and the lower half for fish/shrimp samples to be tested. The dotted samples are air-dried for 4 min. After blocking the dotted membrane with blocking buffer (0.5% BSA in 50 m M PBS, pH7.2) the membrane is continuosly treated with following reagents- i) antibody (rabbit or monoclonal) against the pathogen, ii) second antibody IgG HRP, iii) ready-to-use chromogen substrate (TMB/H_2O_2) without any incubation the steps. Finally the membrane is washed to terminate the reaction with an efficient wash buffer (1%Tween 20 in 50 m M PBS). FTA ends with the development of purple blue colour with-positive and colourless dots with -negative control (Fig 3). The development of purple dots and their intensity with -infected samples are compared with positive and negative controls. The entire procedure in FTA including sample preparation could be completed in 8–10 min at room temperature. Instead of enzyme substrate reagents, second antibody conjugate with colloidal gold can be used ideally for more clarity in colour of dot. An addition of colloidal silver reacting with gold particles can also enhances sensitivity by 20-30 fold. This sensitivity has been evaluated for detection of several antigens and found to be 100 times more than I step polymerase chain reaction(PCR). At college of Fisheries, Mangalore, India MAb based FTA have been developed for detection of four pathogens- WSSV of shrimp, *Aphanomyces invadans* of EUS, *Aeromonas hydrophila* and Vibrio. FTA also developed for detection of antibiotic residues of oxytetracycline and Sulphadimethoxine in fish/shrimp. Later these FTA were developed to field test kits and commercialised for use under field conditions. The kits are ideal for screening brood, seed and routine monitoring of pathogens in farms.

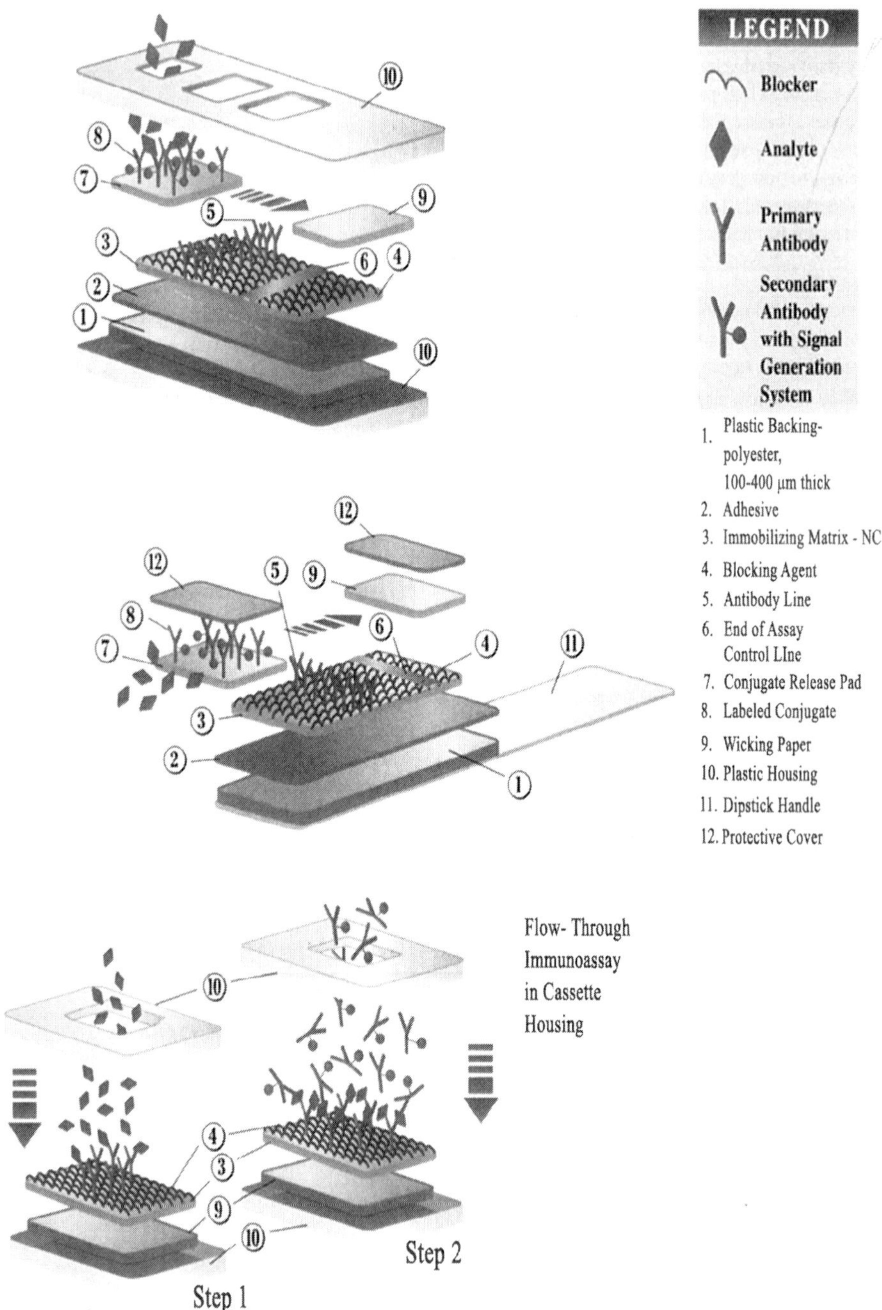

Fig. 1: Schematic diagram of FTA (*Source*: Schleicher & Schuell (Whatman). Products & Protocols. 2005)

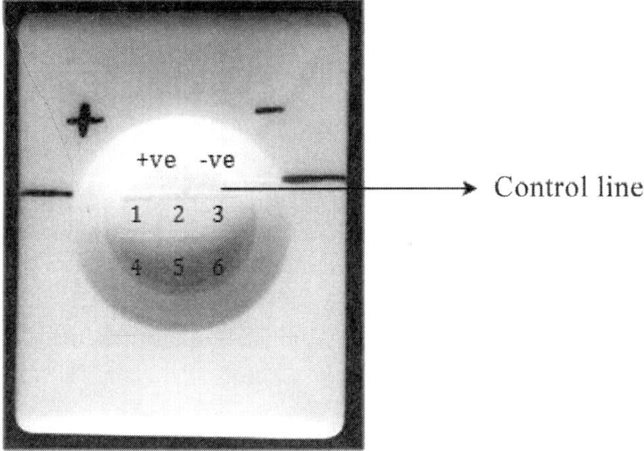

Fig. 2: Ready to use FTA cassette. Test zone and control zone are separated by control line

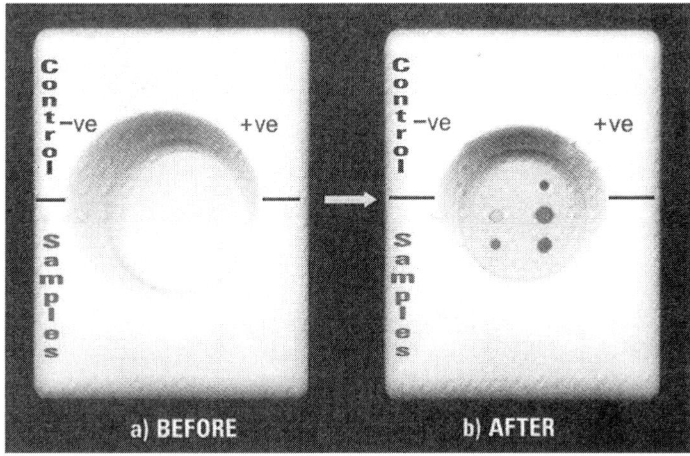

Fig. 3: Developments of dots in FTA

5. Lateral Flow Assay (LFA)

LFA is one of the most popular POCT formats (Fig 4). It is low in cost, simple, rapid, and portable. LFA detection devices have been extensively used in biomedicine, agriculture, food, and environmental sciences. Due to its minimal device size, ease of use, and simple read-out mechanisms, LFA has been increasingly used in aquaculture farms. Researchers are making enormous efforts to improve assay sensitivity and develop user-friendly and highly accurate LFAs for the aquaculture sector. To date, 25 antibody-based LFIAs have been developed. Some of them are frequently used to detect various pathogens in aquaculture farms as they can be performed by unskilled personnel at farm sites within a short time of 10–15 min without any laboratory interventions.

Principle of Lateral Flow Assay (LFA)

LFA is a type of rapid diagnostic test that use immunochromatography to detect the presence of specific analytes in a sample. The principle behind LFA involves the use of antibodies or other binding molecules that can recognize and bind to the target analyte. The basic components of an LFA include a sample pad, a conjugate pad, a nitrocellulose membrane, and an absorbent pad. The sample is applied to the sample pad, which may contain reagents to help prepare the sample for testing. The sample then flows through the conjugate pad, which contains labeled antibodies or other binding molecules that are specific to the target analyte. These labeled molecules can be conjugated into colored particles or enzymes, such as gold or horseradish peroxidase. As the sample flows through the nitrocellulose membrane, the labeled antibodies or other binding molecules will bind to the target analyte if present in the sample. The nitrocellulose membrane may contain a test line that has been coated with antibodies or other binding molecules that are specific to the target analyte. If the target analyte is present in the sample, the labelled antibodies or other binding molecules will bind to both the analyte and the test line, creating a visible signal.

The nitrocellulose membrane may also contain a control line that has been coated with antibodies or other binding molecules that are specific to the labeled antibodies or other binding molecules in the conjugate pad. If the test has been performed correctly and the labeled antibodies or other binding molecules have successfully flowed through the nitrocellulose membrane, the control line will also become visible, indicating that the test has worked properly. Overall, the principle of lateral flow assays involves the use of specific binding molecules to detect the presence of a target analyte in a sample, with the results being displayed in a visual or colorimetric format that is easily interpreted by the user. MAbs can be used to detect viral infections in fish/shrimp by immunological assays such as Western blotting, dot blotting, and immunohistochemistry. Some of the MAbs were used to develop immunochromatographic strip tests for specific detection of white spot syndrome virus, yellow head virus, infectious myonecrosis virus, and *Penaeus monodon* nucleopolyhedrovirus formerly known as monodon baculovirus. The strip test has the advantages of speed, as the result can be obtained within 15 min, and simplicity, as laboratory equipment and specialized skills are not required as mentioned in the flow chart

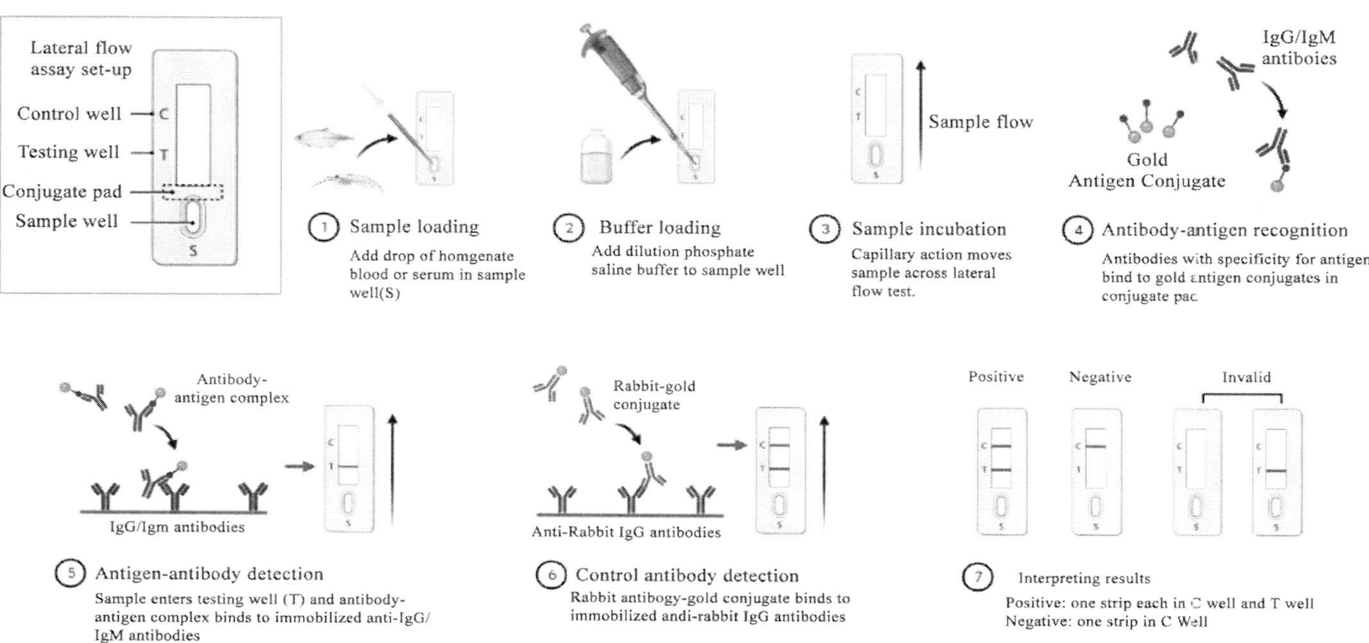

Fig. 4: Flow Chart for Lateral Flow Assay to detect the pathogens. Flow Chart created using Biorender.

It is interesting to note that, to detect pathogens in LFA most of the detection antibodies are either monoclonal antibody (MAb) or antisera (polyclonal antibody, PAb). Most of the reported LFAs for detecting fish and shrimp pathogens have used MAbs as they are more specific in detecting the antigen than PAbs. However, some studies reported the application polyclonal antibody as their detection antibody. Noticeably, most the detecting antibodies have been developed majorly against native or recombinant structural (envelope and capsid) and non-structural proteins of the virus, outer membrane proteins and toxin of bacteria, or spore protein of parasite. For example, as reviewed the anti-LCDV MAb 2D11 targets the surface protein of LCDV, whereas MAb 2E6 and MAb WI-11 target the envelope and non-structural proteins of WSSV, respectively. WSSV LFA was developed using a combination of two MAbs targeting envelope and capsid proteins of WSSV as detection antibodies. LFA was also developed to detect two viruses WSSV and YHV using their respective MAbs in one LFA strip. Moreover, based on reported studies, colloid gold conjugated MAb at concentration at 2–15 μg/ml found to be adequate for detection in the case of fish pathogens. PAb-based LFAs however required higher concentrations of detection antibodies, ranging from 25 μg/mL to 40 μg/ml. However, for detecting shrimp pathogens with LFAs, relatively higher concentrations (0.02–1 mg/ml) of detection antibodies are used.

Comparison of Lateral Flow Through Assay and Flow Through Assay

The main difference between a lateral flow assay and a flow-through assay is the direction of flow of the sample through the assay. In a lateral flow assay, the sample flows laterally across the surface of a nitrocellulose membrane, whereas in a flow-through assay, the sample flows vertically through a porous membrane. In a flow-through assay, the sample is typically applied to a sample well and then flows through the membrane, which contains capture molecules that are specific to the target analyte. As the sample flows through the membrane, the target analyte will bind to the capture molecules, forming a complex that can be detected using labelled antibodies or other detection molecules.

The main advantage of flow-through assays over the lateral flow assays is that they can provide higher sensitivity and specificity. This is because in the flow-through design reagents move by gravity aided by absorbing pad below and pore size of nitrocellulose can be much lower compared that used in LFA. These conditions facilitate longer incubation times and increased closer interaction between the target analyte and the capture molecules, resulting in stronger focused binding leading to better sensitivity than LFA.

Overall, the choice of the assay will depend on the specific needs of the user and the characteristics of the target analyte, including its concentration, complexity, and stability. Both lateral flow assays and flow-through assays have their own advantages and disadvantages and should be evaluated on a case-by-case basis to determine the most appropriate test for a given application.

Antibody Based *In situ* Diagnosis

In situ assays besides diagnosis, are ideal for studies on antigen localisation and distribution. The method followed here is similar to that of ELISA or immunodot, except antigen is detected directly in the tissue. The method does not require homogenisation of tissue with pathogen for coating to a solid phase, as in ELISA. Both fresh and fixed tissue can be employed. However, similar to solid phase assays strict positive and negative controls are required for validating the result.

1. **Immunofluorescence Test:** The test employs histological or croyo cut tissue of a suspected host and reacted with a specific antibody. A second antibody linked to a fluorescing compound usually, fluorescin isothiocyanate (FIT) is reacted with first antibody. Fluorescence is detected by an UV microscope. It is sensitive, rapid, ideal for diagnosis and antigen localisation. However the test requires an UV microscope which is expensive. Further, autofluorescence of some tissue and fading of the FIT are some of the drawbacks of the system. Appropriate positive and negative controls are must for interpretation of results.
2. **Immunoperoxidase Test:** The test employs formaldehyde fixed tissue. Antigen in the suspected host tissue is reacted with first antibody and followed by a second antibody linked to an enzyme usually horse radish peroxidase. The enzyme is detected by reacting with a substrate such as Diaminebenzide (DAB) and observed under an ordinary light microscope. Brick red or grey red coloration in tissue indicates positive reaction.
3. **Immunogold electron microscopy:** This is a highly specific test for intracellular localisation/detection of the pathogen. The tissues are processed for electron microscopy and ultrathin sections of the tissue are obtained. The sections are reacted with specific antibody linked to gold particles. Site of antigen and antibody reaction impart a dark coloration under electron microscope. Appropriate positive and negative controls are necessary for interpretation of specific reaction.

Other Conventional Antibody Based Tests of Importance

1. **Neutralisation Test:** It is a highly specific test ideal for diagnosis and serotyping of viral pathogens. It is based on neutralisation of virus by a specific antibody, which can be determined by inoculation to a cell culture. Virus induced cytopathic effect(CPE) indicates that the virus is not neutralised and hence it is a new virus. Absence of CPE suggest the presence of specific known virus. Besides detection, the test can be used for quantification of antibody or antigen. However, the assay is time consuming and requires cell lines.

2. **Antibody Agglutination Test:** This specific antibody test is used for detection and identification of bacterial pathogens. Antibody binding and specific agglutination of bacterial cells indicates positivity which can be detected visually or using microscope. Extent of agglutination depends on quantity of antibody and antigen. The test is also commonly employed for serotyping of bacterial pathogens.

Summary

Antibody-based disease diagnosis is a commonly used technique in aquaculture for detecting and monitoring diseases in fish and shellfish populations. The technique involves the use of specific antibodies that recognize and bind to antigens, such as proteins or other molecules, that areassociated with the presence of pathogens in fish and shellfish. Antibody-based diagnostic techniques can be classified into two main categories: serological and immunohistochemical assays. Serological assays involve the detection of antibodies produced by the host in response to the pathogen, while immunohistochemical assays involve the direct detection of the pathogen or its antigens in tissue samples.

Examples of antibody-based diagnostic techniques used in aquaculture include enzyme-linked immunosorbent assays (ELISAs), immunofluorescence assays (IFAs), and immunohistochemistry (IHC). These techniques can be used to detect and quantify the presence of pathogens, such as viruses and bacteria, in fish and shellfish. They can also be used to identify specific antigenic markers associated with disease resistance or susceptibility. One of the key advantages of antibody-based diagnostic techniques is their specificity, as antibodies can be designed to target specific antigens associated with the pathogen. They are also relatively easy to use and can be performed in a variety of laboratory settings. However, the technique has some limitations, including the need for prior knowledge of the pathogen and the potential for cross-reactivity with related pathogens. Due to their low-cost, user-friendliness, rapidity, and portability, flow-through assays are the most popular diagnosis systems available for fish/

shrimp farmers. This will be the most suitable technology to detect pathogens in the farm level even for the emerging pathogens in aquaculture.

Detection system employing a combination of silver and gold NPs, a combination of gold NPs and HRP enzymes, gold nano-microspheres for enhancing assay signal enhance the sensitivity to a higher level than I step PCR. Overall, antibody-based diagnostic techniques are valuable tools for disease diagnosis and monitoring in aquaculture, as they provide a rapid, cost-effective, and sensitive method for detecting pathogens and identifying potential disease outbreaks in fish and shellfish populations.

References

Adams, A., & Thompson, K. D. 2011. Development of diagnostics for aquaculture: challenges and opportunities. Aquaculture Research, 42, 93-102.

Chaivisuthangkura, P., Longyant, S., & Sithigorngul, P. 2014. Immunological-based assays for specific detection of shrimp viruses. World Journal of Virology, 3(1), 1.

Di Nardo, F., Chiarello, M., Cavalera, S., Baggiani, C., and Anfossi, L. 2021. Ten years of slateral flow immunoassay technique applications: Trends, challenges and future perspectives. Sensors, 21(15), 5185.

K. M. Shankar and C.V Mohan,2002, Fish and Shellfish Health Management , ISBN 81-7525-328-2, Fish Pathology and Biotechnology Laboratory, Dept of Aquaculture, UAS, College of Fisheries, Mangalore-575002

Pires, N. M., Dong, T., Yang, Z., & da Silva, L. F. 2021. Recent methods and biosensors for food borne pathogen detection in fish: progress and future prospects to sustainable aquaculture systems. Critical Reviews in Food Science and Nutrition, 61(11), 1852-1876

Schleicher & Schuell (Whatman). Products & Protocols. 2005 http://wwwwhatman.com/.

Shyam, K. U., Kim, H. J., Kole, S., Oh, M. J., Kim, C. S., Kim, D. H., & Kim, W. S. 2022. Antibody-based lateral flow chromatographic assays for detecting fish and shrimp pathogens: A technical review. Aquaculture, 738345.

15

Nucleic Acid Based Disease Diagnosis

Omkar Vijay Byadgi, Naveen Kumar B.T. and Raj Reddy

Introduction

Nucleic acid based diagnosis is highly specific, sensitive and rapid method widely used in health management. The detection is mainly based on structure and sequence of bases of nucleic acid which are unique to an organism. In recent years, nucleic acid based diagnosis is becoming common and popular in fish and shell fish health management. Commonly employed nucleic acid based diagnosis are DNA hybridisation and Polymerase chain reaction (PCR).

Nucleic acids are made up of nitrogen bases. The primary sequence of bases in DNA or RNA have regions which are unique for a particular organism or gene. The sequence of DNA is composed of a series of four phosphorylated bases Adenine (A), thymidine (T), guanidine (G) and cytosine (C). Pairs of bases (A=T, or G=C) can form hydrogen bonds, which cause opposing strands of DNA to form a semi-stable bonding (hybridization or annealing) to one another. If the opposing strands of DNA match exactly, then the familiar double helix form of DNA results. In the following example, the opposing strands in (1) will form a double stranded, tightly bonded structure, and the DNA strands are described as complementary. In (2) the strands are non-complementary (opposing A=T and G=C base pairs do not line up) and the two strands will not bond.

1	A	A	A	G	G	T	C	C	G	T	T	T	A	A	C	C	G
	T	T	T	C	C	A	G	G	C	A	A	A	T	T	G	G	C
2	A	A	A	G	G	T	C	C	G	T	T	T	A	A	C	C	G
	G	C	C	A	T	T	C	A	A	A	T	C	C	C	A	T	T

It is this basic hybridisation, which is a function of complementarity of two strands (probe/primers with target) which is exploited in both DNA hybridisation and PCR (Plate 14).

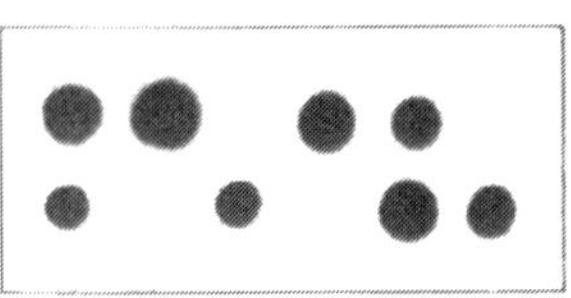

Dot blot DNA hybridization of virus.

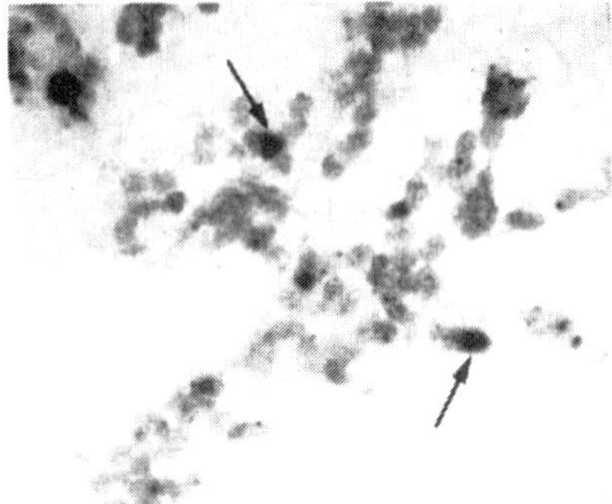

In situ hybridisation of WSV-DNA in gills.

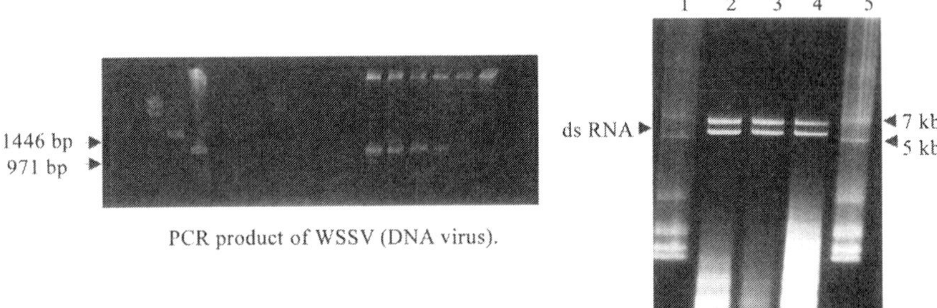

PCR product of WSSV (DNA virus).

Purified nucleic acid (ds RNA) of IPNV for development of diagnostics.

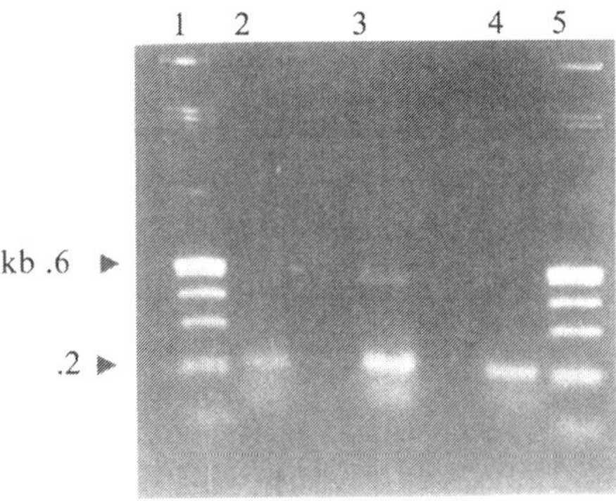

PCR product of IPNV (RNA virus)

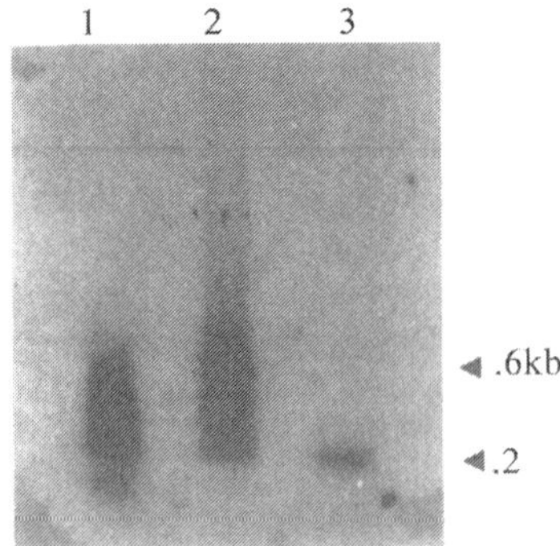

Southern blot DNA hybridisation of IPNV

Plate 14: Nucleic Acid based diagnosis

DNA Hybridisation

The basic principle in DNA hybridization is that, if a strand of double helix of DNA is separated from its complement (by heat) it will bind (reanneal or hybridize) only to its complementary strand (or to a duplicate of its complementary strand). Two complementary RNA structure will hybridise in the same manner. Inaccurate and mismatched bindings will be unstable and fall apart.

The binding of one DNA strand to another is in some respect is analogous to binding of antigen and antibody. However, in the case of nucleic acid bindings, the precise molecular nature of binding (base pairing) is known and can be easily manipulated by choosing DNA strand of appropriate length and by incubating the strands in appropriate salt concentration, temperature. The stringency (faithfulness of the match) of hybridisation can be precisely defined and as a result coincidental cross reaction is virtually eliminated.

Targets and Probes

DNA hybridization, whether done on solid phase, in suspension or *in situ*, requires two strands of DNA – the target and the probe. The target is the sequence of DNA that is to be detected. The probe DNA has several essential features. Firstly, it must be highly complementary to the target strand being sought for. The more nearly complementary (90-100%) the better the bonding, and the better pick of the target from the other masses of DNA in the tissue. Secondly, there must be some way of detecting the probe once it is hybridized. Traditionally, this has been radioisotope labeling of probe with S35, P32, H3 or other radioactive elements. More recently, a number of other detection systems have been used, including biotinylated probes, enzymes and FITC labeled probes. Commercially, the biotinylated probes seems to be the ones of choice, as it is straightforward to design to detect these probes. Thirdly, the probe must be an appropriate length, long enough to specifically detect the target, and short enough that the efficiency of hybridization is high. Probe lengths of 20-100 base pairs are the most common.

Stringency

As a result of the paired bonding, the stability of the double stranded DNA which results is directly proportional to the number of matching base pairs. If we attempt to hybridize two different probes or test strands of DNA to one target, and one probe is 95% complementary and the other is only 15% complementary to the target, then the first will form a stable hybridized double stranded DNA. The second probe may bond, but the bonding will be weak and easily disrupted. This property defines the stringency of the hybridization reaction, and is a critical factor in the specificity of the procedure. The strength of the bonding follows a very well defined formula, which depends on the length of the bonded DNA, the percent of complimentary pairing, the percent of G=C bonds (since the G=C bond is 1½ times stronger than the A=T bond) and the concentration of disruptive reagents such as formamide and high salts. DNA bonding is semistable, and can always be disrupted by heating. The temperature at which the DNA falls apart is determined by the above factors. Thus, in comparing techniques, it is important to look at all the factors listed. All other factors being equal, a technique that uses less formamide in the buffer is less stringent, and may pick up targets which are less well matched with the probe than a technique which uses a higher formamide concentration.

Specificity Vs Sensitivity

In a parallel with the antibody techniques, there is always an uneasy race between specificity and sensitivity in hybridization technique. If comparison of the selection of probe to the selection of antibody is made, then for both cases, the properties of the probe or antibody selected will determine the ultimate specificity and sensitivity. The actual results can then be affected by the tissue processing and the exact protocols used. Table 1 attempts to relate the antibody and nucleic acid methods with respect to those factors which affect specificity and sensitivity.

Table 1: Factors affecting sensitivity and specificity of assays

Antibody	Nucleic Acid
Selection of the best epitope for antibody reactivity	Selection of the best target for probereactivity
Antibody specificity for its epitope	Complementarity of probe and target
Proper fixation of antigen	Proper fixation of nucleic acid
Time, temperature and reaction conditions	Stringency of the procedure
Detection system used	Detection system used

Assuming that the probe selects for the proper target, and that the strands are complimentary, tissue handling is the next most critical issue. As in antibody techniques, DNA may be degraded and physically altered a lot due to improper handling of tissue. For tissues to be processed without embedding, rapid freezing is essential. Tissues may be stored for several months at –70°C or in liquid nitrogen vapor phase. Rapid sectioning and fixation is even better. Tissues to be paraffin embedded should be rapidly fixed, and this means using small tissue blocks, no more than 2-3 mm thick. Formalin or paraformaldehyde fixation appears to be satisfactory in most applications. Over fixation can cause problems of DNA accessibility and should be avoided. To increase accessibility of DNA, a proteolytic digestion is usually essential as a first step in hybridization procedures.

The stringency of the procedure must be considered in any hybridization protocol, and unless a commercial kit is used, this may require the same sort of experimentation that is involved with optimizing an antibody procedure. Examination of different timing, different buffers, etc. may be necessary.

Detection System

The probes are usually linked with radioactive elements such as p32 to enable detection by autoradiography. The current standard for commercial non-radioactive DNA probes is the biotinylated probe. The probe carries a biotin on one or more of the base pairs, and allows the probe to be detected with the very sensitive avidin/biotin or streptavidin/biotin detection systems generally available. The most sensitive system currently available is the ABC-alkaline phosphatase or the streptavidin-alkaline phosphatase system coupled with the blue-black chromogen combination of BCIP and NBT or INT. These chromogens which couple the indole phosphate with a tetrazolium appear to be able to detect in the range of 10-100 copies of the target under the usual conditions for *in situ* hybridization, and may be even more sensitive under optimum conditions. As always, the background-to-positive-signal ratio is the limiting feature in determining a true positive tissue from a true negative tissue, and good negative controls should always accompany a test slide. There are two methods of DNA hybridization: 1. DNA hybridization on solid phase

2. *In situ* DNA hybridisation

1. **DNA Hybridization on Solid Phase:** In this method the suspected fish/shellfish sample is homogenised and dotted on to a nylon membrane or nitrocellular paper and DNA is exposed from protein usually by digesting with trichloroacetic acid. DNA in the sample is then denatured at appropriate temperature and hybridised at appropriate temperature with a probe linked to an isotope or enzyme. Probe hybridization is detected by autoradiography or with a chromogen substrate. This protocol has been adopted to detect DNA in large number of samples to develop a dot blot hybridization. The method is ideal for diagnosis and also for genotyping of pathogens.

2. ***In situ* DNA Hybridization:** Nucleic acid hybridization *in situ* is rapidly being incorporated into the menu of protocols in a number of immunohistochemistry laboratories. Although the procedures for nucleic acid denaturation and hybridization have been known for some

> time, only recently has this knowledge been applied. Commercial availability of sensitive and specific probes, the development of non-radioisotopic detection systems, and numerous improvements in details of the procedures, this technique is beginning to be added to the list of "special stains" in the laboratory repertoire.

In this technique, histological sections of the suspected samples are employed. DNA of the pathogen in the tissues is exposed using trichloroacetic acid and hybridized with isotope or enzyme labeled probe. Hybridization is detected by autoradiography or enzyme – substrate reaction. The technique is ideal for diagnosis and studying site of replication in the target organs of the host.

1. Polymerase Chain Reaction (PCR)

Polymerase chain reaction (PCR) is a novel technique that amplifies specific sequences of DNA with remarkable efficiency. Repeated cycles of denaturation, primer annealing and extension carried out with the heat stable enzyme, Taq DNA polymerase, lead to exponential increase in the target DNA sequences. The technique is increasingly used in diagnosis and molecular biology. Developed originally by Randall Saiki, Henry Erlich and Kary Mullis at Cetus Corporation, USA in 1986 it provides a method for *in vitro* amplification of specific stretches of nucleic acid sequence. This amplification permits rapid analysis or characterization of sequence of interest, even where the starting amounts of sample material are extremely small.

General Principles

The PCR technique was developed to provide highly efficient amplification of DNA sequences of interest. In general, the procedure depends on the availability of sequences that flank regions of interest. Two synthetic oligonucleotide primers are prepared using these flanking sequences, one complementary to each of the strands (Fig. 1). The DNA is denatured at high temperature (95°C) and then reannealed in the presence of a large molar excess of the oligonucleotides. The oligonucleotides, oriented with their 3' ends pointing towards each other, hybridize to opposite strands of the target sequence and prime enzymatic extension along the nuclei acid template in the presence of the four deoxyribonucleoside triphosphates. The end product is then denatured again for another cycle. This leads to the selective enrichment of specific DNA sequences so that they can be readily manipulated or detected. After this three-step cycle is repeated 20-60 times, amplification of up to a factor of 10^{12} can be achieved. The efficiency of this amplification permits the use of very small amounts of starting sequences.

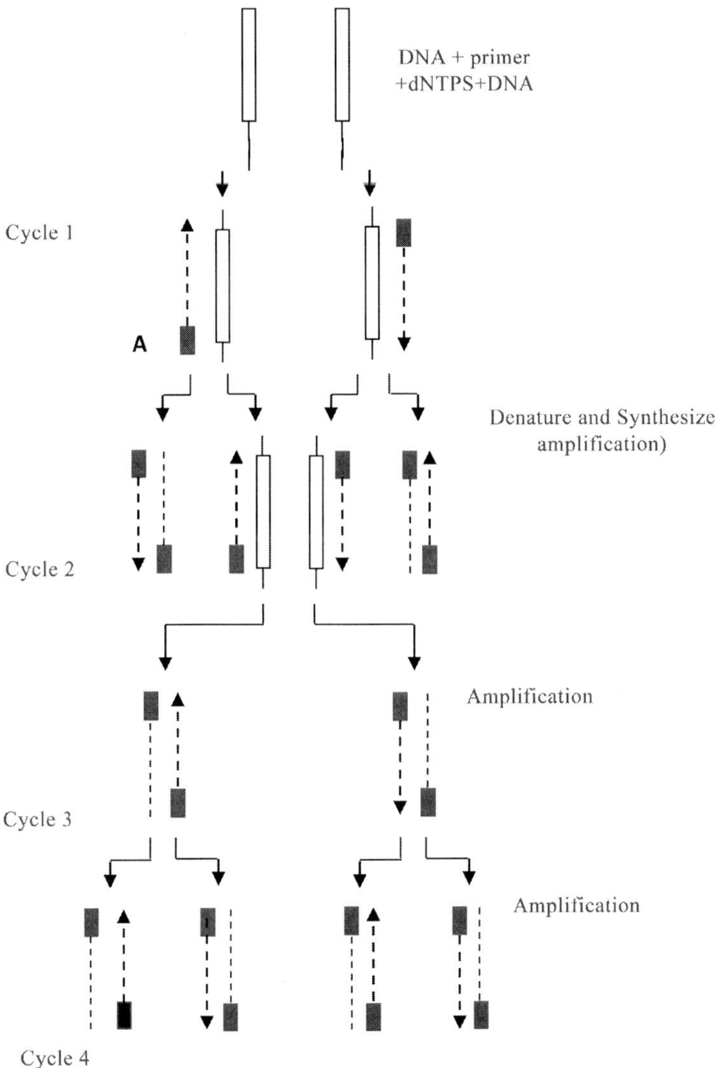

Fig. 1: Polymerase chain reaction: Oligonucleotides A and B hybridized to the denatured DNA, prime synthesis of new DNA in the presence of Taq polymerase and the four dNTPS. This entire event of DNA synthesis is continued in subsequent cycles resulting in amplification of DNA.

Originally, the enzyme used for this procedure was the large fragment DNA polymerase derived from *Escherichia coli* (Klenow). The ingenious identification of an enzyme, Taq polymerase, that is stable at the high denaturing temperature makes it unnecessary to add the enzyme with each cycle and hence has made it possible to automate the technique, resulting in its widespread use. Taq polymerase was obtained from the thermophilic bacterial species, *Thermus aquaticus* has also facilitated the amplification of longer fragments of DNA than was previously possible with the thermolabile enzymes. Both the sensitivity and specificity of the procedure are improved by the high annealling and extension temperatures which reduce secondary structure and increase the specificity of oligonucleotide priming by the high annealing and extension temperatures.

The two priming oligonucleotides dictate the specificity of the reaction. The specificity can also be controlled by varying the reannealing and extension time and temperatures, the magnesium concentration in the buffer, and the enzyme concentration. Each set of oligonucleotide primers has its own optimal conditions. There are no specific rules for the choice of primers except that they are usually 18-22 bases in length, and sequence motifs that might encourage secondary structure should be avoided. Complementary bases at the 3' ends of oligonucleotide pairs should particularly be avoided as they permit the pairing and extension of the oligonucleotides alone (primer-dimer). Oligonucleotides with high GC content generally require higher reannealing and extension temperatures. Inefficient amplification with a given set of primers can be optimized by changing cycle time and temperature or by altering magnesium concentration. Despite this, a small number of oligonucleotide primers remain inefficient and this problem is often best resolved by simply choosing an adjacent new priming site. Several different pairs of oligonucleotides can be used to prime PCR within the same tube. This approach allows a single sample to be simultaneously amplified for several different markers and has important economic implications.

The basic template used for PCR is generally double stranded DNA. RNA can also be detected by PCR after conversion to cDNA by reverse transcription. cDNA can be efficiently used and offer the advantage of fewer sequences being available for specific amplification. In addition, the absence of introns in cDNA avoids the constraints placed on one step amplification of complete expressed sequences by long intervening sequences.

Efficiency, Sensitivity and Specificity

The amplification process is efficient enough to allow the generation of amplified material from a target molecule in a single cell. The efficiency of the technique has been effectively used to amplify DNA sequences from a single cell producing 500 fmol of PCR product from 3.3x10-9 fmol of starting material. This amplification has an average cycle efficiency of 65% and a total amplification of 7.6x10-10 after 50 cycles. Such efficiency readily permits its application to material extracted from hair roots or mouth washings. The efficiency of Taq polymerase enzyme is improved by the presence of non-ionic detergents NP 40 and/or Tween 20 included in the amplification mix to a final concentration of 0.05% each. The length of sequence over which PCR can be used is also greatly increased using this enzyme. Distances of up to 10 kilobases can now be spanned. The magnesium concentration is a critical determinant in amplification; variations of even 100 μM concentration can make a difference to these efficiencies.

Astrategy invoking the use of nested primers has been very effective in improving the efficiency of amplification. This uses one or two oligonucleotide primers designed to hybridize to sequences internal to the original set of primers. Using 1-5% of the original amplification product and fresh deoxyribonucleotide triphosphates, buffer and enzymes, a further set of amplifications can be performed. This provides additional specificity to the reaction and result in dramatic increases in the amplified product. Additionally, specificity of PCR product is checked by Southern blot DNA hybridisation with the primer as probes. In this method PCR product separated on agarose gel is transblotted to nylon membrane and subjected to DNA hybridisation using the primer as probe. Positive hybridisation with one of the primers indicate specificity of PCR. Because of the high efficiencies, a highly purified starting DNA is not needed, as lysed cells can prove sufficient and amplification has even been successfully achieved from tissue sections that have been fixed and paraffin embedded.

Fidelity

The error rate associated with Taq polymerase is a critical factor in determining its general use; this has now been carefully evaluated. The enzyme has little or no 3' to 5' exonuclease activity. In addition, the error rate estimated using DNA synthesis of a Lac Z complementation gene in M13 M2 was 1 in 9000 for single base pair substitution and 1 in 41000 for frame shift errors. This is greatly affected by enzyme concentration. The enzyme is sensitive to low concentration of substrate; when the dNTP concentration falls below 5 μM a high rate of misincorporation is observed. When concentrations of 10 μM are maintained this can be minimized. This contributes to the problems associated with carrying out a large number of consecutive cycles without replenishing the substrate.

Contamination

The high level of amplification possible with PCR create a significant problem with contaminant sequences. This is particularly evident in experiments designed to detect rare sequences of pathogens and may be responsible for the somewhat unexpected results. These problems can best be avoided by using separate positive displacement pipettes, non-reusable containers, gloves when handling reagents and separate work areas for setting up PCR reactions. Aliquoting of PCR mixes in sterile hoods and storage of the mixes at –70°C until target DNA is added, combined with the use of carefully chosen control reactions will also reduce this problem. Apart from the above precautions to avoid contamination leading to false positive, proper collection of samples from aquacultural system need to be considered. Samples should be handled separately to avoid surface contamination.

Different Versions of Polymerase Chain Reaction

There are several versions of the Polymerase Chain Reaction (PCR) technique each with its own advantages and limitations. Here are some common versions of PCR:

1. **Standard PCR:** Standard PCR is the original and most widely used version of the PCR technique. It uses a thermostable DNA polymerase, such as Taq polymerase, to amplify a specific DNA fragment from a template DNA sample. Standard PCR typically involves cycles of denaturation, annealing, and extension, and can be used for a wide range of applications, including DNA sequencing, gene expression analysis and genetic testing.

   The basic protocol of PCR has been further improved to develop several versions; a).Two step PCR-the assay is done in two steps. In the second step 1-5 % of the product developed in the first step is used for application using the same set of primers for further increasing the sensitivity. b).Nested PCR- where the assay is carried out in two steps. In the second step internal primers are used to amplify to increase the specificity and sensitivity. c). *In situ* PCR where DNA in tissue is detected by PCR *in situ*, d). Single tube PCR where several set of primers directed to different regions of pathogen or primers for two pathogens are used for PCR.

2. **Reverse Transcriptase PCR (RT-PCR):** RT-PCR is a version of PCR that is used to amplify RNA instead of DNA. It uses reverse transcriptase to convert RNA to complementary DNA (cDNA), which can then be amplified using standard PCR. RT-PCR is commonly used for gene expression analysis, viral detection, and other applications that require amplification of RNA.

3. **Real-time PCR:** Real-time PCR, also known as quantitative PCR (qPCR), is a variation of PCR that allows for the real-time detection and quantification of PCR products. It uses fluorescent probes or dyes that

bind to PCR products during amplification, allowing for the monitoring of amplification in real-time. Real-time PCR is commonly used for gene expression analysis, viral load quantification, and other quantitative applications.

4. **Digital PCR:** Digital PCR is a newer version of PCR that allows for the absolute quantification of DNA or RNA molecules in a sample. It works by partitioning a sample into thousands or millions of small droplets or wells, each of which undergoes PCR amplification. By counting the number of positive and negative reactions, the number of starting molecules can be determined with high accuracy.

Overall, each version of PCR has its own unique strengths and applications. Choosing the appropriate version of PCR will depend on the specific needs of the experiment or application.

General PCR Protocol Used for Detection of Viral Pathogens in Aquaculture

The protocol has several steps such as collection of sample, extraction of DNA, amplification of DNA and detection of the amplified DNA. Care should be taken in all these steps for successful detection and avoidance of false positive results.

1. **Sampling:** Fresh samples should be collected and fixed immediately in alcohol. Appropriate sample size and sampling strategy are necessary to avoid false negative results.

2. **DNA extraction:** Samples are subjected to DNA extraction by suitable methods (eg. Alkaline lysis method).

3. **DNA amplification:** Target DNA is amplified with a cocktail of Taq polymerase, oligonucleotide primers and nucloetides in a thermal cycler in 20-30 cycles. Each cycle consists of denaturation at 100°C for 1 min, annealing at 55-60°C for 1 min and extension at 70°C for 2 min.

   If required, 2 step PCR is carried out by using 1-5% of amplified DNA from the above 1 step. In 2 step PCR, amplification of DNA can be performed either with the same set of primers or internal primers (Nested PCR).

4. Detection of amplified DNA

The amplified DNA along with DNA markers is subjected to electrophoresis in a 1.5% polyacrylamide gel. The gel is then stained with ethidium dibromide and the fluorescing DNA bands under UV light are recorded. Positive and negative controls should yield appropriate results in each test for validation of the DNA amplification.

Application of PCR in Aquaculture

Polymerase Chain Reaction (PCR) is a powerful molecular biology technique that can be used for a wide range of applications in aquaculture. Here are some common applications of PCR in aquaculture:

1. **Disease diagnosis:** PCR can be used to detect the presence of viral, bacterial, or parasitic pathogens in fish and shellfish. To name a few, PCR has been used to detect the presence of viral pathogens that cause diseases such as infectious salmon anemia (ISA), viral hemorrhagic septicemia (VHS), infectious pancreatic necrosis (IPN) and white spot virus(WSV) of shrimp.
2. **Genetic identification:** PCR can be used to identify fish and shellfish species, as well as to distinguish between different populations or strains. This can be useful for managing fisheries and aquaculture operations, as well as for preventing illegal trade in endangered or protected species.
3. **Quality control:** PCR can be used to detect contaminants or adulterants in fish and shellfish products, such as antibiotics or other chemical residues. This can help ensure the safety and quality of aquaculture products for consumers.
4. **Environmental monitoring:** PCR can be used to detect the presence of harmful algal blooms or other environmental factors that can affect fish and shellfish health. This can be useful for managing aquaculture operations in areas that are prone to environmental hazards or pollution.

Overall, PCR is a versatile and sensitive tool that can be used for a variety of applications in aquaculture, from disease diagnosis to quality control and environmental monitoring. Its applications can help improve the efficiency and sustainability of aquaculture operations, as well as ensure the safety and quality of aquaculture products for consumers.

The sensitivity of PCR allows the detection of pathogens that would be difficult to identify with conventional techniques. PCR is widely used for screening shrimp seed and brood for serious viral pathogens such as WSSV, YHV, IHHNV and TSV. PCR is ideal for studies in epidemiology, genotyping, health certification, quarantine and for screening for development of SPF stocks.

3. Loop-mediated Isothermal Amplification (LAMP): LAMP is a nucleic acid amplification technique that is used to amplify specific DNA sequences under isothermal conditions. Unlike polymerase chain reaction (PCR), which requires temperature cycling, LAMP uses a single temperature for both DNA denaturation and primer annealing. The technique is highly specific, sensitive, and rapid, making it a useful tool for a variety of applications, including disease diagnosis in humans, animals, and plants.

LAMP works by using a set of four to six primers that recognize six to eight distinct regions of the target DNA sequence. The primers hybridize to the target DNA and initiate amplification through a series of strand displacement events. This results in the formation of a stem-loop DNA structure, known as a "dumbbell," which is then amplified by a strand-displacing DNA polymerase. The amplification products can be detected by a variety of methods, including visual inspection, gel electrophoresis, and fluorescence-based detection.

One of the key advantages of LAMP is its ability to amplify DNA at a constant temperature, typically between 60-65°C. This eliminates the need for expensive thermal cycling equipment, making the technique more accessible and cost-effective. In addition, the high specificity of LAMP primers minimizes the risk of false-positive results, while the high amplification efficiency enables the detection of very low amounts of target DNA.

LAMP has been used for the detection of a variety of pathogens in fish and shellfish, including bacteria, viruses, and parasites. The technique has also been used for the diagnosis of genetic diseases in fish and the identification of foodborne pathogens in seafood. LAMP has the potential to be a powerful tool for disease diagnosis and monitoring in aquaculture, as it offers a rapid, sensitive, and cost-effective alternative to traditional diagnostic methods.

4. Microarray Analysis: Microarray analysis involves the use of microarrays, or small chips containing thousands of DNA or RNA probes, to detect and analyze specific nucleic acid sequences. Microarray analysis can be used to diagnose genetic disorders, infectious diseases, and cancers, as well as for gene expression analysis and genotyping.

Microarray analysis is a powerful technique that allows for the simultaneous detection and analysis of thousands to millions of DNA or RNA sequences in a single experiment. Here are some key features and applications of microarray analysis:

High-throughput: Microarrays contain thousands to millions of DNA or RNA probes, allowing for the analysis of large numbers of samples and/or genes in a single experiment.

Versatile: Microarray analysis can be used for a wide range of applications, including gene expression analysis, genotyping, epigenetic analysis, and mutation detection.

Accuracy: Microarrays can provide highly accurate and reproducible data, making them valuable tools for clinical diagnosis, drug development, and disease research.

Speed: Microarray analysis is a relatively fast and efficient allowing for the

rapid analysis of large amounts of genetic data.

Cost-effective: Microarrays can be produced at a relatively low cost, making them a cost-effective tool for large-scale genetic analysis.

Applications of microarray analysis include

1. Gene expression profiling, which can be used to study the expression patterns of thousands of genes simultaneously, and
2. Genotyping, which can be used to identify genetic variations associated with diseases or drug responses.
3. Microarrays can also be used for epigenetic analysis, which involves the study of modifications to DNA that can affect gene expression without changing the DNA sequence itself.
4. In addition, microarray analysis can be used for mutation detection, such as in cancer diagnosis and treatment, and for microbial identification in infectious disease diagnosis.

Overall, microarray analysis is a versatile and powerful tool for genetic analysis, with applications across a wide range of fields including medicine, agriculture, and environmental science.

5. Next-Generation Sequencing (NGS): NGS is a high-throughput technique that allows for the rapid and accurate sequencing of millions of DNA or RNA molecules. NGS can be used for a wide range of applications, including genetic disease diagnosis, cancer diagnosis and treatment, and infectious disease diagnosis. Next-generation sequencing (NGS) is a high-throughput DNA sequencing technology that has revolutionized the field of genomics. NGS allows for the rapid and accurate sequencing of millions to billions of DNA fragments in a single experiment, providing researchers with vast amounts of genetic data in a short period of time. Here are some key features and applications of NGS:

High-throughput: NGS allows for the simultaneous sequencing of millions to billions of DNA fragments, making it possible to analyze entire genome or transcriptomes in a single experiment.

Accuracy: NGS can provide highly accurate and reproducible data, with error rates as low as 0.1%.

Speed: NGS can generate vast amounts of genetic data in a relatively short period of time, making it a valuable tool for large-scale genomic studies.

Cost-effective: NGS has become increasingly affordable over the years with sequencing costs dropping dramatically since the technology was first developed.

Versatile: NGS can be used for a wide range of applications, including whole-genome sequencing, exome sequencing, transcriptome sequencing, epigenetic analysis, and metagenomics.

Applications of NGS Include Genome Sequencing

1. To study the genetic basis of diseases, as well as to identify genetic variations associated with drug responses.
2. NGS can also be used for transcriptome sequencing, which allows for the identification and analysis of all expressed genes in a particular cell or tissue.
3. Epigenetic analysis is another application of NGS, which involves the study of modifications to DNA and histone proteins that can affect gene expression.
4. Metagenomics is another area where NGS has been particularly useful, allowing for the analysis of microbial communities in environmental samples or in the human microbiome.

Overall, NGS has revolutionized the field of genomics, providing researchers with a powerful tool for studying genetic variation, gene expression, and epigenetic modifications. The technology continues to evolve, with new sequencing platforms and methods being developed that promise to further increase the speed and accuracy of DNA sequencing.

Summary

Nucleic acid-based disease diagnosis is a powerful and rapidly advancing field in aquaculture that involves the detection and identification of pathogens using nucleic acid amplification techniques. These techniques can detect the presence of pathogen-specific DNA or RNA in fish and shellfish samples, allowing for the rapid and sensitive diagnosis of diseases in aquaculture.

Examples of nucleic acid-based diagnostic techniques used in aquaculture include polymerase chain reaction (PCR), real-time PCR, loop-mediated isothermal amplification (LAMP), microarray and next-generation sequencing (NGS). These techniques offer a high degree of sensitivity and specificity, as they can detect very low levels of pathogen DNA or RNA in a sample and can differentiate between closely related pathogens.

Nucleic acid-based diagnostic techniques are particularly useful for detecting emerging and exotic pathogens in aquaculture, as well as for monitoring the spread of established pathogens. They are also valuable tools for identifying the genetic basis of disease resistance or susceptibility in fish and shellfish populations. One of the main advantages of nucleic acid-based diagnostic techniques is their speed and sensitivity. Many of these techniques can produce results within hours, allowing for rapid response to potential disease outbreaks in aquaculture. They are also relatively easy to use and can be performed in a variety of laboratory settings.

However, there are some limitations to nucleic acid-based diagnostic techniques, including the need for specialized equipment and expertise, and the potential for false positives or negatives due to sample contamination or other factors. Overall, nucleic acid-based diagnostic techniques are a valuable tool for disease diagnosis and monitoring in aquaculture, as they offer a rapid, sensitive, and specific method for detecting and identifying pathogens in fish and shellfish populations.

References

Abdelsalam, M., Elgendy, M.Y., Elfadadny, M.R.(2023. A review of molecular diagnoses of bacterial fish diseases. Aquacult Int 31, 417–434 https://doi.org/10.1007/s10499-022-00983-8

Adams A, Thompson KD ,2011 Development of diagnostics for aquaculture: challenges and or references REFEREN opportunities. Asquac Res 42:93–102

Ador MAN, Haque MDS, Paul SI, Chakma J, Ehsan R, Rahman A.2021).Potential application of PCR based molecular methods in fish pathogen identification: a review. Aquac Studies 22(1):621. https://doi.org/10.4194/2618-6381/AQUAST621

Altinok, İ., & Kurt, İ.2003. Molecular diagnosis of fish diseases: a review. Turkish Journal of Fisheries and Aquatic Sciences, 3(2).

Austin, B. 2019, Methods for the diagnosis of bacterial fish diseases. Mar Life Sci Technol 1, 41–49. https://doi.org/10.1007/s42995-019-00002-5

Biswas, G., & Sakai, M. 2014. Loop-mediated isothermal amplification (LAMP) assays for detection and identification of aquaculture pathogens: current state and perspectives. Applied Microbiology and Biotechnology, 98(7), 2881–2895. https://doi.org/10.1007/s00253-014-5531-

Chakravorty S, Helb D, Burday M, Connell N, Alland D. 2007, A detailed analysis of 16S ribosomal RNA gene segments for the diagnosis of pathogenic bacteria. J Microbiol Methods 69:330–339

Cunningham CO.2002, Molecular diagnosis of fish and shellfish diseases: present status and potential use in disease control. Aquaculture 206:19–55

16

Aquatic Medicine

K.M. Shankar and Prakash Patil

Introduction

Scope and importance

Aquaculture is breeding, rearing and husbandry of aquatic animals such as fishes, crustaceans and molluscs for food, ornament and sports. It is a husbandry involving special animals, the cold blooded vertebrates and invertebrates, whose physiology, growth, metabolic activities and overall performance depend heavily on ambient aquatic temperature and environment. Therefore, husbandry and the health management of aquatic animals is quite different from that of the other land based warm blooded animals.

Aquaculture is one of the fastest growing food production sectors in the world including India. Three decades ago, few species were cultured in India mainly in inland waters and then aquaculture was contributing less than 10 % of total fish production in the country. In the last 25 years, there is a rapid growth in aquaculture with annual rate of 7-8% with significant contribution (above 60%)to total fish production today. The number of species cultured today in India including ornamentals is close to 40. Besides, a good number of plant and animal fish food organisms are cultured to support the hatcheries and nurseries in aquaculture. The area under aquaculture is increasing steadily in inland and coastal brackish water while marine water is also being added up for aquaculture slowly and steadily. Very interestingly, non-traditional environment such as manmade saline soils are also explored successfully for aquaculture. Intensification of culture activity with high stocking density artificial feed and fertilizers is another dimension of aquaculture noteworthy. Large scale movement of aquatic animals and their raw or processed products across national and international borders is important from aquaculture production and health point of view. In general, aquaculture package of practices alter the very water in terms of its chemistry and microbial load and has significant negative impact on the health of fish in culture system.

Due to the increase in aquaculture activities, incidences of diseases are increasing. Non-communicable diseases related to water chemistry and nutrition are one part of it. Communicable diseases due to viruses, bacteria, fungi and parasites are taking heavy toll with significant economic loss. In India, it is estimated that annual loss due to one viral disease, the white spot virus disease in shrimp is more than Rs. 1000 crores. Zoonotic diseases from the growing aquaculture is another important aspect needs to be given attention

Health management of aquatic animals with medicine is a new era. Besides its positive impact on the aquatic animal health, the aquatic medicines have also a negative impact on aquatic life, environment and fish consumers at the end. Over the years, there is significant increase in gathering of information particularly on the biochemistry and immunology of fish, crustaceans and molluscs. In view of poor immune system of aquatic animals, it is better to prevent the diseases by avoidance (biosecurity) and at the best, manage the diseases with judicial application of drugs and chemicals.Against this background health managers in aquaculture should ideally be a fishery professionals with adequate knowledge and training in fish biology, environment and aquaculture package of practices. Health management of aquatic animals by the veterinary professionals is inefficient as they have limited knowledge and exposure to fisheries and aquaculture and the new trends accompanied with the vast expansion of aquaculture in the last 30 years. In the Western world, where aquaculture hardly contributes 10% to total fish productionaquatic medicine is practiced by veterinarians who also study fisheries/aquaculture, which is covered under the wild life. However, aquaculture in the Asian countries which constitute 80% of world aquaculture production, is quite diverse involving several species cultural practices and hence aquatic medicine should be handled appropriately by fisheries professional provided with adequate training and knowledge in biology, immunology, pharmacology, chemotherapy and toxicology.

Development of Aquatic Medicine

First and foremost, the development of therapy and drug evaluation in coldblooded aquatic animals should be based on time and temperature. Mode of action of drugs in different species of fishes, crustaceans and molluscs should be initiated on a war-foot basis. It is a herculean and time requiring task yet very crucial for the growing aquaculture. Information on the chemicals or drugs used in fish farming, run-off of chemicals from agriculture and allied activities, kinetics of chemicals, residues and their effect on fishes and ultimate consumers - the humans is very important.

At the present aquatic medicine is practiced based on the empirical knowledge and research findings in human and livestock health management. Chemicals pesticides, antibiotics and hormones are widely used in chemotherapy in aquaculture. In aquaculture these products are mostly applied to water or incorporated in feeds. When applied to water their effects depends on soil and water chemistry. Also often fish and other non target organisms get affected. Besides, the use of chemicals is associated with problems such as emergence of resistant pathogens, residues in cultured animal, destruction of useful microorganisms and deterioration of aquatic environment.

Overall, it is important to design and develop aquatic medicine considering immune system, benefits and hazards of drugs and chemicals. It is very necessary and important to know that drugs and chemicals used in aquaculture are based on research on warm blooded animals. Effect, kinetics, excretion residual effect of drugs and chemicals depends on time and temperature in the coldblooded fish and shellfish which is different from that of warm blooded animals. As there are good number of fish and shellfish varieties in tropics from fresh, brackish and marine environment, effect of these drugs and chemicals need to be researched on each species appropriately. Furthermore several products with proper research on fish of cold waters are used in fish in the tropics where the efficacy may vary. Otherwise in the years to come lot of these products and other spurious products will be used in aquaculture to the detriment of both fish and consumers. Further regulatory organisations should act appropriately to curb the misuse of drugs and chemicals.

While considering the above basic facts in pharmacology and toxicology for aquaculture, the scope of aquatic medicine should be explored keeping in view the following facts.

a) Aquaculture is practiced in fresh, brackish and marine environment. Aquatic medicine should be developed appropriately to suit water chemistry of these environments and associated changes due to aquaculture activities.

b) Similarly, chemistry of soils in fresh and brackish/salt water has greater influence on water and further on the drugs and chemicals applied. In simple, aquatic medicine should be viewed with soil chemistry.

c) Physiology and immunology of aquatic species is highly variable. Therefore, aquatic medicine should be custom made to species - fishes crustaceans, mollusc from fresh, brackish and marine water.

d) Aquatic medicine should be viewed from different package of practices – feed-based aquaculture and fertilizer-based aquaculture.

e) Physiology, enzyme and immune profile of aquaculture species at different life stages is highly variable. Therefore, aquatic medicine should be fine-tuned to different life stage of fishes and crustaceans - larvae, fry, fingerlings, sub-adults and adults.

f) Aquatic medicine should be viewed from medicine application point of view- to water, to feed, and to soil.

Overall, it is important to design and develop an aquatic medicine considering immune system (prophylaxes) vs benefits and hazards of drugs and chemicals vs avoidance of pathogens through bio-security measures.

As stated earlier, aquatic medicine is a growing science still at the infancy. Therefore, basic knowledge and experience gained in the medicine of human and livestock in the area of 1. source of drugs, 2. drug discovery and development, 3.pharmacokinetics, 4. pharmacodynamics, 5. dose -response relationships 6. metabolism and excretion of drugs and 7chemotherapy and drug interaction need to be understood properly for development and adoption in medicine of cold blooded vertebrates and invertebrates in aquaculture.

Pharmacology and Modern Medicine

Medicines are part of our life. From birth to death drugs and chemicals are invariably involved in our life due to various reasons. Ancient use of drugs arose by the experience of using various plants containing chemicals known to alter biological functions. While many terrestrial and marine organisms remain valuable sources of naturally occurring compounds with various pharmacological activities, especially including lethal effects on both microorganisms and eukaryotic cells, drug invention became more allied with synthetic organic chemistry as that discipline flourished over the past 150 years. Drugs are used to cure the disease, to suppress the disease, to prevent the disease and to diagnose the disease. Drugs or medicines are substances that changes biological system by interacting with it.

WHO describes drug as any substance used to modify or explore physiological system or pathological status for benefit of the recipient. Medicine could be a substance or a mixture of substances used for preserving health. Drug is a single chemical substance which is active component of a medicine, where as medicine contains several substances to deliver a drug in a stable, acceptable and convenient form to the patient.

Brief History of Modern Medicine

Development of vaccination against small pox is a great step and weapon during middle of the 19th century. Invention of anesthetics and antiseptics during the period also contributed significantly to health management. Identification of causative agent of malaria, plague and cholera during the last quarter of 19th century was first step in developing strategies later for managing these diseases. Beginning 20th century pharmacological science revolutionised the industry with synthetic chemistry. This period also witnessed synthesis of new drugs such as barbirates, anesthetiscs and discovery of antimicrobial chemotherapy, for example arsenical compounds for treatment of syphilis. Other landmark discoveries during the middle of 20th century were sulfonamides (1935) and antibiotics (1945). Later additional drugs were added rapidly during the later half of 20th century leading to several pharmacological developments over the years catering to modern medicine. Pharmacology is defined as an experimental science dealing with property of drugs and their effect on living system. It is a comprehensive science pursuing historical knowledge, compounding, sources, physical and chemical properties, biochemical and physical effects, mechanism of action, absorption, distribution, biotransformation and excretion of drugs.

Sources of Drugs

Active principles of drug are a constituent of a drug, upon the presence of which the characteristic therapeutic action of the substance largely depends. Every drug has three names viz. i. Chemical name, ii. Non-proprietary name or generic name, and iii. Proprietary name (Patent or Brand name). Chemical name is given according to the chemical constitution of the drug. It gives exact chemical composition of the drug with molecular structure.Non-proprietary or generic name is given by regulatory bodies. It is uniform all over the world. This generic name is derived from chemical name and usually small. Proprietary or Brand or Patent name is given by the manufacturing and marketing company. This cannot be used by other manufacturers or sellers. This may be fancy name or given by the manufacturer only and cannot be used by others.

Major Sources of Drugs

Natural sources can be divided into four groups- plant, animal, micro-organisms and minerals. Modified natural drugs are termed as semi-synthetic sources. Major source of drugs comes under synthetic source. In recent years genetic engineering plays important role in synthesizing drugs for human and animal health management.

Plant Source

Active principles in plants are key to their pharmacological activities. A long list of active components in plants such as alkaloids ,glycosides, fats, oils, tannins, saponins, and gums are pharmacologically important.

Alkaloids

Alkaloids are nitrogenous waste substances of plant metabolism are of two types 1. solid and non volatile alkaloids containing oxygen, 2. liquid and volatile alkaloids without oxygen. Alkaloids are bitter to taste and are insoluble in water except their salts. They are incompatible with alkalies, tannic acid and heavy metals and have their names usually end with 'ine "eg atropine, morphine, quinine, nicotine, strychnine.

Glycosides

Glycosides are non nitrogenous waste from plants, which upon hydrolysis give a sugar and a genine or aglycone. The pharmocological action is due to aglycone, while sugar helps in dissolution of preparations.

Oils

Available in two forms 1. fixed (castor oil, coconut oil) and 2 volatile or essential (turpentine and eucalyptus oil). Fixed oils are obtained by extraction and volatile oils by distillation of plant parts. Besides there are mineral oils such as liquid paraffin sourced from earth having pharmacological importance.

Saponins

These are water soluble non nitrogenous substances similar to glycosides. Upon hydrolysis saponin splits into a sugar and aglycone (sapogenin), the latter is responsible for pharmacological action. Saponin after dissolving in water gives characteristic foam upon shaking. There are also toxic saponine termed sapotoxin.Eg Ginseng, Fenugreek

Resins

Oxidation of terpenes gives rise to water insoluble solid brittle resins. There are several types of resins such as oleoresins, gum resins or balsum. Eg Asfoetida, camphor, storax

Gums

These are exudates obtained by incision on stems of plants. Gums usually form gelly with water. Eg gum acasia.

Mineral Sources

There are several metallic and non metallic inorganic substances sourced from nature having pharmacological potential. These minerals are used as such after purification or combined with other ingredients. Long list of metals and their salts , non metals, metalloids, acids, alcohol, coal tar, have potential pharmacological applications.Eg Copper sulphate, sodium chloride, magnesium sulphate, potassium permanganate.

Animal Sources

In olden days dried toad skin was used for treatment of tooth ache, gum bleeding (China). Liver oil of cod fish was used as source of vitamin A & D. Currently insulin from pork (porcine), cow and buffalo (bovine), heparin from leech, IgG (antisera)from different animals, gonadotropins such as FSH and L H, human chorionic gonadotropins are used for treatment in human /animal medicine.

Microorganism Sources

Bacteria and Fungi, are important source of many lifesaving drugs. These drugs are obtained mainly from micro-organisms and are also used to neutralise/ kill microorganisms. Some of these drugs are purified naturally or sometimes expressed/produced synthetically in micro-organisms.Eg. penicillin, chloromphenicol, streptomycin, neomysin

Synthetic Sources

Presently most of the drugs are obtained synthetically. Some of the drugs that are earlier obtained from plants, are synthesized in laboratory today. These drugs are more potent and effective and also free of impurities. Some of the other advantages of drugs produced synthetically include, controlled quality, easier process, cheaper, more potent and safer, and scope for large scale production. E.g. Hepatitis-B, Insulin production by rDNA technology.

Semi-synthetic Sources

Some of the drugs are altered structurally in the laboratory to get newer agents with less adverse effects, prolonged duration of action and better pharmacological actions. Eg.: Ampicillin, Amoxycillin -a semi-synthetic Penicillin and Heroin from Morphine

Drug Discovery and Development

A general idea about drug discovery and development is essential for adoption in animal and aquaculture treatment. Drug development processes is long and tedious involving identification of a drug molecule for particular use by an organic chemist for synthesis, purification and formulation, short term and long term studies on animal pharmacology, toxicity and approval from regulatory bodies till reaching the market for end use. It is essential in drug discovery and development for good understanding of anatomy, physiology, biochemistry, microbiology, parasitology, pathology, mathematics, and statistics. The drug discovery has three major steps.

Step I: Discovery and Development: New drugs are discovered by understanding disease process of a disease which helps to design a product to stop or reverse the disease effect. Against this requirement, thousands of compounds appear to be potential candidates for treatment. However, a small number of promising ones are chosen after initial tests for further studies. Following selection of few compounds, detail studies initiated on absorption, distribution, metabolism, excretion, dosage, potential benefits, mechanism of action and important side effects or toxicity. Simultaneously, effect of compounds on different age group, species and interaction with other drugs are obtained

Step 2: Pre-clinical Research: Before testing a drug in human beings/animals, researchers must find out whether it has the potential to cause serious harm, also called toxicity. Preclinical research is performed *in vitro* and *in vivo*. Good laboratory practices(GLP) designed for a product development with regulation are used by researchers for preclinical laboratory studies. GLP set minimum basic requirement for studies such as personnel, facilities, equipment, written protocols, operating procedures, study reports and system of quality assurance aimed at safety of a product. Preclinical studies are small but must provide detail information on dose, toxicity levels, overall efficiency of drug to decide for further test on animals/fish.

Step 3: Clinical Research: Preclinical studies although provide additional basic answers on safety but certainly are not a substitute study providing information on actual interaction of a drug with fish/animal. Clinical research refers to trials on fish conducted in different phases each with a target by beginning investigational new drug process (IND). IND provides information for designing clinical trials.

Designing Clinical Trials

In clinical trials, research/manufacture designed protocol get specific answer for a product by developing a protocol for the study. Information on the product/drug is obtained developing specific research questions and objectives. Following this, decision is made on number of animals, quality of members for participation, duration, control, form and dose of the drug, type of assessment, reviving data obtained. Clinical trials follow typical series for early small scale phase 1 studies to late stage large scale phase 3 studies.

Pharmacokinetics

Pharmacokinetics deals with absorption, distribution, biotransformation and excretion of drug in the body over a defined period of time. In simple, pharmacokinetics is what the body does to the drug. A good understanding of the following transport mechanisms of drug molecules across the biological membrane is the first step in the study of pharmacokinetics.

Passive Diffusion

A random movement of drug molecules from a site of higher concentration to a site of lower concentration is termed passive diffusion. And in this process cell does not spend energy. Many drugs enter the cells through this mechanism. The drugs then eventually enter blood capillaries facilitating systemic circulation. cell membrane is made up of phospholipids and a drug molecule must dissolve in the membrane to diffuse and reach an equilibrium concentration on either side of the membrane. Diffusion process is important for majority of drugs, particularly lipid soluble ones, which diffuse by dissolving in the lipoidal matrix of the membrane. The rate of diffusion of a drug is proportional to lipid : water partition coefficient and depends on several factors such as, a), concentration of gradient on either side of the membrane, b) molecule size of the drug c) lipophilic nature d) temperature with direct relation with diffusion and e). thickness of the membrane.

Carrier Mediated Transport

In this transport drug molecule forms a complex with carrier present in the membrane to facilitate transfer across the membrane. Polar molecules get a hydrophobic coat over the hydrophilic groups to pass through the membrane. Carrrier transport is specific, saturable, and inhibited by analogues using the same carrier. Examples of carrier transport are intestinal absorption of iron and calcium. Facilitated diffusion and active transport are the two types of carrier mediated transports.

a. **Facilitated Diffusion:** In this diffusion, a 'special carrier molecule' is involved in the movement of drug molecule without spending energy where direction of movement is determined by concentration. This transport is rapid than simple diffusion because of the carrier molecule. However, many other compounds can compete and inhibit the transport. Absorption of Vit B 12 from the gut, transport of sugars and amino acids across the membrane are of the some examples.

b. **Active Transport:** This transport also involving a special carrier molecule like facilitated diffusion needs energy to move the molecule across the membrane. Advantage in the active transport is, drug molecule can move against concentration gradient ie from lower to higher concentration. However, this process leading to high concentration may create toxicity by disruption of energy production of a cell and subsequent active transport. Some examples of this transport are, transport of glucose across gastric mucosa, non specific excretion by proximal renal tubules, transport of certain drugs to brain, secretion of some drugs in bile and secretion of many organic acids and bases by renal tubules.

Pinocytosis and Phagocytosis

Pinocytosis (cell drinking) and phagocytosis (cell eating) are processes where molecules are engulped by cells. Cell membrane form vesicles surrounding a molecule to carry it across the membrane. This is an active transport, where cell expends energy. These two processes are particularly important to carry large molecule such as antibodies or protein complexes. Generally these processes are of less significance in pharmacology.

Filtration

Movement of the drugs or other molecules through the aqueous pores of the membranes or paracellular spaces is termed filtration. Usually small lipid insoluble drugs pass across membrane by filtration. Large molecules unable to pass through small pores by the process, except in the case of capillary cells with large pores. Hence filtration of drug molecules depends on rate of blood flow, rather than lipid solubilty or pH of the medium. Filtration has a minor role in drug transfer except glomerule filtration, drug removal from CSF and movement across hepatic cell membrane.

Pharmacodynamics

Study of action of drugs based on physiological and biochemical effects is termed pharmacodynamics. In simple it is response of organism to action of a drug also referred to as ‘ what a drug does to body’

Principles and Mechanisms of Drug Action

Basic types of drug actions area). stimulation- boosting existing level of action eg adrenaline on heart b). depression- selective inhibition of cell activity eg barbitrate on CNS, c). irritation, a non selective stimulation of some cells for a benefecial effect, d) replacement of natural substance eg. hormone therapy e), cytotoxicity- toxicity to cells eg. antibacterial, antiviral and anticancer agents.

Mechanisms of Drug Action

Drug action depends on physical and chemical properties of a drug. Physical activity of a drug depends on physical properties which cause action. Eg charcoal adsorption. Chemical action depends on chemical properties of a drug. Eg antacids, acidifiers, alkanizers, oxidising and chelating agents.

Enzyme Mediated Drug Action

Enzymes are ubiquitous biological catalysts are main targets of drug actions. Drugs either increase or decrease enzyme mediated reactions. Enzyme stimulation is favored by endogenous substances (co-factors), vitamins and minerals

Receptor Mediated Drug Action

An agent that can bind to a receptor and elicit a biologic response is called as agonist. It is the most important type of drug action. Here, receptor is a binding site within biological tissues for either endogenous or exogenous substances. The macromolecules situated on the surface of the cell or inside the cell regulate critical functions like enzyme activity, permeability and structural features. Many drugs show structural specificity, minor substitution alters the effect. Stereo-specificity of drugs d-propranolol less active than l-isomer, l-noradrenaline – more potent than d- noradrenaline. Intensity of response depends on fraction of receptors occupied; maximal response when all receptors are occupied, 'all or none' action of drug on each receptor – each receptor is either fully activated or not at all. Drug and receptor have structurally 'lock and key' arrangement". The interactions involved in the drug-receptor complex are the same forces experienced by all interacting organic molecules and include, Covalent bonding, Ionic interactions, Hydrogen bonding, Hydrophobic interactions and Electrostatic interactions

Types of Receptors

Type I – Channel linked receptors (ligand-gated ion channels or inotropic receptor).

Receptors coupled directly with ion channel. Fast neurotransmiters such as nicotinic acetylcholine act through these receptors.

Type II – G-Protein coupled receptors.

These are G protein membrane receptors coupled to intracellular effector system, usually serving as receptors for hormones Eg adrenaline receptor

Type III – Kinase-linked receptors

These are membrane receptors possessing intracellular protein kinases within their structures. Eg receptor for insulin, cytokines and other growth promoters.

Type IV – Nuclear receptors

Nuclear receptors regulate gene transcription. They are present in cytosol and serves as receptors for steroid hormones and thyroid hormones.

Signal Transduction

It is the process involving the conversion of a signal from outside the cell to a functional change within the cell. It converts a mechanical/chemical stimulus to a cell into a specific cellular response. There are two stages in this process: 1. A signaling molecule activates a specific receptor protein on the cell membrane, 2. A second messenger transmits the signal into the cell, eliciting a physiological response.

Second Messengers System

There are two types of receptors 1. specific membrane receptors(first messenger) 2. cytoplasmic component of receptors. Action of many hormones, neurotransmitters on specific membrane receptors depends on cytoplasmic component of receptors, which acts as second messenger that actually carry forward stimulus from the receptors. Eg cAMP, cGMP, Ca2, G proteins.

Dose- Response Relationships Graded dose–response relations

As the concentration of a drug increases, the magnitude of its pharmacologic effect also increases. The relationship between dose and response is a continuous one, and it can be mathematically described for many systems by application of the law of mass action, assuming the simplest model of drug binding. The response is a graded effect, meaning that the response is continuous and gradual. This contrasts with a quantal response, which describes an all-or-nothing response.

Potency

From a graded dose -response curve, two important properties of a drug a). the potency, b).the efficacy can be determined. Potency is the amount of drug required to produce an effect of given magnitude. A concentration producing a 50 % of the maximum (EC 50) effect is used to determine potency.

Efficacy

It is ability of a drug to elicit a physiological response upon interaction with a receptor which is thus dependent on number of drug -receptor complexes formed and its efficiency of coupling of receptors activation to cellular responses. Efficacy or maximum responses (E max) of a drug is very important than its potency, as greater efficacy is more therapeutically advantageous.

Therapeutic Window

A range of doses producing therapeutic response is a ratio between minimum effective concentration or minimum inhibiting concentration(MIC) to minimum toxic concentration (MTC). For a risk free effect ,a range of doses between MEC and MTC is required as beyond MTC a drug is toxic, and further if it fails to surpass MEC it will lead to failure. A drug with a wide range of therapeutic window is safer compared to narrow window which is very difficult to dose and administer. Therapeutic window or safety window is quantified by therapeutic index

Therapeutic Index (TI)

It is a ratio of dose at lethal/toxic effect to dose that causes therapeutic effect. Lethal dose or toxic dose 50(LD 50/TD 50)is minimum amount of a drug causing adverse effect in 50 % of the animals. On the other hand, ED 50 is quantity of a drug producing desired therapeutic in 50 % of the animal population.

Ideally any drug that requires more amount to produce toxic or adverse effects in 50% of population will have wider therapeutic index and vice versa. Drugs having wider therapeutic index are safer in comparison to those having low therapeutic index because minor modification in the dose of such drugs (aspirin, acetaminophen) will not produce toxic effects

Therapeutic Threshold

Minimum amount of drug that is required to produce the desired response. It is the difference between therapeutic threshold and the amount of the drug considered to produce unwanted and possibly dangerous side effects. It is used to understand the margin of safety that exists for the use of a drug.

Half-Life

It is the amount of time that a drug remains at a therapeutic level to produce a desired effect and the time it takes the body to metabolize or reduce a drug's concentration by 50%. It is used to determine when more of the drug shall be given.

Metabolism and Excretion of Drugs

Metabolism

Metabolic processes convert lipid soluble, non polar compounds to water soluble polar compounds for easy excretion by the body. Water soluble substances are excreted while lipid soluble (lipophilic) substances are reabsorbed from renal excretory sites into blood.

Biotransformation

It is chemical transformation of a drug in the body by a series of enzyme catalysed processes. These processes alter the physical property of a drug from absorption through cell membrane to elimination in urine or bile.

Chemical alteration of a drug in the body is to convert non -polar lipid soluble complexes to polar water soluble compounds to facilitate excretion to avoid toxicity. Liver, kidney, intestine are the major sites of chemical alteration.

Biotransformation reactions can lead to any one of the following: 1.Inactivation – most drugs are rendered pharmacologically inactive. Example: Drugs such as morphine, chloramphenicol, procaine,2.Formation of active metabolite – the metabolite formed possesses biological activity. Example: Ciprofloxacin from enrofloxacin, Oxyphenbutazone from Phenylbutazone, Salicylic acid from aspirin,3.Inactive drug to active drug (Prodrug) An inactive drug gets converted into an active metabolite once it undergoes biotransformation inside the body. Example: Hetacillin to ampicillin, Prontosil to sulfanilamide,4.Less active or inactive drug is converted into a more toxic drug (Lethal synthesis). Example: Parathion to paraoxon, Fluoroacetic acid to fluorocitrate

Drug Metabolism in Fish

The detoxification of drugs and chemicals in fish is carried out by the liver. The rate of metabolism is about one tenth of that of the mammals. The optimum temperature for many of the phase I reaction (oxidation, reduction and hydrolysis) and phase II reactions (conjugate formation and subsequent excretion) in fish is close to the temperature of the natural environment of the fish.

Types of biotransformation reactions-Phase I reaction or non-synthetic phase includes oxidation, reduction and hydrolysis, Phase II reaction or synthetic phase includes glucoronidation, sulfation, methylation, glutathione conjugation, acetylation, amino acid conjugation, thiosulpahte conjugation

Phase-I: Biotransformation Reactions: Liver is the main organ of drug metabolism, besides considerable activity of other organs such as gastrointestine, gills, skin and kidney. Majority of xenobiotics undergo biotransformation by catalytic action of specific cellular enzymes, which are located in the endoplasmic reticulum, cytosol, lysosome envelope of cells and nucleus. These enzymes called mixed function oxidases or cytochome P450 enzymes, catalyse oxidation and glucuronide conjugation reaction.

Phase-II: Biotransformation Reactions: This phase involves conjugation of a functional group to the drug or its metabolite to make it water soluble. The presence of –OH, -COOH, -NH2, -SH groups favour formation of conjugates (these groups are added to (or) exposed in Phase I)

Glucuronidation

Most drugs including endogenous steroids and bile pigments undergo this important reaction, which takes place in microsomes. Source of glucuronide is UDPGA through glucuronyl transferases which is highly soluble and excreted through urine if mol wt<300-500 or via bile if mol wt>500.

There are also several non microsomal conjugations such as a).sulphate conjugation, b) acytylation and c) methylation. Besides there are also minor pathway conjugations such as glycine conjugation and glutathione conjugation.

Enzyme Induction and Inhibition Enzyme inhibition

Repeated administration of some drugs inhibit cytochrome p-450 enzyme activity, which can reduce metabolism of other drugs administered subsequently. The direct inhibition of enzymes or due to competition by drugs for the enzyme site is an immediate action and as a result drug concentration increases leading to increased drug effect. Inhibition besides prolonging the activity of co administered drugs, can increase toxic effect of unchanged drugs.

Enzyme Induction or Stimulation

Repeated administration of some drugs increase synthesis or reduce degradation of cytochrome p-450 enzyme leading to 'induction' of the enzyme. Accelerated metabolism of drugs due to induction results usually in decreased pharmacological action of inducer and also co administered drug. These inducers take long time to set in and to return to original values upon their withdrawl.

Factors Affecting Biotransformation

Several factors affect biotransformation of drugs such as a). age-biotransformation is poor in both young and old animals, b) diseases of liver with low blood flow, c). species variation, d). co administration of other drug inducers or inhibitors

First Pass Metabolism

Metabolism of a drug before it reaches systemic circulation is termed as first pass metabolism. Normally it happens with oral drugs, besides topical applications, however at low level. This process leads to increased oral dose, unpredictable absorption and variation among individuals in drug action.

Enterohepatic Recirculation

After biotransformation by glucuronide conjugation, some drugs undergo biliary excretion. Interestingly, after biliary excretion they get detached from the conjugation in large intestine leading to their release, reabsorption and rentry to liver. The recirculated drug has a long duration of action.

Excretion of the Drugs

Excretion or elimination means drugs are transferred from the internal to the external environment. The principal organs involved are kidneys, gills, biliary system and intestines. Drugs are excreted mainly by the kidneys into the urine and by the liver into the bile and subsequently into the feces. More details on excretion of drugs/chemicals in fish is covered in the chapter 10 Toxicology under the detoxification/biotransformation section.

Main function of kidneys include maintenance of volume and composition of body fluids, production and elimination of metabolites, elimination of drugs and maintenance of acid base balance. Kidneys receive nearly 25% of cardiac output. Besides absorption and reabsorption also takes place in nephrons. Liver excretes drugs into bile which further conveys to intestine. Usually these drugs from biliary excretion include unchanged drugs and other endogenous substances.

Concept of Chemotherapy and Drug Interactions

Apart from diseases arising as a result of malfunction of some body system or the disturbance of a homeostatic control mechanism, they can also be caused by invading micro-organisms or by aberrant tissue growth. Treatment of diseases caused by invading organisms with chemicals is referred to as chemotherapy. Drugs in this class differ from all others in that they are designed to inhibit/ kill infecting organisms and to have no or minimal effect on the recipient. This type of therapy is generally called chemotherapy, which has come to mean treatment of systemic infections with specific drugs that selectively suppress the infecting microorganism without significantly affecting the host. Treatment of neoplastic diseases with drugs is also called chemotherapy. Antimicrobial drugs are the greatest contribution of the present century and they are also one of the most frequently used as well as misused drugs.

Sulfonamides

Sulfonamides are non antibiotic antibacterials are synthetic drugs from sulfanilamide. These drugs are insoluble in water and soluble in alkaline pH. Sulfonamide is a functional group which is the basis of several groups of sulfonamides also called sulfa drugs. Important drugs of this group are sulfadiazine, sulfadoxine, sulfalene, sulfamerazine, sulfamethizole and sulfamethoxazole.

Mode of Action

Sulfonamides (SN) or sulfanilamides are broad spectrum antibacterials for the treatment of human and animal infections. The drugs get rapidly absorbed from gastrointestinal tract. Sulfonamides have similar structure as that of *p*-aminobenzoic acid(PABA), required by bacteria for folic acid synthesis essential for DNA synthesis and cell division. The sulfonamides are thus are bacteriostatic than bactericidal. Sulfonamides are competitive antagonists competing and replacing PABA in the production of bacterial folic acid required for DNA in the bacteria. But Sulfa drugs do not affect animal cells because they consume folate in the form of dietary requirement. A free amino group at position 4 and a sulfonamide group at position 1 are involved in antibacterial activity. Sulfonamides have a wide range of activity against gram-positive and gram-negative bacteria except *Pseudomonas aeruginosa* and *Serratia* spp.

Bacterial Resistance to Sulfonamides

Sulfonamides were the first selective antibacterial drugs used systemically but are used less today due to widespread resistance from bacteria. Overuse of these medications has led to antimicrobial resistance. Bacteria can resist the medicines in two ways 1. by endogenous vertical evolution 2.by exogenous horizontal evolution. In vertical evolution resistance acquired through spontaneous mutation within the bacterial genome is transferred to its offspring. In horizontal evolution resistance genes are transferred between non-related bacteria, which are commonly found among pathogens.

Antibiotics

These are substances produced by microorganisms which can suppress the growth or kill other microorganisms at very low concentrations. Besides production from microorganisms several antibiotics and their analogues are synthesized and hence antimicrobial agents (AMA) include both synthetic and naturally sourced antibiotics.

Penicillin

The essential penicillin molecule contains a fused ring system, the β-lactam thiazolidine. The physical and chemical properties, especially solubilities of penicillins are related to the structure of the acyl side chain and the cat ions used to form salts. Penicillin is incompatible with heavy metal ions, oxidizing agents, and high concentrations of alcohol.

Mechanism of Action

B -lactum antibiotics by preventing cell wall synthesis and damaging cell integrity are bactericidal. Penicillin binding proteins located on bacterial cell membrane are involved in the catalytic reaction leading to damaged cell wall and osmoticaly unstable cells, leading to cell death by lysis mediated by autolysins.

Microbial resistance: The three independent factors that determine the bacterial susceptibility to β-lactam antibiotics includes the production of β-lactamases, permeability of cell wall, and the sensitivity of the penicillin-binding protein. The two important mechanisms of β-lactamase production include- a).chromosomally derived β-lactamases which are species and genus specific and can be induced by the presence of any β-lactam compound, b). Plasmid- derived β-lactamases that can be transferred between bacteria, increasing the number of bacteria resistant to the antibiotic.

Toxicity: Penicillins are very safe drugs, with relative few adverse effects reported. Acute allergic reactions are the most common untoward effect in people. Acute anaphylaxis, collapse, hypersalivation, shaking, vomiting, urticaria, fever, eosinophilia, neutropenia, agranulocytosis, thrombocytopenia, leukopenia, anemia, and lymphadenopathy can occur in sensitized animals. Coomb's-positive hemolytic anemia has been reported in horses following penicillin administration.

Aminoglycosides

The group of natural and semi-synthetic antibiotics discovered as a result of systematic search for drug active against gram-negative organisms.

Aminoglycosides means amino sugar joined by a glycosidic linkage, whereas aminocyclitols means amino group is on cyclitol rather than sugar ring eg. Spectinomycin, Apramycin.Aminoglycosides are relatively broad spectrum in terms of type of microorganism – active against bacteria, mycoplasma and mycobacteria. They are bactericidal with activity against gram-negative aerobes and newer groups especially against pseudomonas, but not active against anaerobes.

These inhibit protein synthesis by binding to 30s ribosome, leading to misreading of codon and incorporation of faulty amino acid. Acute anaphylactic response due to these antibiotics are nausea, vomition , respiratory distress, coma and death

Chloremphenicol

First obtained from *Streptomyces venezuelae*in 1947 is now commercially synthesised. A derivative of dichloroacetic acid with a nitrobenzene moiety is a white/grey/yellow crystalline powder, bitter in taste. It is slightly soluble in water while soluble in most of the organic solvents

Mechanism of Action

A bacteriostatic drug that interferes with protein synthesis. It irreversibly binds to 50 S ribosome and inhibit peptidyl transferase preventing amino acid transfer for the growing polypeptide chain. It is a broad spectrum antibiotic inhibitive to both gram positive and gram negative bacteria Ricketsiella, Chlamydia and Mycoplasma.

Toxicity and adverse effects: Dose dependent reversible suppression of bone marrow activity; more common in cats and dogs. It can be reversed on withdrawal of the drug. Dose independent irreversible bone marrow depression –aplastic and fatal anemia. This is rare and is due to idiosyncratic reaction. Reports of aplastic anemia after exposure to chloramphenicol in the form of ophthalmic ointment restricted its use in food producing animals. Human beings exposed to small doses through residues in food animals develop aplastic anemia. Hence, it is banned by FDA for human use.

Fluoroquinolones

Fluoroquinolones are the first entirely man made synthetic antibiotics or antibacterial agents. They were developed from structural modification of chloroquine. Reasons for their limited use are- a). Narrow spectrum - only Enterobacteriaceae, b). Low potency - Short half life, c). High plasma protein binding - low tissue level, Toxicity related to CNS

Mechanism of Action

Fluoroquinolones specifically inhibit microbial growth by inhibiting activity of DNA gyrase, which is responsible for DNA supercoiling. Fluoroquinolones are classified a).First generation Quinolones, Eg: Nalidixic acid, Oxolinic acid, Flumequine,b).Second generation Quinolones Eg: Ciprofloxacin, Danofloxacin, Enrofloxacin, Norfloxacin.

Antimicrobial action and resistance: They have high activity towards gram-negative and gram-positive aerobes. The second generation quinolones are active against mycobacteria (Eg.Ciprofloxain and Ofloxacin), mycoplasma, rickettsia and inactive against anaerobes. Relatively safe drugs in terms of toxicity and adverse effects.

Macrolides

Macrolides consists of macrocyclic lactone ring (usually 14-16 atoms) attached to one or more deoxy sugars. Some of the macrolides are prepared semi synthetically. Macrolides are bacteriostatic drugs. Weak bases (dimethylamine group) and lipid soluble, used as esters (estolate) form to increase oral bioavailability. Macrolides are characterized by macro cyclic lactone ring attached to two or more sugar moieties. Eg. Erythromycin, Tylosin, Spiramycin, Tilmicosin, Roxithromycin. The spectrum of activity of Erythromycin is similar to penicillin but wider including mycoplasma, mycobacteria, chlamydia and rickettsia to some extent and can be readily used in penicillin allergic animals. Erythromycin has been used in aquaculture the treatment of bacterial kidney disease caused by *Renibacterium salmoninarum* and Streptococcosis in yellowtail in Japan.

Mechanism of action: They reversibly bind to 50s ribosome and prevent the translocation of peptidyl tRNA to 'p' site and inhibit protein synthesis. Bacteriostatic at therapeutic concentration but bactericidal at higher concentration. The source is *Streptomyces erythreus.*

Tetracyclines

The tetracyclines are broad-spectrum antibiotics with similar antimicrobial features, but differing somewhat from one another in terms of their spectrum and pharmacokinetics. Oxytetracycline, chlortetracycline, and Demethylchlortetracycline are the three naturally occuring tettracylines, while semi-synthetically derived ones are rolitetracycline, methacycline, minocycline and doxycycline. Based on elimination time, Tetracyclines are further classified into short-acting (tetracycline, oxytetracycline, chlortetracycline), intermediate-acting (dimethyl chlortetracycline and methacycline), and long-acting (doxycycline and minocycline) types.

Mode of Action

These antibiotics inhibit protein synthesis in bacteria by reversibly binding to 30 s ribosome. These are very effective in the pH range of 6-6.5 against multiplying microorganisms.

Antimicrobial Spectrum: Tetracyclines are about equally active and typically have about the same broad spectrum, which comprises both aerobic and anaerobic gram positive and gram negative bacteria, mycoplasmas, rickettsiae, chlamydiae, and even some protozoa (amoebae). After usual oral dosage, tetracyclines are absorbed primarily in the upper small intestine and effective blood levels are reached in 2-4 hr.

Tetracycline Used in Aquaculture

Oxytetracycline HCL is a FDA approved antibiotic drug for use in aquaculture feed, is marketed as "Terramycin®10" by Pfizer, Inc. The application of oxytetracycline through feed for different species in aquaculture has been studied to control various bacterial diseases in fish and shellfish.The antibiotic has been effectively used for treatment of bacterial hemorrhagic septicemia in salmonids and catfish culture at 2.5 to 3.75 g/100 pound of fish for 10 days, with a with drawl period of 20 days. Shellfish such as lobsters treated for bacterial diseases at 1g/pound body wt for 5 days with withdrawl period of 20 days.

Antimicrobial Resistance (AMR)

Antimicrobial Resistance(AMR) is the ability of microorganisms to resists the effect of antimicrobials, to which they were sensitive earlier. It is not a new phenomenon and evolves naturally through genetic changes. Indiscriminate, excessive and inappropriate use of antimicrobials in human, animal and plant health management is one of the causes of AMR

Increasing incidences of microbial diseases in aquaculture today is causing estimated annual loss of USD1.6 billion in India. Antimicrobials are employed for prophylaxis and treatment in health management in animal husbandry including aquaculture. However, irrational use of antimicrobials in animal husbandry has led to AMR which has two dimensions, ineffectiveness of the antimicrobials in i)animal health management, 2) human health management. Safety of aquatic food from residues of antimicrobials to source of resistant microbes has raised concern in both domestic and international market. Therefore, aquaculture and animal husbandry entrepreneurs, R and D organisations, health managers, manufactures and traders of drugs and meat processors should be aware of the irrational use of antimicrobials leading to AMR

Success of antimicrobial therapy depends on appropriate concentration at the site of infection, which must be sufficient to inhibit growth and multiplication of microorganisms while remaining below level of toxicity to human/animal cells.

Antimicrobials based on mechanism of action are classified as those i) inhibit bacterial cell wall synthesis- *B* lactum antibiotics, ii) act on cell membrane iii). disrupt or inhibit protein synthesis-Tetracyclines, aminoglycosides and microlide antibiotics iv). inhibit nucleic acid metabolism- rifamycin, quinalones, v). antimetabolites to bacterial cells-sulphonamides.

AMR is bigger threat to global health, food security and development. AMR is substantial economic burden world over, causing deaths and long treatment causing billions of USD. Problem gets further aggravated in the absence of any new antibiotics in the pipeline. A large threat looming with worst scenario in near future is where world will have no antimicrobials for treating bacterial diseases. The threat to poor countries is much more than to developed countries. Permanent resistance can not be prevented in the long run unless use of antimicrobials is completely stopped.

Bacterial pathogens of human and animal origins are becoming increasingly resistant to most of front line antimicrobials. More than 80 % of bacteria associated with hospital acquired infections are resistant to one or more drugs previously used to treat them. Wide use of antibacterials in animal production system including aquaculture is contributing significantly to AMR. AMR has been increasing in zoonotic and commensal bacteria, which has increased risk of transmission from animals to humans or vice versa. Emergence of superbugs ranging from multidrug resistant(MDR), extensive drug resistant(XDR) and totally drug resistant (TDR) pathogens is a grave concern. In addition, cross resistance to closely related antibiotics also recorded eg sulfonamides.

Microorganisms develop AMR through i. natural resistance, ii. acquired resistance. a). biochemical mechanisms b) mutation with passing of the trait. Irrational use (misuse and overuse) of antimicrobials in farm animals is directly associated with the transmission of resistance as well as resistant microbes to human via animal source foods. Antimicrobials are heavily used in animal husbandry (70 % of total antimicrobial consumption)than in human health management(30%). Important, nearly 75 % of antibiotics fed to animals excreted un metabolised in urine and feces and enter sewage, thus animal waste besides resistant bacteria also source of antibiotics to nature. The role of sewage, waste from pharmacological industry also promote AMR. Other important factors of drivers of AMR : poor sanitation, poor water hygiene, spurious drug quality, non availability of diagnostics and vaccines, and non implementation of travel or migration quarantine. And food borne route is the major transmission pathway for AMR, therefore surveillance of use in non human use is important. Antibacterials are used in animal production including aquaculture for treatment of infection as growth promoters and for prophylaxis to control infection. Use of biocides and heavy metals in animal production including aquaculture also cause AMR, which may be cross resistant to antibiotics. The mechanism of development of AMR to metal are similar to that of for antimicrobials and as such microbial communities develop cross resistance and co resistance between heavy metals and antibiotics. In addition to antibiotics, disinfectants and heavy metals, biocides also play important role in AMR. AMR can be lost but slowly after several decades provided antibiotics applications are completely stopped.

Global and national action plans to curb AMR: Global action plan(GAP) and National action plan(NAP) have been developed in collaboration with FAO,WHO and OIE(World organisation animal health). OIE recommends which align with the principle of Global action plan (GAP) on AMR, that antibacterials which are WHO category of highest critically important for human health are not to be used as preventive treatment in animals. EU bans the preventive and prophylactic use of antimicrobials in animal production. Further EU restricts use of human important antimicrobials in animal production and reservation of some antibacterials only for human use. FSSAI in India has banned use of antibiotics and several pharmacologically active substances in aquaculture, poultry and dairy ((NAP-AMR 2017). Overall there is greater need for "One health approach" which can regulate use of antibacterials.

Several strategies have been developed for managing AMR: a). Antibiotic usage policy i.: policy on prudent and judicious use of antibiotics in human and animal medicine, ii strict implementation of drug with drawl period, iii Strict regulation on industrial antibacterial pollution. b). surveillance and monitoring of pathogens and their susceptibility, c). Robust immunising programme and use of alternatives such as pre and probiotics, antimicrobial peptides, bacteriocins, d). Curbing production and sale of poor quality and sub standard antimicrobials, e). Educational and awareness programme

Anesthetics in Aquaculture

Often, fish seed and brood need to be transported from faraway places to hatchery or grow-out sites. This may exert greater stress and physical damage to the fish. Hence, anaesthesia is recommended to lower the metabolic rate and oxygen consumption and reduce handling stress. List of anesthetics used in aquaculture is given in Table 1.

Anesthetics by controlling excitability, reduce injury and time for handling fish. Selecting a sedative and its dosage is important, so to avoid total suppressing of the escape reaction and for promoting quick revival of fish.

Mechanism

Anesthetics are causing a low state of sensation through depression of nervous system is termed anesthesia. Most anesthetics have a neurophysiological action to depress both spontaneous and evoked neuronal activity in several sites of the brain by acting on hydrophilic sites on excitable membrane.

Anesthetics after absorption through the gills, rapidly enter the blood to reach the central nervous system for action. They are excreted via the gills upon the fish's return to fresh water. Anesthetics induce a calming effect followed by a successive loss of equilibrium, mobility, consciousness and reflex action. Determining appropriate dose is very important, as excess dose may lead to respiration and cardiac failure.

Stages of anesthesia: Three stages of anesthesia can be identified.Stage I, Loss of equilibrium, total sedation with loss of reactivity to external stimuli with slight reduction in opercular rate. Stage II, Loss of equilibrium with belly up and loss of gross body movements but with continued opercular movements, Stage III, Total anesthesia–loss of reflex activity, failing to respond to strong external stimuli with almost complete stop of opercular beating.

Stages of recovery: Similar to stages of anesthesia, three stages in recovery can be identified when fish are returned to fresh water. Stage I. Opercular movements just starting although body is immobilized, Stage II. Regular opercular movements started and gross body movements beginning, Stage III Original equilibrium regained with pre anesthetic appearance.

Factors affecting the efficacy of anesthetics are: i). Size of fish - younger fish is more susceptible than older fish, ii). Lipid content of fish - oily fish takes more time than lean fish, iii). Sexual maturity - sexually mature fish is more susceptible than immature fish, iv). Condition of fish - exhausted and diseased fish is more susceptible, v). Temperature - higher dose is required at higher temperature, vi). Salinity - higher dose is required at higher salinity.

Anesthetics in aquaculture are mainly used for i).Transportation of fish seed and brood, ii). Injection of brood fish for induced breeding, iii). Treatment of fish for diseases and parasites, iv). Tagging and marking of fishes. Greatly facilitates in procedures including induction of breeding, handling during stripping and transport of brood stock. It is essential to minimize stress and physical damage in handling the fish for routine operations and also for holding fish immobile while the animal is being handled for sampling and conducting painful procedures (e.g. tagging or injection vaccination).

Rules for Fish Anesthesia

i).Before sedation or general anesthesia, fish should be fasted for at least 12 to 24 hrs or until one can ensure that the stomach is empty of food to prevent regurgitation. ii).Anesthesia and recovery (without anesthetic agent) tanks are prepared ahead of time. iii)Anesthesia should be carried out in water from which the animals originate and at the temperature to which the animals are acclimated iv)Recovery tanks should be aerated at a level required by the particular species of fish v) Temperature, dissolved oxygen concentration, ammonia levels should be monitored

An ideal anesthetic should have the following properties: Produce rapid anesthesia effect(1 - 5 mins),have a quick recovery (5 min) time, large safety factor with no/low tissue residue, easy to handle and affordable, no persistent effects on fish physiology and behaviour, rapidly excreted or metabolised and excreted and zero or short withdrawal time.

Table 1: Anesthetics used in fishes

Drug	Route	Fish taxa	Dose	Comments
Tricaine methane-sulfonate (MS-222)	IN	General (Non-elasmobranch)	75-200 mg/l	Anesthesia induction
			50-150 mg/l	Anesthesia maintenance
	IN	Freshwater eels	75 mg/l	Sedation
	IN	Cyprinids (carp, koi, goldfish, roach, dace, minnows)	60-300 mg/l	Dose-dependent anesthesia
Benzocaine	IN	General (Non-elasmobranch)	25-200 mg/l	Dose-dependent anesthesia
	IN	Freshwater eel	60-80 mg/l	Anesthesia
	IN	Pacus and other characins	50-150 mg/l	Light to deep anesthesia
Clove Oil	IN	General (Non-elasmobranch)	20-100 mg/l	Anesthesia
	IN	Freshwater eels	100 mg/l	Anesthesia
	IN	Cyprinids (carp, koi, goldfish, roach, dace, shark)	25-100 mg/l 4 mg/l	Anesthesia Sedation

Drug	Route	Fish taxa	Dose	Comments
Eugenol	IN	General (Non-elasmobranch)	100-120 mg/l	Anesthesia induction
			40 mg/l	Anesthesia maintenance
	IN	Pacus and other characins	50-200 mg/l	Anesthesia
Isoeugenol	IN	General (Non-elasmobranch)	6-17 mg/l	Anesthesia
Metomidate	IN	General (Non-elasmobranch)	5-10 mg/l 2.5-5 mg/l 0.06-0.2 mg/l	Anesthesia Heavy sedation Light sedation
		Catfish	6 mg/l	Anesthesia
2-PE	IN	General (Non-elasmobranch)	0.1-0.5 mg/l	Anesthesia
	IN	Cyprinids (carp, koi, goldfish)	0.1-0.5 mg/l	Dose-dependent sedation to light anesthesia
Quinaldine	IN	Cyprinids	10-50 mg/l	Anesthesia
Quinaldine sulphate	IN	General (Non-elasmobranch)	50- 100 mg/l	Anesthesia induction
			15-60 mg/l	Anesthesia maintenance
Ketamine	IM	General (Non-elasmobranch)	66-88 mg/l	Anesthesia
		Elasmobranchs	13-30 mg/l	Anesthesia

IN: immersion, IM: intramuscular

Source: Aquatic medicine manual K. M. Shankar (Editor Patil, 2014)

Hormones in Aquaculture

Hormones are chemicals that target specific gland to exert their physiological role, thereby bringing about desired effect in the target organs. The chemical nature of hormones could be proteins or amino acid derivative or steroid. Although produced in minute quantity, they carry out several vital physiological activities such as reproduction, growth, metabolism, osmoregulation, colour change and behavior. In aquaculture, hormones are primarily used in induced breeding, promotion of growth, production of monosex/sterile seed and enhancement of colouration. Of late, use of hormones to produce quality fish seed or promote growth or obtain monosex seed is gaining increased attention of aquaculturists.

I. Induced Breeding

Fish pituitary gonadotropins

The gonadotropins (GtH) are key hypophyseal hormones for gonadal maturation and spawning. Hormonal manipulation of maturation and spawning of fish is about eight decades old wherein exogenous fish pituitary extract was first employed in the 1930s to produce fish seed in captivity. The other hypophyseal hormones are - adrenocorticotropic hormone (ACTH), melanophore stimulating hormone (MSH), thyroid stimulating hormone (TSH), prolactin and growth hormone (GH). Among these, the GtH regulate spawning. Two isoforms of GtH occur in fish, viz. luteinizing hormone (LH) and follicle stimulating hormone (FSH), both of which perform specific function in male and female individuals.

Purified Gonadotropins

In view of inconsistent results obtained with crude pituitary hormones, the gonadotropin is purified partially or completely and successfully used for induced breeding of fish. The partially purified gonadotropin (SG-G100) from salmon is used as a substitute to fish pituitary gland extract (PGE), but is not available on commercial scale and is more expensive. Alternatively, human, bovine, porcine and ovine GtH were tested of which, human GtH was proved to be more effective.

Human Chorionic Gonadotropin (HCG)

The HCG, produced by the chorion, finds increasing use in aquaculture because of its easy availability, uniform potency and low cost. It is very effective in freshwater as well as marine species. The HCG could be used both for enhancing maturity and inducing ovulation. Subsequently, several hormone combinations are employed for induced reproduction. They are: HCG, HCG + PGE, Synahorin (HCG + mammalian pituitary), HCG + SG-G100, mammalian LH-RH and dopamine antagonist.

Other Pituitary Hormones

The other pituitary hormones such as TSH, prolactin and GH are also known to be involved in regulating spawning are seldom used.

LH-RH/Gn-RH and Its Analogues

Among various hypothalamic peptides, gonadotropin releasing hormone (GnRH/LHRH) is the best studied peptide hormone in induced breeding of fish. Mammalian (LHRH) and fish (GnRH) and their analogues are used on a wide range of species (carps, catfishes, sea bass, groupers, milk fish, grey mullets, etc.) as they are not species-specific. In particular, the analogues (synthetic) are more widely used than their natural forms due to their more efficacy, when used in ready-to-inject form, e.g. ovaprim, ovatide and WOVA-FH. GnRHa/ LHRHa in combination with dopamine antagonists such as domperidone, pimozide.have been successfully used in induced spawning. This is because GnRHa/LHRHa is a small molecule (mol. wt. 1180), while GtH is large peptide (mol. wt. ranges from 38,000 to 42,000) and hence hypothalamic peptides find wider applications in aquaculture.

With the use of ready-to-inject spawning agents, the fish seed production of India increased from 409 million fry in 1973-74 to more than 30,000 million fry in 2010. Such a tremendous increase in major carp seed production over a period of 3-4 decades was possible through the application of traditional as well as synthetic hormones. Similar hormone-induced seed production programmes are also prevalent in a number of other countries. Several spawning agents are available in the Indian market, Viz ovaprims(salmon)GnRHa + domperidone, Ovatide (Synthetic LHRHa + dopamine antagonist),WOVA-FH(Synthetic GnRHa) and gonadoprim(sGnRHa + domperidone). Some of these spawning agents are commonly used in carps and other fishes in India.

Steroids

Steroids not only mediate the actions of GtH, but also regulate the GtH secretions through negative as well as positive feedback. Androgens, estrogens and progesterones also play major role in maturation and spawning. A number of natural and synthetic sex steroids have been used for advancement of maturation and spawning in a number of cultivable fishes. Among these, natural hormones such as testosterone (T), 17β-estradiol (E2) and progesterone, and a synthetic steroid, 17α-methyl testosterone (17α -MT) are more effective in induced breeding.

II. Growth Promotion

A fairly good amount of information is now available on the growth promoting substances in poultry, pig and cattle. Application of growth-promoting substances, in particular hormones, is relatively recent in aquaculture. Hormones increase the efficiency of feed utilization at very low levels of inclusion which leads to increased production, thereby bringing down the cost of fish production. Among the four groups of steroid hormones, androgens and estrogens are very useful as growth promoters in fish. HCG and thyroxin are known to induce better growth in carps. Growth hormone gene for increased growth has been demonstrated in several cultivable fishes such as trout, salmon, carps and tilapia. In *Oreochromis hornorum*, recombinant tilapia GH gene has shown positive growth response. Thyroxin is also known to promote growth in tilapia.

III. Production of Monosex/Sterile Fish

Sex hormones play an important role in sex differentiation and sex control in fish. Androgens and estrogens when used during sex differentiation can reverse the sex in male or female direction or suppress gonad development. The technique of hormonal sex reversal is common in the farming of warm water fishes. The technology to produce all-male tilapia by means of hormones (sex hormones) is one of the most important keys to successful and widespread commercialization of tilapia culture. Among steroids, 17α -MT and E2 are the key hormones in the sex inversion programmes of tilapia aquaculture. In India, stocking of sterile common carp, produced through steroid hormone (17α -MT) treatment not only helped preventing undesirable spawning in culture ponds but also leads to higher yield (up to 40-45%). The super-male tilapia (YY genotype) produced through estrogen sex reversal and progeny testing is widely used to obtain an all-male tilapia progeny which is known as genetically modified tilapia (GMT).

IV. Enhancement of Colouration

All-male ornamental fish population (males are preferred due to brighter colour and attractive finnage) can be produced through hormonal manipulation.

Safety of Hormone Treated fish to Human Health

Pituitary hormones (glycoproteins) and GnRH/LHRH(peptides) used in induced breeding of fish may not pose health problem to consumers. However, various steroids employed in aquaculture are a concern from human safety point of view. But research has demonstrated that steroid get metabolised and eliminated in urine after a reasonable with drawl period and hence safe for human consumption. In one study tilapia treated with 17 alphaethyltestosterone (ET) at 60 mg/kg fish for 11 weeks, the steroid could not be detected 7 days post termination of feeding compared to that in the untreated control.

The genetic modification technology (GMT) is safe as it avoids direct consumption of steroid-fed fish. Researchers in Mexico have found that three common species of bacteria have voracious appetites for 17 α -MT, a potentially harmful steroid that fish farmers use to change the sex of tilapia. Some bacteria in ponds can reduce steroid residue for safe. *Pseudomonas fluorescens*, *P. aeruginosa* and *Bacillus cereus* can remove 97-99% of the hormone after 16-20 days of their presence in culture tanks.

Principles and Methods of Chemotherapy in Aquaculture

It is well known that prevention is always better than cure. However, in intensive culture systems it is impossible to prevent disease outbreaks and chemotherapy becomes inevitable. Chemotherapy refers to the use of chemicals to treat the disease. It is very essential to resort to rational, environment friendly chemotherapy practice. Indiscriminate use of chemicals in aquatic systems can have adverse effects on other aquatic fauna and flora, other water users and can lead to environmental disasters. Indiscriminate use of anti bacterials can lead to problems like drug resistance in fish and human pathogens and tissue residues. The four K's of chemotherapy are: knowledge about the pathogen, knowledge about the host, knowledge about the environment and knowledge about the chemical. Only with a sound knowledge of the pathogen, host, environment and the chemical it will be possible to resort to scientific chemotherapy which is effective, safe, cheap and environment friendly. The choice of chemical for therapy should take into account the toxicity of the chemical, efficacy of the chemical at therapeutic levels and long-term effects on the ecosystem. The chemical selected for treatment should be effective at very low levels and should be environment friendly. It is not always possible to carry out precise diagnosis before resorting to chemotherapy. Many a times, chemotherapy is decided based on case history and clinical sign information and rational presumptive diagnosis. It is very essential to determine the nature of disease and the primary etiology, if possible, before resorting to any chemical treatment. Chemical treatments should be suitably modified after proper diagnosis.

The major chemotherapeutic methods available for aquaculturists are immersion methods, oral medication and injection. Immersion methods involve immersing the sick fish/shrimp in a chemical solution of desired concentration. Immersion methods are ideal to tackle disease conditions confined largely to the surface of the fish/shrimp. Depending on the number of fish, area of the system and infrastructure, various methods like dip, short bath, long or indefinite bath can be used to treat fishes. All these methods are same in principle but differ with regard to exposure time and concentration. These methods have their own advantages and disadvantages. In large grow-out ponds, indefinite bath is the method of choice. Chemicals can be broadcasted or solutions can be sprayed over the pond surface.

Systemic disease conditions can be best tackled by oral therapy or injection. Injection though very efficient is not a practical method as it suffers from many disadvantages. The ideal method for mass treatment of systemic diseases is medicated feed. Medicated feeds can be prepared at the farm site. It is necessary to take into account the total biomass, feeding percentage, concentration of the antibacterial, etc while formulating the medicated feed at the farm site. Medicated therapy is normally given for a duration of 1-2 weeks.

Treatment Tips

Fish disease treatments are seldom straight forward. An understanding of the volume and condition of the water, the species and age of fish, the activity of the chemical, the intensity of the disease outbreak and the rate of treatment as well as other factors are important in obtaining consistently beneficial treatment results.

It is essential to have a proper diagnosis before resorting to any treatment. After a diagnosis is made, several other factors should be considered. Are the fish too near death to be saved? Does the extent of the mortality justify the expense of treatment and labour involved? Is the infection likely to spread to other fish stocks if treatment is delayed or withheld? Does the nature of the disease and size of fish preclude the immediate processing of the fish? These and other similar questions should be answered before any treatment is given. If benefit is to be gained from a treatment, the chemical used must be effective against the pathogen and nontoxic to the fish. Water quality and temperature greatly influence the effectiveness and toxicity of certain chemicals.

Treatment Units

Fry and small fingerlings often cannot tolerate chemical dosages that are safe for larger fish, it may be necessary to use lower dosage in treating young fish. Chemical tolerances also differ among fish/shrimp species. Treatment concentration applied to the water are usually expressed as parts per million (ppm). One part per million equivalents for different units of volume used are1.0 milligram per liter or1.0 g per m^3 (1000 liter) or 10 kg. per ha-meter

Determination of Volume of Ponds

Determination of the water volume in a pond, or tank is critical to safe and effective treatments. One should never rely on an estimate by guess. In many cases fish have been killed by over doses resulting from such procedures.

The volume of water in a tank is usually expressed in liters or cubic meter. Volume in cubic meter is computed by multiplying the length by the width by the depth (all measurements in meter). Thus, a tank 40 m long and 6 m wide with water 1.5 m deep contains 360 cubic meters of water. The volume in liters can be determined by multiplying the cubic meters by 1,000.

Conversion Tables

Acre	=	4046.9 square meters
Acre meter	=	4046.9 cubic meters
Hectare (ha)	=	10,000 square meters
Ha-meter	=	1 ha surface of water 1m. deep
	=	10,000m3, 1m3 =1000 liters.

Active Ingredient

Some chemicals are not pure compounds, that is, they contain less than 100% active ingredient. The percent of active ingredient in a product is shown on the label. Treatment dose, unless otherwise stated, is based on the active ingredient only. If 10 kg of a pure chemical per ha/metre gives a 1 ppm treatment, more than 10 kg will of course be needed if a product contains less than 100% active ingredient.

Formaldehyde solution, also known as formalin, contains 37 to 40% formaldehyde for fish cultural purposes, however, it is considered a pure compound. Dose of formalin required for treating external parasites at a rate of 25 ppm in a 1 meter deep 1 ha pond is computed as follows.

Volume of water 1x10,000 = 10,000m3

Dose 25 ppm = 25 mg/l=25 ml/1000l=25 ml/m3

25 ml x10,000=2,50,000 ml=250 liters

Medicated Feed Formulation

For treatments incorporated into the daily ration of the fish, the treatment rate is based on body weight, because medication is prescribed according to the weight of the fish. If 100 kg of fish are to be treated with Terramycin at the rate of 2.5 grams per 100 kg of fish per day, the amount of feed offered each day must contain 2.5 grams of pure Terramycin. Fish are generally fed at the rate of 3% of their body weight per day. At this feeding rate, 3 kg of feed is fed each day to 100 kg of fish. If properly prepared, the 3 kg of feed should contain 2.5 grams of Terramycin; and 100 kg of feed should contain 83.3 grams of Terramycin.

Premixes used in feed formulations usually contain 25 grams of activity per Kg of material. To incorporate 2.5 grams of active ingredient into a quantity of feed when using such a premix, one must use 0.1 kg of premix.

Treatment Methods

The method of application chosen for a particular treatment depends on the disease organism, the fish, the therapeutant, and the size and type of aquacultural unit. Chemicals applied to the water are usually not absorbed by fish and so cannot usually be used for treating internal parasitic and bacterial infections.

Internal or systemic diseases caused by bacteria, tapeworms, flukes, and roundworms are usually treated with medicated feeds. Two considerations related to medicated feed are that; (1) the fish may be too sick to feed, and thus, will not receive the required medication, and (2) the disease agents are sometimes resistant to the drugs.

For external bacterial and parasitic infections, the chemical is added to the water. If the fish are held in small tanks (less than 1,000 cubic meter) that can be rapidly flushed, bath treatments of few minutes to a few hours can be given. However, if fish are held in large ponds where rapid water exchange is not possible, an indefinite treatment is the only practical procedure.

One of the best ways to apply chemicals to ponds is to use a boat and sprayer. Dry chemicals can be placed in a tightly woven sack and pulled behind the motor of the boat. The chemical dissolves in the water and the propeller aids in its distribution. Chemical solutions can also be spread from a tank in the boat by using a boat bailer or a siphon hose that empties behind the propeller. Large ponds can be treated by aerial application. Small ponds can be treated with hand-held sprayers.

Examples of Computing Treatments

If the volume of an aquacultural unit is known, the weight of chemical to be added can be computed from the following formula:

Volume x ppm desired x (100% active ingredient)x desired ppm equivalent for unit of volume used = weight of chemical needed.

Example: A pond holding 14,350 cubic meter is to be treated with 0.5 ppm of a 25% active ingredient chemical. How much chemical should be added to the pond?

It is calculated as follows:

14,350 cubic meter x 0.5 ppm x 4x1 g per cubic meter =28,700 g =28.7 Kg.

Treatment Frequency

The treatment frequency can be decided only by determining a pathogen's susceptibility to the compound used, the effect of the first treatment, and the potential for recurrent infection. Some pathogens, such as the anchor parasite or Ich, require several treatments for effective control because parts of the life cycle is spent in or below the skin, where external treatments cannot penetrate. Others, such as the protozoan parasites *Chilodonella* and *Trichodina*, usually require only one treatment. To be sure that a treatment has been effective, however, one must make a follow-up examination of the fish about 24 hours after treatment, either by watching for the presence of additional sick and dying fish, or by examining some fish microscopically to check for the presence of the pathogen.

Some of the commonly used chemicals and their dosage for treating important diseases and addressing environmental problems in fish/shrimp culture are given in Table 2.

Table 2: Treatment Chart for Common Diseases of Cultured Fish (Source: Fish and shellfish Health management, K. M. Shankar and C.V. Mohan 2002)

1. **Parasitic Diseases**

Disease Agent	Treatment Chemical	Method of Treatment	Concentration/Time	Remarks
A. Ectoparasites I. Protozoans i. ciliates *Ichthyophthirius m ultifiliis*	Malachite green Malachite green + Formalin	Prolonged bath Prolonged bath (Slow turnover-6 hr or more)	0.15-0.20 mg/L for 4-6 hr, repeated 4 times at intervals of one-two days 0.1 ppm malachite green +15-25 ppm formalin –3 treatments on alternate days or 1 ppm malachite green + 160 – 250 ppm Formalin for 1 hr –3 treatment on alternate days.	Only malachite green oxalate should be used, chloride and sulphate of malachite green are toxic to fish These baths are fully effective only if the water is changed after every 24-48 hrs and the therapeutic concentration recorrected.
Chilodinella	Formalin	Short bath	250-300ml/m3 for 30 min.	Formalin is a reducing agent causes O2 depletion, especially at high temperature. Use at maximum O2 level or aerate the water. If stock is old, formation of ineffective white precipitate of paraformaldehyde -toxic to fish.
Trichodina	Formalin Formalin	Short bath Permanent bath	250-330 ml/m3 for 30 min. 15-25 ppm	See precautions for formalin
ii. Flagellates *Ichthyobodo*	Formalin Formalin	Short bath Permanent bath	0.25-0.33 ml/L for 30 min. 15-25 ppm	See precautions for formalin
II. Monogenetic Treamatodes *Doctylogyrus and Gyrodactylus*	Dipterex Trichlorphon	Bath (permanent)Dip	0.25-0.5 ppm 1%, 2-3 min	Toxic to human beings, use rubber gloves and protective clothing, rapid breakdown in warm water, use 0.5 ppm at over 27°C. Much slower breakdown at cold temperature especially at pH below neutral, therefore potentially toxic in soft fresh water.

Disease Agent	Treatment Chemical	Method of Treatment	Concentration/Time	Remarks
III. Crustaceans i. Copepods *Ergasilus* and*Lernaea*	Trichlorphon Trichlorphon Dipterex NaCl Formalin $KMnO_4$	BathDip Bath (permanent) Bath BathBath	5 mg/L 101 for 10 min. 1%, 2-3 min; 4 treatment at 5-7 day intervals 0-25-0.5 ppm 0.8-1.1% for 3 days 250 ppm for 30-50 min. for 3 weeks 25 ppm – Repeat after 90 min	For controlling larval stages of Lernaea and Ergasilus. Only for fish of >25 g size. Concentration of dips and baths lies close to the lethal dose for fishes also. Organic load in water reduces the efficacy.
ii. *BranchiuraArgulus*	$KMnO_4$ Trichlorphon Trichlorphon Trichlorphon Reverse salinity	Short bath Prolonged bathProlonged bath Short bath Bath	10 mg/L for 30 min 0.2 mg/L for 24 hrs 0.4 mg/L for 6 hrs. 5 mg/L for 30 min. 1 % NaCl for 3 days	Only for NaCl tolerant fishes
IV. *Leeches*	TrichlorphonNaCl Quicklime	Bath Immersion Immersion	0.25 g/m3 2.5% for 15 min. 1 gm/2 lit of water	Kills all free-living leeches
B. Endoparasites I. Protozoans i. *Sporozoans* *Myxobolus sp*	Furazolidone	Feed	152-194 mg/kg fish	No known effective treatments. For all sporozoans soil treatment with lime to kill the spores is the best option. Furazolidone is a nitrofuran antibiotic. Only slightly soluble in water. No toxicity and secondary side effects.
ii. Flagellates *Trypanosoma* *Hexamita*	Metronidazole Emtryl (dimetridazole) Furazolidone	FeedFeedFeed	50 mg/kg fish/day for 6 days 40 mg/kg fish/day for 3-5 days 50 mg/ kg fish/day	Flagyl (Metronidazole) may be used at similar rate

Disease Agent	Treatment Chemical	Method of Treatment	Concentration/Time	Remarks
II. Digenetic trematodes *Diplostomum (Eye fluke)*	Paraziquantel	Feed	300 mg/kg fish every 4th day for 1 month, or 6 consecutive days	No known treatment In general control of intermediate host (snail) and prevention of fecal contamination from birds is the best approach. Experimental and expensive, but may be useful for a variety of metazoans. Single dose may be beneficial
III. *Cestodes (adult tape worms in gut)*	Paraziquantel	Feed	300 mg/kg fish for every 4th day for 1 month, or 6 consecutive days	Not used widely in aquaculturepractices
IV. Acanthocephalans	Di-n-butyl tin oxide	Feed	250 mg/kg fish for 10 days	Control of intermediate host is the best option
2. Fungal Diseases				
I. *External Fungi: Saprolegnia* sp. *Achlya* sp.	Malachite green (Zn-free) Malachite green(Zn-free) Formalin Formalin Malachite green Methylene blue	Dip Bath Bath Bath Prolonged bath Bath	67 ppm; 1 min. 1-2 ppm; 1 hr. 1:500 or 1:1000 for 15 min 150-300 ml/m3 for 30-40 min 0.15-0.20 mg/L 10-20 mg/L for 15 min.	Highly toxic to juveniles of fish and some eggs are sensitiveFor fry For adult fish CuSO4, KMnO4, also may be used against external fungi

Disease Agent	Treatment Chemical	Method of Treatment	Concentration/Time	Remarks
II. Systemic fungi: *Branchiomyces* *Ichthyophonushoferi*		No known treatment No known treatment		It is reported that fungicides may be effective. Prevention achieved by strict hygiene. Prevention- infective stages in water destroyed by chlorine (20 mg/L) or Malachite green (290 mg/L).
3. Bacterial Diseases				
I. External Surface Ulcerative type (*Aeromonas* sp. *Pseudomonas* sp. *Vibrio* sp.)	$CuSO_4$ Proflavine Hemisulphate Oxytetracycline or Terramycin Gentamycin	Dip Bath or flowing Bath or flowing In feed Injection (intraperitoneal)	500 ppm; 1 min. 20 ppm for ½ hour 20 ppm for ½ hour 75 mg/kg fish for 7-21 days. 20-40 mg/kg fish.	Systemic therapy with antibiotics also should be considered simultaneously Oxytetracyline hydrochloride is a broad spectrum antibiotic aqueous solution which is unstable and destroys on heating.
Myxobacteria (environmental gill disease)	Quaternary NH+4 compounds (Roccal, Hyamine, Benzalkonium chloride) Trypaflavine Furance	BathBathBath	1 ppm (soft water) for 1 hr. to 4 ppm (hard water) for 1 min 3-6g/m3 for 6-12 hrs. for 1-3 consecutive days. 1 mg/L for 1 hour	Non-corrosive, inactivated by normal soaps. More toxic in soft water.
II. SystemicInfections Vibriosis (*V. anguillarum*)	Sulphonamides	Feed	200 mg/kg fish/day initially, later maintain at 100 mg/kg fish/day	All treatments through feed for minimum of 10 days. Sulphonamides may be toxic in seawater (crystallise in fish's urine)
Furunculosis (*Aeromomonas salmoncida*) and Other *Aeromonads*, *Pseudomonas*	Potentiated sulphonamides Nitrofurans (eg: Furazolidone)	FeedFeed	30 mg/kg fish/day of total trimethoprim/sulphonamide combination 50 mg/kg/day	Expensive Should be made by the feed mill. They do not taste well. Use of fish oil better than vegetable oil

Disease Agent	Treatment Chemical	Method of Treatment	Concentration/Time	Remarks
	Oxolinic acid Oxytetracycline	FeedFeed i/p injection	10 mg/kg fish day75 mg/kg fish/ day 100 mg/kg (continued in feed)	4 week withdrawal period before marketing, especially important with oxytetracycline
Edwardsiella tarda	Oxytetracycline (or chloramphenicol or Kanamycin)	Feed	50 mg/Kg/fish day for 10 days	All strains are resistant to penicillin, erythromycin some strains resistant to tetracyclines.
*Yersinia ruckeri (*enteric red mouth disease*)*	Suplhamethazone Oxytetracycline Chloramphenicol Sulphamethrizine + Furazolidone	FeedFeedFeedFeed	200 mg/kg/day for 5 days 50 mg/kg/day for 3 days 50 mg/kg/day for 3 days 66 mg+ 44 mg/kg/day for 5 days	
Mycobacterium sp	Oxytetracycline Kanamycin sulphate	Feed	10 mg/kg feed for 4 days	No proven therapy, only experimental
4. Environmental Problems				
1. Algal blooms: i. *Prymnesium parvum*	Ammonium sulphate 'Aqua ammonia' Copper sulphate	Spread over pond surface Distribute into water by pipe Spray	10 ppm (pH> 9, temp > 200C). Up to 25 ppm (pH < 8.6, temp. <170C) 10 ppm (pH 9, temp. >200C) upto 14 ppm (pH 8.6, temp. <170C) 2 ppm (pH 9, 200C) to 3 ppm (pH <8.6 temp. <170C)	Aqua ammonia preferred below 170C Great care in handling toxic ammonia compounds Not recommended at temp. > 200C.
Microcystis and *Oscillatoria* sp.	Copper sulphate Ammonium sulphate	Treat empty pond bottom Spread over pond surface	0.1 kg/m220kg/ha	Effective upto 1 year, but expensive for large pond Induces oxygen production from bloom starved of nutrients

Common Treatments Followed in Shrimp and Freshwater Prawn Culture

1. Epicommensal organisms (surface fouling)Formalin – 25-30 ppm for 24 hr. (75 ppm formalin can reduce DO to 0 ppm within 48 hrs).
2. Larval Mycosis
   a. Disinfecting spawners with 5 ppm Treflan for 1 hr. before spawning.
   b. Treating affected shrimp with 0.2 ppm Treflan for 24 hr.
   c. Malachite green 0.01 ppm – 24 hr.
3. Microsporideans:
   a. Destroying affected individuals.
   b. Disinfecting culture systems with bleach, lime or iodine compounds.
4. Cuticular or surface vibriosis

   Use of antibiotics and sulphadrugs for immersion therapy. Furazolidone 100 ppm for 12 hr for 4-5 days.

   Furanace at 0.1 ppm. Sulfamethazine at 3 ppm.
5. Enteric and systemic vibriosis: Systemic therapy through medicated feed. Use of antibacterials in feed for 7-10 days.
6. Viral disease: No effective chemotherapy to inactivate virus inside the host cells. Free virus particles outside the host can be inactivated by UV, chlorine, iodine and formalin.

Other Compounds Used in Shrimp Farming

Chelated copper compounds are used as algicides, parasiticides, bactericides and "forced moulting" agents at 0.25 ppm for 24 hr.

Quaternary ammonium compounds (Benzalkonium chloride and Hyamine) used at 1-2 ppm baths for shell and gill necrosis and bacterial diseases.

Other Chemicals used in Aquaculture

i) Weedicides

Several weedicides are recommended and used for the control of aquatic weeds in aquaculture ponds. The recommended weedicides and their dosages are shown in Table. 3.

Table 3: Other chemicals used in aquaculture

Type of weeds	**Chemical**	**Dosage**
Planktonic algae Filamentous algae	$CuSO4$	0.25-1.0 ppm
	Simazine	0.5-1.0 ppm
Floating weeds	2,4 D	2-4 kg/ha
	Diquat	1-2 kg/ha
Submerged weeds	2,4 D	1-2 kg/ha
	Diquat	3-5 kg/ha
Emergent weeds	2,4 D	2-4 kg/ha
Marginal weeds	Diquat	2 kg/ha

ii) Insecticides

Chlorinated hydrocarbons, organo-phosphorous compounds, bleaching powder are used for different purposes in aquaculture.

1. **Chlorinated hydrocarbons:** Aldrin at 0.2 ppm kills weed fishes and predatory fishes, without affecting zooplankton. Dieldrin at 0.01 ppm effectively kills fishes and 0.5 ppm kills prawns and insects. Endrin (0.001 ppm) is the most poisonous and effective chemical for eradication of fishes, prawn and insects.

2. **Organophosphates:** Three organophosphates, viz., Thiometon, DDVP and Phosphamidon have been found to be effective in killing fish. DDVP at 3-30 ppm is recommended for eradication of trash fishes in nursery ponds.

3. Gammexaneat 06-1.0 ppm and Sumithionat 0.25-3.0 ppm are effective in controlling aquatic insects.

4. Bleaching powder with 30% chlorine content, when applied at 25-30 ppm, kills all varieties of fishes, including catfishes, murrels, weed fishes and carps.

5. Butox (Deltamethrin, a veterinary drug used for healing wounds on live-stock) is used for the control of predatory aquatic insects in fish nursery pond at 1.0-1.50 ml/25 m2 area

iii) Antiseptics

1. Recommended dose of antiseptics in the transport medium or short-term bath prior to transport. Acriflavin (10 ppm), Methylene blue (2 ppm), CuSoo4 (5 ppm), $KMnO_4$ (3 ppm), NaCl (3 %) and Formalin (15 ppm)

2. $KMnO_4$at 5 ppm is used for foot dips (hatcheries, processing hall, etc.), treating plant substrates for egg collection, site of injection, spent brooders.
3. Malachite green is used at 2 ppm for treating trout eggs to control fungi.
4. Iodophorehaving 100 ppm available Iodine is used as an anti-viral agent.

Common drug related terms for information

The word drug is derived from a French word 'Drogue' meaning a dry herb. Drug is any chemical that is used to promote or safeguard the health of human beings or animals.

Over-the-counter drugs are those preparations that are sold without any prescription.

Prescription drugs or Legend drugs are those prescribed by a licensed veterinarian/physician/dentist/surgeon.

Essential drugs are agents that satisfy the healthcare needs of majority of the population. They should therefore be available at all times in adequate amounts and in appropriate dosage form.

Orphan drugs are drugs or biological products useful for diagnosis/treatment/ prevention of a rare disease condition. These drugs may be life saving for some patients, but are not commercially available.

Placebo is a vehicle for cure by suggestion and is surprisingly often successful though only temporarily. It can be used as a control in scientific evaluation of drugs and to benefit or please a patient not by pharmacological actions but by psychological means (Latin: Placebo – I shall be pleasing or acceptable). Placebo reactor is an individual who report changes of physical and mental state after taking a pharmacologically inert substance.

References

Antimicrobial Resistance, 2021, Policy paper 103,National Academy of Agricultural Sciences, New Delhi 110012

K. M. Shankar, 2016, Importance and Scope for Aquatic medicine in India, In Aquatic medicine-a training manual, Special publication No 01/2016, Editor Prakash Patil, Aquatic Animal Health Management Laboratory, College of Fisheries, Karnataka Veterinary, Animal and Fisheries Sciences University, Mangalore-575002

K. M. Shankar and C.V Mohan, 2002, Fish and Shellfish Health Management, ISBN 81-7525-328-2, Fish Pathology and Biotechnology Laboratory, Dept of Aquaculture, UAS, College of Fisheries, Mangalore-575002

17

Defense Mechanisms in Fishes and Crustaceans

Prakash Patil and K.M. Shankar

Farming of fish and shellfish has gained significant grounds in several parts of the world. However, the major bottleneck to successful farming is disease in farmed aquatic animals. In recent years, lot of attention is being given to health management using various forms of immunoprophylactic measures such as vaccination and immunostimulation. To rationalise immunoprophylactic applications, it is vital to have insight into the specific and non-specific defense mechanisms of farmed aquatic animals. Through disease process studies it is very well known that a pathogen can cause disease only if it can overcome the non-specific and specific defence barriers of the host and successfully establish and proliferate. Non-specific and specific immune systems are the two main branches of the defense system. The immunity which an organism derives against a pathogen is the result of a delicate interaction and cooperation between the two systems.

Defense System of Fish

Surface Barriers as Local Immune System

The immune systems present in the surface structures of fish have their potential importance in local immune responses. Important surface barriers are skin, gills and gastrointestinal tract.

Skin

The skin surface of fish, the epidermis is composed of non-keratinised living cells. Epidermal integrity is vital to fish in maintaining osmotic balance and excluding microorganisms. The epidermal healing response in fish is extraordinarily rapid, even at low temperatures. It involves a migration of malphigian cells from the periphery of a wound over the wound surface, rapidly closing the lesion. The epidermis in fish may also respond to nonspecific irritation by a thickening of the cuticle or a hyperplasia of the malphigian cells, thus minimising the chances of epidermal disruption.

Colonisation of skin by bacteria and parasites is prevented by active immune responses in the skin. There are several reports of lymphocytes, plasma cells and macrophages present within the epidermis of fish for mounting a local adaptive response. Immunoglobulin is present at low concentration in the skin mucus, with few antibody-secreting cells (ASC). ASC number can increase some 20-fold post-vaccination indicating they are the source of the cutaneous immunoglobulin (Ig). Recent discovery of IgT in skin mucus, suggests its role in mucosal immunity. Expression of cytokine genes is also known to occur in the epidermis of fish with various ectoparasitic infections.

Gills

Gills comprising a large surface area of delicate epithelium, is considered to be an important route of entry of microorganisms. The organ is protected by mucus production and a highly responsive epithelium resulting in hyperplasia frequently seen in many gill bacterial and parasitic infections. The gills also contain phagocytic cells which line the branchial capillaries, and lymphoid cells on the caudal edge of the interbranchial septum.

Phagocytosis by reticuloendothelial macrophages lining blood vessels in the secondary lamellae and the accumulation of lymphocytes on the caudal edge of the interbranchial septum has been reported. The gills are considered an important organ of antigen uptake, particularly of particulate antigens. The gills contain quite high numbers of resident lymphocytes, macrophages and plasma cells and local antibody production may play an important defense especially against bacterial diseases.

Gastrointestinal Tract

Mucous membrane lining the gastrointestinal tract is similar in several ways to that on the skin.

The digestive action of the gut provides an extremely hostile environment to pathogens by virtue of the low pH and the secretion of digestive enzymes and bile. The mucosa of the gut and the lamina propria are well populated by a variety of leucocytes including macrophages, NCC, T and B cells and plasma cells. Immunoglobulin (Ig) is present in the gut which may have its origin from mucosal secretions and the bile. The presence of specific Ig has been observed in gut after oral vaccination of fish with relative high IgT+ B cells at this site. The mucosal epithelial cells themselves, especially in the second half of the intestine, can endocytose antigens and transport them to macrophages and lymphocytes present in the mucosal tissue.

Fish have local mucosal defense in the gut and produce local immunoglobulin responses. Leucocytes are abundantly present in the fish gut's lamina propria and intestinal epithelium. After systemic intestinal or immersion immunization specific antibody secretion in the fish intestine has been recorded. Immunoglobulins (Ig) produced in the intestine are a result of local synthesis has been confirmed. Ig isotype, the IgT is specialized for mucosal immunity and Ig + B cells and Ig-T cells are abundantly present in fish's gut, but limited data is available regarding their functional relevance. Posterior segment of fish intestine housing various immune cells including B cells, macrophages and T cells is very immunologically active.

Mucus and Mucosal Barriers

The integument of fish (skin, gills and gut) is covered by a layer of mucus which, by entrapping microorganisms and continuously sloughing, inhibits colonisation of the integument. The mucus production increases in fish in response to infection or by physical or chemical irritants. The mucus of fish is toxic to certain microorganisms due to a variety of humoral factors.

Mucosal barriers are complex tissues which besides maintaining homeostasis perform several physiological functions such as gas exchange and nutrition including immunity. The gut associated lymhoid tissues (GALT), gill associated lymhoid tissue (GIALT), skin associated lymhoid tissue (SALT) and nasopharyngeal associated lymhoid tissue (NALT) are the main mucus associated lymhoid tissues (MALT). These systems serve as very active immune system protecting fish from the pathogens.

Several secretary cells mainly goblet cells(GCs), sacciform cells and club cells can produce mucous. Mucous produced by goblet cells contain high molecular weight glycoproteins, are prevalent on external surfaces of several organs serving as first line of defense against pathogens. Mucin, a major molecule of mucus is responsible for its viscosity, entrapping pathogens and physical protection of skin surfaces. Mucins contribute significantly to innate immunity in two ways. As first line of defense, mucin prevent adhesion and colonisation of bacteria and parasites by continuos production and frequent shedding. Furthermore, mucins contain large kinds of immune related proteins such as lysozomes, esterases, lectins, complement proteins, c-reactive proteins(CRP) transferrin, immunoglobulins, proteases, antimicrobial peptides which collectively inactivate pathogens and subsequent infection.

Oral or immersion vaccination can stimulate antibody responses in mucous membranes without inducing serum antibody titres. Several deliveries of antigen via the oral or immersion routes are usually required to stimulate serum antibody production, with oral boosting following immersion vaccination proving to be an effective way to induce protective immunity. Immunisation by the injection route can stimulate antibody producing cells at mucosal sites indicating a two-compartment model of humoral immune responses in fish systemic and mucosal.

Lymphoid Organs of Fish Thymus

Evidences to indicate that the role of the thymus in fish is similar to that in mammals, as a primary lymphoid organ in which T lymphocytes differentiate and mature to leave the organ and participate in immune responses in the peripheral lymphoid organs and tissues. There is considerable evidence to indicate that the thymus of fish functions as a primary non-executive lymphoid organ, as in mammals. Thus the available evidence indicates that the thymus is involved in the production of large numbers of lymphocytes for export to other tissues, while having little involvement in the execution of immune responses. It is also likely that at this site cells that can react with self antigens are deleted from the T cell repertoire, to give tolerance to self. In these respects the thymus can be considered a primary lymphoid organ in fish.

Kidney

The hemopoitic tissue of the kidney bears a close resemblance to the bone marrow of higher vertebrates but differs in having a highly active reticuloendothelial and antibody producing cell component. In primitive teleosts (e.g. salmonids), there is little tissue specialisation, but in higher teleosts (e.g. the cyprinids and pleuronectids), there is a degree of specialisation and organisation of the haemopoietic tissue. Throughout the hemopoitic tissue is a system of sinusoids lined by reticuloendothelial cells which are phagocytic for carbon particles and immune complexes. Within the hemopoietic tissue are spherical aggregates of pigment containing cells which are of two types: melanocytes and melanomacrophages, essential for melanin production and confirms that at least some leucocytes are capable of melanogenesis and melanosome secretion in fish. In higher teleosts these aggregates, melanomacrophage centers (MMC), possess a delicate reticulin capsule and contain other cell types including lymphocytes and pyroninophilic cells. Lymphoid system in kidney with lymphoid cells of all developmental stages (small and large lymphocytes blast cells and plasma cells) are present within the hemopoietic tissue.

Spleen

The lymphoid tissue in the teleost spleen is more organised in the advanced species although it does reach a high degree of extensive organisation. Generally it forms a cuff of tissue surrounding the ellipsoid system and the associated MMC. Macrophages which are enmeshed in the densely packed reticulin fibres in the wall of ellipsoids are active in uptake of bacteria and foreign particles. Upon antigen stimulation, antibody-producing cells appear with similar kinetics to those in the kidney. Furthermore, following antigen stimulation in the carp, clusters of pyroninophilic cells appear in the ellipsoid walls. They expand into structures similar to those in the kidney and are thought to develop into MMC. The reticulin fibres in the ellipsoids are important in trapping immune complexes which are retained for prolonged periods of time considered important in the development of immune memory.

Immunoglobulins of Teleosts

Antibodies are a class of proteins called immunoglobulins (Ig). In teleost fish there are three classes of Ig; IgM, IgD and IgT (IgZ), compared with five classes in mammals. The B cells producing these Igs appear to be mutually exclusive at least for IgM and IgT production, and hence different B cell subsets exist in fish. The basic structure of Ig is composed of two heavy (H) chains and two light (L) chains, as measured by molecular weight, with both a heavy and light chain required to form an antigen-binding site (complementarity determining region, CDR). It is the physicochemical properties of the H chains which determine classification of the Ig class. In mammals IgM is a pentamer of this basic unit, while in teleosts it is a tetramer, though in some species a monomeric form of IgM can be present. IgT is monomeric in the blood. Ig also exists in mucus secretions of the skin and gut, and in the bile. IgT/IgZ are important, specialised for mucosal immunity are produced by the cells in the mucosal membrane. Polymeric immunoglobulins receptors (pIgR) in the gut and skin of teleosts carrry IgM and IgT through mucosal barriers. The IgM of fish which is a tetramer with eight antigenic combining sites is predominant and most efficient. Functional specialisation of the different Ig classes in mammals is a property of the structure of the H chains which determine the class or subclass. Ig has also been found in the eggs of some species of teleosts (e.g. plaice and carp), suggesting transfer of maternal antibodies and possibly also maternal Ig transcripts. The health status of brood stock fish may impact on the type and quantity of immune molecules transferred to their offspring. Important protective roles of antibody include neutralisation of viruses, toxins and bacterial adhesins, activation of the complement system and opsonisation of particles (e.g. bacteria and viruses).

Cell-Mediated Immunity (CMI)

Different populations of T-lymphocytes namely T-helper cells, T-Suppressor cells, T-killer cells and T-lymphokine producing cells are involved in the execution of the CMI. This aspect of the immune response is wide ranging and also recruits the macrophages which constitute the body's main line of non-specific defense by phagocytosing and digesting invading pathogens. On primary antigen stimulation, the T-lymphocyte clones differentiate into several different functional types involved in CMI. The CMI is generally measured by allograft rejection. *In vitro* assays include mixed leucocyte reaction and macrophage migration inhibition test (MIF). These responses are mediated by T-cytotoxic cells and T-lymphokine producing cells. The different functional subpopulations of T-lymphocytes are;

T-helper cells: These cells cooperate with B-lymphocytes in recognition of T-dependent antigens. On stimulation, the clone proliferates to T-helper memory cells which are long lived. T-helper memory cells cooperate with increased number of B-memory cells so that antibody production to a thymus dependent antigen on secondary exposure is more rapid. T-helper clones also possess clones responsible for CMI.

T-Killer cells: These cytotoxic T cells are capable of lysing foreign cells by a mechanism depending upon physical contact between the T cell and the target cell.

T-Lymphokine producing cells: Upon antigen stimulation a clone of responsive T cells release humoral factors called lymphokines (interleukins) which enhance the non-specific defense capacity of macrophages.

T-Suppressor cells: Suppressor cells regulate the production of antibodies and lymphokines during the course of an immune response and this is performed by the proliferation of suppressor cells.

Immunological Memory in Fishes

An important feature of the immune system is specificity and the capacity to develop immunological memory. A first contact with an antigen usually induces relatively short-lived effector cells (activated Th, plasma cells or cytotoxic T-cells). There are also long-lived memory cells among the progeny of the original non-primed lymphocytes. These memory cells retain the capacity to be stimulated by the antigen. The development of immunological memory is often measured indirectly by monitoring the secondary response. In the case of positive memory, this response will be faster and more vigorous than the primary response. The height of the secondary response is dependent on the amount and antigenicity of the priming antigen. A relatively low priming dose is usually optimal for memory development in carp. In carp, the ratio between secondary and primary antibody responses never reaches the high levels of that in mammals (5–20-fold in carp and up to 100-fold in mammals). Imunological memory has also been demonstrated in both vaccine challenged and infected–recovered fish. Several differences between the secondary responses of mammals and teleosts have been found. One distinction that can be made is that the ratio between the secondary and the primary response is much higher in mammals than in teleosts (which can be expressed as the 'memory factor' (MF)). During the secondary response the dominant Ig isotype in mammals is IgG. IgG is absent in fish, since teleosts possess only the IgM isotype. Isotype switching is triggered in mammals during the secondary response, in contrast to teleosts, where this phenomenon has not been demonstrated. A temperature dependence of the secondary response in carp has been observed

Immune memory can be either positive or negative. The latter form is termed tolerance. Positive memory is expressed in a secondary response and is typified by a shorter lag phase and higher magnitude of reaction (higher antibody titre or shorter graft rejection time) as compared with the primary response. Memory is antigen-specific and in tetrapods there is a shift of high-molecular-weight Ig to low-molecular-weight Ig with higher affinity. This is called isotype switching and in mammals IgM becomes replaced with IgG in blood. Tolerance is a state of antigen specific nonresponsiveness induced by exposure to antigen. It can also be induced by administration of very small or very large doses of antigen. Tolerance to self-antigens is an obvious example. Memory and tolerance may be long or short lasting. The mechanism of memory resides in the differentiation of long-lived T and B memory cells and changes in the nature of the reactive cells, so that clones with high affinity for the antigen are selected. Memory has been demonstrated in fish for both humoral and cell-mediated immunity. Establishment of memory in carp appears to take longer (maximum 3–6 months) than in mammals and remains for 8–12 months after primary stimulation. The data for secondary humoral responses in other fish species are similar.

Overall Immune Response in Fish

Immune system of fish is similar to that of higher vertebrates. However, fish depends more on non specific immune system. The specific and non specific immune systems are interdependent, acting in concert with each other. This is clear when macrophage activity depends on coating with specific antibody and antigen processing by macrophage is crucial for response and uptake by lymphoidal system for antibody production. Overall some of the defense factors for immunity involves both specific and non specific components and is very complex.

Factors Affecting Immune Response

Factors affecting the immune response may be classified into three broad categories: extrinsic factors which relate to the environment and the nature of the antigen, intrinsic factors which relate to the immunoregulatory mechanisms within the immune system and the physiological state of the animal, and ontogenetic factors which relate to the maturation of the immune system in young animals. Extrinsic factors are temperature, antigen dose, nature of antigen, route of administration, adjuvants, and environmental factors such as photoperiod, dissolved oxygen, salinity/pH, pollution.

Immune Cells—Nerve Interaction

Nervous system and immune system work in concert to deal with infections and inflammation. Brain and peripheral nervous system can stimulate or inhibit both innate and adaptive immune system. Antimicrobial peptides significant in microbial immunity are synthesised by neuropeptide precursors. Several peptides involved in neuronal or neuro endocrinal signaling also possess antimicrobial activities. These peptides are widely distributed across immunological, endocrine, neuroendocrine and neuronal cells. The classical example is reciprocal interaction dependency of microbial-gut- brain axis(MGB axis) on CNS and gut bacteria with involvement of endocrine immunological and neurological pathways. First, lymhocytes stimulation in the gut releases cytokines, which in turn are activated by gut peptides from enteroendocrine cells from sensory nerve terminals. Communication pathways involving neurotransmitters, neuropeptides, cytokines, hormones serve as link between immune system and CNS leading to homeostasis.

Defense systems of Crustaceans

Crustaceans encounter a variety of pathogens in the environment they live in. Though the specificity of crustacean defense mechanism is not comparable to that of higher vertebrates or even fish, the efficiency with which their defense system overcomes the pathogenic hostility is highly remarkable. In crustaceans non-specific immune system has a greater role to play due to a poorly developed specific immune system with a very weak or no memory function in it. The defense in crustaceans is largely dependent on the blood cells and hemolymph activities, where the hemocytes involve in clotting melanisation and phagocytosis.

Crustaceans completely depend on the innate immune system with overlap of cellular and humoral components. Their defense system is activated by recognition of a pathogen associated molecule pattern(PAMP). PAMP includes soluble or cell surface host proteins such as lectins, antimicrobial, clotting and pattern recogntion proteins which in turn activate cellular or humoral components. Special recognition of carbohydrate containing molecule such as glycans, glycopeptides, glycoproteins, pepetidoglycan, or lipopolysaccarides (LPS) from bacteria, virus or fungi is major part of immune defense in crustaceans. Defense processes like agglutination, encapsulation, phagocytosis clotting proteins are induced by the carbohydrate driven recognition patterns. Thus the innate defense mechanisms is very efficient for protection against the invading pathogens. These mechanism in invertebrates indicate that they are precursors of vertebrate immunity. Although there is little evidence, the process are antigen-specific, however some cellular and cell free hemolymph factors show high specificity for non self or damaged cells.

Circulatory system in crustaceans is open where cells, nutrients, oxygen hormones are distrtibuted in the hemolymph. The circulatory cells (hemocytes) involves in non self matter recognition and elimination of pathogens. Depending on pathogen's characteristics several defense mechanisms of crustaceans such as prophenol oxidase (proPO), phagocytosis and encapsulation are activated. Pathogens are recognised by Pattern Recognition Receptors(PRRs) followed by deployment of cellular and humoral immune response protection. Important cellular responses include phagocytsis, nodulation, encapsulation leading to melanisation. These responses are also complemented with humoral responses such as secretion of clotting proteins, agglutinins, antimicrobial peptides proteinase inhitbitors, and reaction intermediates of oxygen or nitrogen. The innate immunity of crustaceans is simple, but very potent that can defend them against a variety of pathogens from diverse habitat.

Types of Defense Systems in Crustaceans

It is very well known in crustaceans that the non-specific defense system has a greater role to play because the specific immune system is very poorly developed with no specificity and memory. The major defense system of crustaceans may be classified as follows:1. Fixed defense 2. Mobile defense

1. Fixed Defenses

Blessed with hard cuticular exoskeleton, the first line of defense serving as structural and chemical barriers to attachment and penetration of pathogens. The oral tract also a potential route of infection is provided with gut enzymes and acids to destroy pathogens. Furthermore, the intestinal epithelium is protected with chitinous lining preventing entry of pathogen from the gut. However, pathogens can enter the body by breaching the first line of defense through damaged cuticle or during intermolt period, which is considered very vulnerable phase in the life cycle of crustaceans. During this stage parasites find it easy for attachment and penetration which also leads to other infections and complexities. Branchial podocytes act as fixed phagocytic cells and remove pathogens from hemolymph. Crustaceans have remarkable ability for autotomy and regeneration of appendages. Clotting and wound repair is rapid and well developed in crustaceans.

2. Mobile Defenses

The mobile innate defense system is regulated by humoral (antisomes) and cellular (hemocytes) factors. Main humoral components are prophenol oxidase activating system, agglutination, protease inhibition, AMPs, phospotases and lysozyme. Mechanism directly mediated by cells or hemocytes are phagocytosis, encapsulation and nodule formation.

a. Humoral Immune Responses

Pattern Recognition Proteins (PRPs) of Crustaceans

Receptors on the hemocytes termed " pattern recognition receptors" can recognise the highly conserved and widely distributed signature molecules on the cells of microorganisms. These molecules are unique to microorganisms such as lypopolysaccarides (LPS), peptidoglycans (PGNs) of cell wall of bacteria and b-1-3 glucans of fungal cell wall. The innate system thus can prevent initial or early stage of infection by inducing the chemical oxidative modification on their molecule by the proPO system, through proteolytic processes and phagocytosis. In crustaceans and other arthropods, immunity is based on hemocytes able to recognize foreign material through proteins found in the cell-free hemolymph. These proteins are able to interact with specific carbohydrate residues, such as lectins, and act on pathogens. This interaction allows for the phagocytosis or encapsulation of the pathogens, or the lectin action is exerted through the recognition of specific molecular patterns of pathogens. Based on the aforementioned, it can be implied that crustaceans possess efficient defense mechanisms in spite of lacking immunological memory and antibodies, which are found in vertebrates. It has been considered that certain physiological factors are importantly involved in the immunity of these species. Among these factors are developmental conditions, the stage close to ecdysis, their adaptation to the aquatic environment in which they develop (saline, stereo saline or freshwater), and the stress conditions to which the organisms are subjected when being cultivated extensively. Response against pathogens seems to be different for each crustacean species, since, for example, *L. setiferus*, *L. vannamei*, and *F. chinensis* are susceptible to the Taura syndrome virus, whereas *P. aztecus* and *P. duorarum* are not.

Pathogen Recognition

Crustaceans have a well articulated defense mechanisms, employing Pattern Recognition Receptors (PRR) on their hemocytes. This non self recognition although non specific is a highly effective response to pathogens. These PRRs expressed ubiquitously in all tissue cells (both immune active and non immune active cells) recognise well conserved pathogen associated molecules pattern (PAMP) and damage associated molecular pattern (DAMP). The individual members of receptor family identifies through their ligands specificity, cellular localisation, and activation of downstream signaling pathways. Crustacean cells possessing several family of PRR is a well developed strategy designed to react to a wide range of microrganisms. PAMPs include LPS of gram negative, PGN of gram positive, ds RNA of viruses and beta glucans (GLU) of fungi. Binding of PRR with specific PAMPS activate signal transduction pathways. Three major types of signal pathways: Toll pathways and immune deficiency(IMN)pathway against gram negative and JAK/STAT pathway in antiviral defense have been identified in shrimp.

Lectins

Lectins are proteins or glycoproteins which can bind to carbohydrates with or without catalytic activity. They are known to bring about agglutination of foreign cells, containing carbohydrate in their membrane. Lectins, by their abilities of recognizing the carbohydrate membrane surfaces, common to many of the pathogens, represent a very primitive immune response. In crustaceans lectins known to have at least two roles to play. Firstly, they can bring about agglutination of the foreign cells and secondly can help in adhesion of haemocytes to foreign cells and thus function as opsonins.

An important and essential feature of crustacean immunity is its ability to distinguish between non self and self particles. There are molecules mostly proteins/ group of proteins capable of recognising considered as precursors of antibodies. Lectins recognise specifically carbohydrate from membrane or surface of cells and induce agglutination of the cells or diverse cellular events such as phagocytosis acting as opsonins. Lectins have been identified in hemolymph, on hemocyte membrane and in cytoplasmic grannules.

Lectin could be produced in hemocytes, hepatopancreas, be expressed in several organs or tissues, including the hepatopancreas, muscle, eyestalk, and cuticle. Some of the expressed lectins remain on the membrane surface and participate in foreign recognition or in carbohydrate transport. Lectin inducible acute-phase proteins might be involved in immune responses to microbial infections and in virus resistance by inducing agglutination and opsonization processes. *In vitro* assays allowed identification of the carbohydrate specificity of crustacean lectins. The specific recognition of carbohydrate structures allows for the identification of several crustacean pathogens, which are agglutinated by cell free hemolymph or membrane lectins.

Prophenol Oxidase System (PPO System)

An important component of shrimp immunity, the PPO system is comparable to complement system of higher animals. A series of enzymes take part in the PPO system. Granules in the semigranular and granular hemocytes contain these enzymes, which upon exocytosis activate PPO enzyme system. Exocytosis (degranulation) is induced by beta 1,3, glucans of fungi, LPS endotoxins of gram negative bacteria, peptidoglycans of gram positive bacteria, other lipids detergents and damaged host tissue. In crustaceans, this enzymatic system could be activated by LPS or by b-glucans, leading to hemocyte degranulation. In hemocytes, the enzymatic precursor (proPO) is present as an inactive precursor; after its release to the hemocoel, the proPO is activated in the presence of Ca. The phenoloxidase (PO) catalyzes the reaction in which tyrosine is converted to DOPA and DOPA to DOPA-quinone, the precursor of melanin. Melanin is a brownish-black pigment with diverse biological properties, such as inhibition of fungal and bacterial enzymes activity.

Melanin is responsible for inhibition of the growth of microorganisms by suppressing the activity of extracellular proteinases and chitinases. In addition PPO system also increases phagocytosis, nodule formation, encapsulation and promote haemolymph clotting. Two endogenous proteins associated with PPO system are 76 kD protein and Beta –1,3-glucan binding protein. These proteins are multifunctional in nature and act as a cell adhesion factor for SG and G haemocytes, causing degranulation by a regulated exocytosis and thereby amplify further the PPO system. The defense mechanism in crustaceans like that of finfish gets weakened due to the stress such as accumulated metabolites poor water quality and nutritional deficiencies.

Clottable Protein

An efficient clotting system is very essential for crustaceans possessing open circulatory system to avoid excess loss of hemolymph during injury and preventing and spreading of pathogens. Clotting is mediated by both cellular and humoral components of hemolymph. Enzyme transglutaminase, stored in hemocyte, catalyse polymerisation of plasma clotting protein in presence of calcium and iron. Clotting proteins besides helping clotting, prevent hemolymph loss during injury and in recognition and neutralisation of non self cells and particles. Clottable proteins are found in the hemolymph plasma rather than in hemocytes, and the transglutaminase is released from hemocytes when Gram-negative bacteria invade the hemolymph and hemocytes detect bacterial LPS molecules on their surfaces through membrane specific LPS receptors.

Antimicrobial Peptides (AMPs)

These small molecular weight proteins have broad spectrum antimicrobial activity against bacteria, fungi, yeast and even protozoans. A large number of AMPs such as penaedins, lysozyme, crustin, antipolysaccaride factors are produced and stored by hemocytes. Penaedin PmPEN4 is identified as most powerful bactericidal AMP in shrimp. AMPs also promote phagocytosis of bacteria through opsonisation.

Penaedins comprise a family of peptides with antibacterial and antifungal activity, and are unique to penaeid shrimp. Penaeidins may be involved in local defense reactions through their release by hemocytes and then binding to shrimp cuticle surfaces. Through their antimicrobial activities against Gram-positive bacteria and fungi, penaeidins may protect tissues from infections and/ or participate in wound healing processes; they can also contribute to bacterial elimination by phagocytic cells through a potential opsonic function. The diversity, distribution, and abundance of penaeidins in shrimp reveal that this protein family constitutes a major component of the shrimp immune system.

Scavenger Receptors

Scavenger receptors (SRs)are located on cell membrane of crustaceans serving as PRRs. They are a large family of diverse transmembrane glycoproteins possessing ligand binding properties capable of recognising microorganisms as well as their components. Mainly SRs work as phagocytic receptors mediating direct non opsonic phagocytosis of pathogenss. Besides some SRs also involve in pro-inflammatory cytokine responses to various PAMPS. Based on their multidomain structure eight different classes of SRs (A-H) have been identified. Croquemort a class B SR considered a homologue of human CD36 has been identified , characterised and its responses against different microbes investigated in crustaceans. A scavenger receptor C (SRC) having antiviral function has been identified and elucidated in shrimp. In particular, a Mj SRC interacting with VP19 and serving as WSSV phagocytic receptor has been reported in *M japonicus*

Cell Adhesion Proteins

Wound healing in crustaceans is an important event, is dependent on cell adhesion proteins. Besides, this protein also has role in nodulation and encapsulation to sequester larger pathogens. Peroxinectin, a cell adhesion protein has a putative function for activating proP enzyme cascade. Antibacterial activity of Mas-like protein, a cell adhesion protein capable of binding *Vibrio harvey* and its LPS has been reported.

Hemocyanin

Hemocyanin synthesised by hepatopancreas being main protein component (95%) of hemolymph, mainly involve in transport of molecular oxygen. Hemocyanin is also reported to mediate wound healing and induce PO leading to AMP activity. It has different role against viral, bacterial and fungal infection. A hemocyanin protein complex binding WSSV is reported to delay the virus infection *in vivo.*

Interferon (IFN)

Shrimp also possess an interferon system-like antiviral regulatory mechanisms. A novel interferon regulatory factor (IRF) has been reported in shrimp which regulates the expression of Vago gene, an antiviral cytokine in arthropods, which activates the JAKSTAT pathway for limiting viral infection. Upon binding with PAMP such as ds RNA some Toll receptors can activate expression of IRF. DNA sensing pathways and its interaction with IFN activator through activation of IFN-b signaling pathways leading to antiviral response demonstrated in shrimp. Involvement of IFN activation in antibacterial immunity has also been shown in shrimp.

Antiviral Proteins

Several studies have demonstrated over expressed genes and proteins in response to viral infection. These proteins include virus recognition proteins, proteins involved in virus endocytosis, hemolymph associated proteins, transcriptive factors, cell signalling proteins, AMPs-penaedin, isoforms of crustin and ALF. However, the exact antiviral mechanism has not been demonstrated yet.

RNAi

RNAi are produced during replication of most positive sense RNA, ds RNA or DNA viruses. These gene fragments of size more than 30 base pairs act as RNA interference named RNAi. They are the key activation of post transcriptional gene silencing, possessing antiviral properties which has been demonstrated in shrimp.

b. Cellular Immune Responses

Crustaceans have three types of hemocytes namely hyaline cells, semigranulocytes (SG) and granulocytes (G). Hyaline cells, active phagocytes primarily responsible for phagocytosis, are agranular and are the smallest among three types of hemocytes. Semigranulocytes and granulocytes involves in encapsulation and nodule formation, play a major role in activation of prophenol oxidase system (PPO system). The semigranulocytes (SGCs), are the most abundant (up to 65%) cell type, contain many small eosinophilic granules. SGs are responsible for recognition of microorganisms leading to encapsulation, coagulation and occasional phagocytosis. Some cell adhesion proteins, such as peroxinectin, is localized in SGC. Encapsulation of microorganisms is accompanied by melanization due to proPO which is stored in an inactive form within granules of the GCs. GCs are known to contain AMPs, protease inhibitors and cell adhesion/degranulating factor called peroxinectin. The degranulated haemocytes are reinforced with new hemocytes in minimum time from the hematopoietic tissue.

i. Phagocytosis

Phagocytosis eliminates both miocrorganisms and apoptotic cells. The agranular hemocytes, the hyaline cells are mainly involved in pahagocytosis which begins with recognition and binding of target particles followed by their internalisation and intracellular vesicle formation. The phagocytosed particles are digested by enzymes and eliminated by burst of reactive oxygen or nitrogen. Phagocytes are found free in hemocoel or on the surface of arterioles of hepatopancreas and or gills.

Hyaline cells involve in phagocytosis following stimulation from peroxinectin and masquerade-like protein (mas-like protein) that are involved in cell adhesion and opsonic activities. The phagocytic process is regulated by many small GTPase-binding proteins (Rab), Ras related nuclear protein (Ran), ADP ribosylation factors(ARFs) and PRRs.

Phagocytosis is the immediate mode of cellular reaction the defense system exhibits when a foreign body enters the host after crossing the external barrier.

The phagocytic cell number in crustacean haemocytes vary from 2-28%. Opsonin, compounds that help in enhanced phagocytosis have been recorded in shrimps and other crustaceans. There are indirect evidences that the PPO system may provide factors that enhance phagocytic activity. Hayaline cells are primarily phagocytic whereas, the semi-granular haemocytes help in phagocytosis through the activation of PPO system.

ii. Encapsulation

Generally, when a parasite is too large to be engulped or phagocytosed, it is encapsulated by haemocytes and thus it is sealed off from circulation. Encapsulation is the most easily observed defense response in crustaceans. Semi-granular hemocytes are the first to react to a foreign particle or organism by encapsulation. It is reported that a haemolymph protein 76 KD, acts as a binding factor in the encapsulation process. Encapsulation is a multicellular response to eliminate foreign particles that cannot be destroyed by humoral mechanisms. This process kills pathogens or, at least, restricts their movement and growth in the hemocoel cavity. Invertebrate haemocytes are known to sequester nonself-molecules by either encapsulation or nodulation.

Semi-granular hemocyte aggregation and the presence of adhesive factors surrounding particles larger than 10 mm in diameter, such as helminths and fungal spores has been described. A similar reaction has been observed during damage in the cuticle. In this case, encapsulation works as a protective barrier, which prevents pathogens from entering the muscle or the hemocoel. It has been observed that a typical capsule consists of 5–30 compact layers of hemocytes, without intercellular spaces. The morphological aspect of the capsule can present a slight variation in cell extension, cellular necrosis, presence or absence of melanin, and the amount of extracellular material according to different crustacean species. Through histochemical analyses, it has been demonstrated that hemocytes, which participate in the encapsulating process, show acid or neutral mucopolysaccharides and glycoproteins.

iii. Nodule Formation

Nodule formation is known to occur in several invertebrates including the crustaceans when microbial invasion is far in excess of the phagocytic capabilities of the host. Microbes become entrapped in several layers of hemocytes. These nodules become heavily melanised because of the phenol oxidase activity. However, mechanism of clearance of entrapped microbes in the nodule is not clearly understood.

iv. Cytotoxicity

Cytotoxicity is the mechanism by which some cells of the haemocytes specialised to kill the target cells or organism by their toxic properties. Though importance of cytotoxicity is not established in crustaceans the process has been observed in cray fish although, the molecules responsible for guiding the cytotoxicity process in crustaceans remains to be isolated and characterized.

v. Apoptosis or programmed cell death

Apoptosis is an immune response against viral infection studied extensively in metazoans. In this process infected cells are killed during viral replication preventing further spread of virus infection. Several apoptosis related genes and large number of apoptosis inducing factors have been identified in different groups of crustaceans. In shrimp apoptosis mediated disruption or thwarting of WSSV and Taura syndrome virus has been reported. Interestingly, several viruses are also known to possess antiapoptoic proteins that can suppress or delay apoptosis which may help spreading virus to other healthy cells. WSSV antiapoptoic proteins include WSSV 449, WSSV 222, Vp 38, WSSV134 and WSSV 322. Therefore, apoptosis mechanism to disrupt virus infection is debatable.

References

Amod Kulkarni, Sreedharan Krishnan, Deepika Anand, Shyam Kokkattunivarthil Uthaman, Subhendu Kumar Otta, Indrani Karunasagar and Rajendran Kooloth Valappil1 2020, Immune responses and immunoprotection in crustaceans with special reference to shrimp Reviews in Aquaculture, 1–29

K. M. Shankar and C.V Mohan,2002, Fish and Shellfish Health Management , ISBN 81-7525-328-2, Fish Pathology and Biotechnology Laboratory, Dept of Aquaculture, UAS, College of Fisheries, Mangalore-575002

Lage Cerenius, Pikul Jiravanichpaisal, Hai-peng Liu3Lage Cerenius,, Pikul Jiravanichpaisal, Hai-peng Liu and Irene Söderhäl, 2010 Crustacean immunity, In Invertebrate Immunity, Editor Kenneth Söderhäll.©2010 Landes Bioscience and Springer Science+Business Media.

Lorena Vazque, Juan Alpuche, Guadalupe Maldonado, Concepcio´n Agundis, Ali Pereyra-Morales, Edgar Zenten, 2009, Review- Immunity mechanisms in crustaceans , Innate immunity 15(3) 179–188

Roberts, R. J. (2012). Fish Pathology: Fourth Edition. Fish Pathology: Fourth Edition, 1–581. https://doi.org/10.1002/9781118222942

Willem B. Van Muiswinkel1 and Brenda Vervorn-Van Der Wal1, The Immune System of Fish 2006,. Fish Diseases and Disorders Vol. 1 678 (ed. P.T.K. Woo) CAB International 2006

18

Prophylaxis in Aquaculture

M.N. Venugopal

Introduction

Aquaculture is given major thrust in many countries including India not only as a source of good quality food of animal origin but also because of economic gains it brings to aqua-farmers, creation of employment and export trade. It has emerged as one of the fastest growing food producing sector sowing to intensification and diversification of culture systems. However, the success of the intensive aquaculture system depends on several factors including species cultured, type of feed, water quality, management practices and strategies adopted to prevent/ control disease problems. Further, the intensive aquaculture operation results in accumulation of wastes leading to water quality deterioration, oxygen depletion, build-up of toxic metabolites such as hydrogen sulphide, methane, ammonia and nitrites that cause stress to culture animals leading to susceptibility to diseases. Among the several challenges that seriously affect the growth and sustainability of aquaculture, the incidence of microbial diseases are considered as major threat. Thus, keeping cultured aquatic animals in good health becomes major requirement for the success of aquaculture.

Though application of chemicals and drugs are often considered necessary to maintain water quality and animal health, use of these, in the long run are known to cause harmful effects both to the animal and the environment. Further, accumulation of drugs and chemicals in tissues, development of resistance to drugs by pathogens, pathogen build up, destruction of useful microorganisms, deterioration of pond environment and toxicity issues have limited their use in aquaculture system. Under these circumstances several alternative strategies have been proposed as viable option for the success of aquaculture in the form of prophylactic measuressuch as use of probiotics, prebiotics, synbiotics, immunostimulants and bacterial vaccines. The term prophylaxis refers to all preventive steps taken including better aquaculture management practices that are adopted during a hatchery and farming operation to improve growth, minimize the load of pathogens and prevent occurrence of diseases.

Probiotics

The term probiotic is derived from the Greek words 'Pro' (favour) and 'bios' (life) meaning 'for life'. Probiotics generally refer to include healthy microbiota associated with the host and the use of which will have beneficial effect on host. It includes organisms and substances which contribute to the intestinal microbial balance. The traditional knowledge of beneficial microbial cultures on humans was well documented and the same was extended for use in farm reared animals. Thus the probiotic can be defined as 'a live microbial feed supplement which beneficially affects the host animal by improving its intestinal microbial balance'. The concept of probiotics was introduced by Dr. Elie Metchnikoff, a Russian scientist in 1905 for the first time and used it to describe the positive role played by some bacteria among farmers who consumed pathogen containing milk.

The term probiotic was first used way back in 1965 by Lilly and Stillwell to describe 'substances secreted by one microorganism which stimulates the growth of another. 'The Joint Food and Agriculture Organization of the United Nations and World Health Organization (FAO/WHO) have defined probiotics as 'live microorganisms, which when consumed in adequate amounts confer a health benefit on the host' (FAO/WHO 2001). However, the application of probiotics in aquaculture is not limited to the intestinal tract, but also can improve the health of the host by controlling pathogens and improving water quality by modifying the microbial community composition of the water and sediment. Thus, to encompass various effects of probiotics when used in aquaculture, Verschuere et al (2000) have defined probiotics as a 'live microbial adjunct which has a beneficial effect on the host by modifying the host-associated or ambient microbial community by ensuring improved use of the feed or enhancing its nutritional value, enhancing the host response towards disease, or by improving the quality of its ambient environment.'

Gut Microbiota

The microbiota of gastrointestinal tract of fish and shellfish is dependent on ambient aquatic medium and in fact microbiota of gut is reflection of microflora of external environment. But, due to continuous flow of water and food, most of the gut associated microorganisms are transient in nature. It has been well established that by supplementing right kind of probiotics it is possible to maintain healthy gut microflora thereby improving the overall health status of the animal. Therefore, use of probiotic is merely for restoring desired gut microorganisms to their full protective capacity. Probiotics may contain one or several strains of microorganisms and may be presented to the animal in the form of powder (loose or capsule), tablets, granules or paste. It may be administered through food or water. In spite of careful selection of strains, it seems unlikely that it would be possible to establish permanently the probiotic organisms in intestinal tract. Hence to realize full potential of probiotic organisms multiple administration is required during the culture/ hatchery operation.

Mechanism of Action of Probiotics

Probiotics offer beneficial effect to the host in several ways. These include, creation of hostile environment for pathogens by producing inhibitory compounds (organic acids, antibiotic substances, bacteriocins, lysozymes, proteases, hydrogen peroxide, siderophores etc), competition with pathogens for essential nutrients, competition for adhesion sites, enhancement of immune response, production of digestive enzymes enabling feed utilization, providing growth promoting factors, neutralisation of bacterial toxins by metabolites produced by probiotic bacteria, and improvement of water quality.

Characteristics of Probiotics

The selection of right kind of probiotic bacteria is a fundamental requirement for aquaculture use. For a bacterial isolate to qualify as probiotic it has to possess certain characters. These include;

- Able to survive, colonize and proliferate in the host gut.
- Useful to host in growth promotion, food utilization and improved health.
- Safe to target and non-target animals at the site of application.
- Antagonistic activity to pathogens.
- Competitive exclusion of pathogens by competing for nutrients and adhesion sites.
- Able to utilize nutrients and substances of normal diet.
- Non –pathogenic to the host.
- Exert beneficial effect to the host.
- Absence of virulence and antibiotic resistant genes.
- Remain stable and active at the site of action and under any storage conditions.
- Able to be produced in large scale as viable product.

- Retain desired characters during processing, storage and transportation.
- Possess anti-inflammatory, anti-mutagenic and immunostimulatory properties.

The isolation and selection of probiotic bacteria is a challenging task. Several stages are involved before an isolate gets qualified as a probiotic. The stages involved are, selection of potential microorganism,and evaluation of microbe under *in-vitro* and *in-vivo* systems, standardization of dose, schedule and route of application, optimizing the production in bulk industrial scale, development of effective formulations and shelf life evaluation, and performance of the final product under different culture systems.

Use of Probiotics in Aquaculture

Realising their beneficial effects, the use of probiotics has become an integral part of present dayaquaculture practice. Feeding probiotics to cultured animals leads to improvement in intestinal tract health by means of regulation of microbiota, stimulation of immune system, synthesis and enhancement of bioavailability of nutrients, and reduced risk to certain diseases. Though the action of probiotics in aquaculture is considered to be similar to that of the mechanisms observed in terrestrial animals, the relationship of aquatic organisms with the culture environment is highly complex. The microorganisms present in the surrounding aquatic medium and those present in the ingested food have an easy access to digestive tract of aquatic animals. The pathogens of aquatic animals are generally opportunistic in nature and have potential to cause infection and subsequent mortality under stressed conditions caused due to high stocking density, poor nutrition and poor water quality. The effectiveness of the probiotic bacteria is known to be affected during dominance of native gut microbiota caused due to variations in water temperature, dissolved oxygen and salinity.

Use of probiotics in fish larval rearing system has demonstrated the establishment of probiotic organisms on egg surface and in gut resulting in significant increase in larval survival. In grow out systems, use of beneficial probiotic microorganisms has shown their establishment and production of antibacterial compounds that inhibit opportunistic pathogens. Several probiotic products are commercially available for use in fish and shrimp culture systems, however the results are not always consistent. Scientifically validated and certified probiotics need to be made available to protect the farmers from unscrupulous vendors selling spurious products. Further, instead of use of single strain, a mixed probiotic culture is known to be more effective may be because of synergistic effect, differences in their ability to establish in host gut environment and mechanism of action. A probiotic product can be considered successful and ready for field use only when it meets the economics of production on commercial scale and better shelf-life under varied environmental conditions.

The species of microrganisms normally used as probiotics and supplemented through feed are usually non pathogenic and are normal microflora of the host. The microorganisms used as probiotics in aquaculture are generally members of Gram-positive and Gram-negative bacteria and yeasts. Some of the microorganisms having probiotic potential include members of the genus Bacillus (*B. subtilis*, *B.cereus*, *B. pumilus*, *B. licheniformis*, *B. megaterium*, *B. polymyxa*, *B. coagulans*, *B.amyloliqifaciens)*,Pseudomonas (*P. aeruginosa*, *P.* synxantha),

Lactobacillus (*L.plantarum*, *L.lactis*, *L. acidophilus*, *L.casai*, *L.sporogenes*, *L. brevis*, *L. bulgaricus*, *L. plantarum*, *L. curvatus* , *L. reuterii)*,Streptococcus *(S. diacetylactis, S. thermophyllus)*, Bifidobacterium *(B. animalis, B. bifidum, B. longun, B. thermophyllum), Pediococcus (P. acidilacticii, P. cervisae, P. pentosaceus)*, Saccharomyces (*S. cerevisiae*, *S.boulardi*, *S.cerisia*, *S. faeium*, *S. intermedius*) and *Aspergillus*(*A.oryzae*, *A. niger).*

Prebiotics

The term prebiotics mainly refers to useful food ingredients that enhance intestinal commensal bacterial activity thereby resulting in improved health of the host. The concept of prebiotics was introduced for the first time by Gibson and Roberfroid in 1995 and they defined prebiotics as 'an indigestible fibre that can enhance the growth and activity of health-promoting bacteria in the intestine and beneficially affect the host'. Later others described prebiotics as 'non-digestible food ingredients that are broken down into simpler substances, enhancing the growth and activity of preferred microorganisms found in the gastrointestinal tract and benefiting the health of the host'.

To overcome the drawbacks associated with probiotic such as their inability to survive or grow during processing stages of probiotic based feed and reduced efficacy due to adverse conditions, use of prebiotics has emerged as an alternative strategy in aquaculture. The application of prebiotics in aquaculture is considered beneficial as it improves growth performance, feed utilization, enhances immune system and improves resistance to diseases. The beneficial effects of prebiotics are mainly due to fermentation or degradation of prebiotic substances by gut bacteria. Any compound, substrate, long chain sugar, nutrient or fibre which can serve as food for beneficial gut microorganisms are considered potential prebiotics. However, any compound to be considered as prebiotic has to meet criteria such as, must offer resistance to gastric/abdominal acidity, initiate various enzymatic processes, enhance gastrointestinal digestion, support microbial fermentation and stimulate the activity of desirable microbes that can confer health benefits. But, all prebiotics may not meet all these characters.

Prebiotics are mostly plant based products and are naturally found. The use of prebiotics is useful to aquaculture in many ways. It helps in the elimination of pathogenic microorganisms by preventing attachment to gut epithelial cells, modulation of host immune system, controls inflammation, and promotes growth and general wellbeing of host. Some of the compounds having potential as prebiotics include inulin, fructooligosaccharides (FOS), short-chain fructooligosaccharides (scFOS), mannan oligosaccharides (MOS), galactooligosaccharides (GOS), galactogluco-mannans (GGM), xylooligosaccharides (XOS), arabinoxylooligosaccharides (AXOS) and isomaltooligosaccharides (IMO). Prebiotics are metabolized by beneficial gut bacteria such as *Lactobacillus* and *Bifidobacterium* yielding short-chain fatty acids and lactate. The short chain fatty acids help in lowering colon pH thereby promote beneficial gut bacteria and prevent proliferation of pathogens. The gut bacteria inhibit pathogens also by competing with glycoconjugates on the epithelial lining that results in mucus production, increased short chain fatty acids and cytokines. Further, the use of prebiotics is known to enhance growth through improved weight gain, specific growth rate, feed conversion ratio, feed efficiency ratio and protein efficiency ratio.

Use of prebiotics in combination / mixture rather than singly is found to be more effective. Growth promoting effect was observed when FOS was used in combination with GOS, MOS, and GGM in red drum (*Sciaenopsocellatu*s), inulin along with FOS and GOS in *Litopenaeus vannamei* post larvae, and mixture of β-glucan , inulin and MOS in Chinese mitten crab (*Eriocheir sinensis*). A commercially available prebiotic mixtures (immunogen) containing β-glucan and MOS has shown growth promoting effect in common carp, Nile tilapia, Caspian trout, snake head, and Chinese mitten crab. The prebiotics are effective for increasing disease resistance in aquaculture. The prebiotic mixture containing β-glucan and MOS is reported to increase disease resistance of several cultured animals such as sea cucumber (*Apostichopus*) against *Vibrio splendidus*, Atlantic cod (*Gadusmorhua*) against *Vibrio anguillarum*, common carp (*Cyprinus carpio*) against *Aeromonas hydrophila*, juvenile tiger shrimp (*Penaeus monodon*) against White spot syndrome virus, Nile tilapia (*Oreochromisniloticus*) against *Yersinia ruckeri*, *Aeromonas hydrophila*, *Lactococcus gravieae and Pseudomonas fluorescens and Pseudomonas aeruginosa*.

Synbiotics

Synbiotics refers to simultaneous dietary supplementation of both probiotics and prebiotics which act synergistically on gastrointestinal tract. The nutrient present in prebiotics promote adherence, survival and growth of live microorganisms supplied as probiotics in the gut of cultured animals. Realising the advantages, both probiotics and prebiotics have gained importance for potential application in aquaculture for promoting growth and pathogen control.

For a suitable synbiotic formulations, the choice of prebiotics is based on their ability to promote growth and survival of beneficial bacteria in the gut while, probiotics are chosen depending on their overall beneficial effects on the host. However, prebiotic with low degree of polymerization are preferred for the development of synbiotics than the ones with higher degree of polymerization as they are easily hydrolysed by the beneficial strains which produce primary and secondary metabolites that are beneficial to the host. Further, administration of synbiotics is reported to be more effective than the use of probiotics alone, as it ensures better survival of probiotic organisms when used in combination with prebiotics. The effectiveness of synbiotic use in shrimp culture has emphasised its importance in promoting growth and survival caused due to secretion of pancreatic enzymes resulting in improved enzymatic digestion, modulation of intestinal microflora, reduced mortality rate due to enhanced immune response, and antagonism to pathogens.

Immunostimulants

The cultured aquatic animals are exposed to invasion by pathogens from the surrounding environment and combat pathogen attack by exhibiting a variety of immune responses. The fish and shellfish depend on non-specific or innate immune responses which function as first line of defense against invading pathogens. In this context, use of immunostimulants as dietary additives is known to enhance innate defense mechanism and there by increase resistance to specific pathogens.

The term immunostimulants refer to a group of biological and synthetic compounds that directly interact with non-specific cellular and humoral defense mechanism and enhance immune response. These may be administered alone as dietary additives to activate non-specific defense mechanisms as well as for elevating a specific immune response. Thus, an immunostimulant may be defined as an 'naturally occurring compound that modulates the immune system by increasing the host's resistance against diseases that in most circumstances are caused by pathogens. The immunostimulatory effects of various compounds studied has revealed their immune enhancing potential by elevating nonspecific immune responses in fish and shellfish. Use of immunostimulants thereby considerably improves growth, survival rates and resistance to disease in aquaculture systems. However, the immunostimulatory effect depends on the structure and function of different immunostimulants. Realising their immune enhancing potential, immunostimulants are considered as alternative for antibiotics and chemotherapeutics.

Mechanism of Action of Immunostimulants

Immunostimulants activate the immune system of animals and make them more resistant to microbial infections by enhancing non-specific immune factors. However, no memory component is developed and duration of the immune response is for over a short period thereby requiring priming of immune factors through multiple applications of immunostimulants. Their mechanism of action involves activation of immune components such as phagocytic cells natural killer cells, complement, lysozyme, activity of T cells and B cells inflammatory agents and activity of macrophages. In case of invertebrates such as shrimps, immunostimulants help in enhancing the phagocyte activity phenol oxidase activity, sodium dismutase activity, total haemocyte count respiratory burst activity, etc. Thus, immunostimulants are considered as useful alternatives for commonly used antibiotics and chemotherapeuticsin controlling disease problems in aquaculture.

Immunostimulants use in Aquaculture

The immunostimulants that have shown potential for use in aquaculture are included under different categories based on their source as synthetic chemicals, bacterial derivatives, animal and plant extracts, nutritional factors antimicrobial components and nucleic acids. Use of these different types of immunostimulants is considered as an effective means to increase the immune competency and disease resistance of fish and shellfish. Particularly immunostimulants are very useful to boost immunity of fish and shellfish which depend largely on non specific immunity.

Synthetic chemicals: Among the synthetic chemicals, levamisole immunoactive peptide(FK-565) and muramyl dipeptide (MDP) have shown proven immunostimulatory activity.

Levamisole

Levamisole is a synthetic phenylimidathiazole compound used as antihelmenthic drug for the treatment of nematodes in humans and animals. This has been approved for use by the US Food and Drug Administration

(FDA) for the treatment of helminths infection. In aquaculture, levamisole has shown considerable immunostimulatory effect against bacterial and parasitic infestations. This compound activates non-specific defence mechanism by enhancing the cell mediated cytotoxicity, lymphokine production, suppression of cell function and stimulation of phagocytic activity of macrophages and neutrophils. Levimasole use in carp culture showed upregulation of non-specific immune response mediated through the enhancement of phagocytic activity, myeloperoxidase activity in neutrophils, increase in leukocytes number and serum lysozyme levels. Application of levamisole through oral and immersion methods has resulted in improved resistance to diseases.

Immunoactive Peptide (FK-565)

FK-565 is chemically 'heptanoyl-y-glutamyl(L)meso-diaminopimelyl(D)-alanine'related to lactoyltetrapeptide (FK-156) isolated from cultures of *Streptomyces olivaceogriseus*. It was found effective against microbial infection and by activating the phagocytes it induced resistance in rainbow trout against *Aeromons salmonicida* infection.

Muramyl Dipeptide (MDP)

Muramyl dipeptide (MDP) is derived from Mycobacterium and chemically it is 'N-acetylmuramyl-L-alanyl-D-isoglutamine. MDP activates immune system by enhancing the activity of macrophages, B lymphocytes and alternative pathway of complement. Use of MDP through injection showed increase in phagocytic activities, respiratory burst and migration activities of kidney leucocytes as well as resistance to *A. salmonicida* challenge in coho salmon and rainbow trout.

Glucan

Glucan is a high molecular weight polysachharide derived from yeasts having a chemical name 'peptide-glucan-ß-1,3, glucan'. Immunostimulatory effect of glucan is due to their ability to enhance lysozyme and macrophage activity. Treating fish and shellfish with glucan has shown increased phagocytic activity and protection against some bacterial pathogens.Use of glucans in fishes offered increased protection against *Vibrio anguillarum*, *V. salmonicida* and furunculosis.

Lipopolysaccharide (LPS)

LPS forms the major component of Gram negative bacterial cell wall and the immunostimulatory effect is due to its ability to stimulate B cell proliferation and enhance phagocytic activity of macrophages. It increases disease resistance and acts as a prophylactic agent when used in low doses. Feeding LPS incorporated feed to shrimps showed higher survival rate when challenged with *Vibrio harveyi*. Stimulation of macrophage activating factor and the production of interleukin-1 like molecules was observed in goldfish and catfish upon LPS administration.

Freund's Complete Adjuvant (FCA)

This bacteria based immunostimulant is a mineral oil adjuvant containing killed cells of *Mycobacterium butyricum*. FCA improves immune responses and enhances the efficacy of vaccination in fish. It has shown activation of macrophages in fish and increase in respiratory burst, phagocytic and NK cell activity of leucocytes and protection against *V. anguillarium* infection. FCA use has resulted in increased protection in Coho salmon challenged with *Aeromonas salmonicida*, *A. hydrophila* and *Vibrio oradalii*.

Vibrio Bacterin

Vibrio bacterin is a suspension of killed or attenuated cells of pathogenic vibrios having immunostimulatory effect. Administration of vibrio bacterin to fish as well as shrimps has resulted in increased protection against vibrio infection.

Chitin and Chitosan

These animal derived polysaccharides have immunostimulatory effect on a short time basis and play an important role as immunostimulants in aquaculture. Chitin is one of the most abundant polysaccharide in nature and forms a major component of exoskeleton of insects and crustaceans and cell wall of fungi. Chitin activates defense system by increasing macrophage activities, haemolytic complement activity, leukocyte respiratory burst activity and cytotoxicity. Chitin supplementation showed increased resistance to *A. salmonicida* in brook trout, *V. anguillarum* in rainbow trout and *Pseudomonas piscicida* in yellowtail. Chitin use resulted in increased growth and survival rates in *Macrobrachium rosenbergii* culture and improved disease resistance in shrimps against *V. alginolyticus*. Chitosan is a deacetylation product of chitin and a linear homopolymer of ß-(1, 4)-2-amino-deoxy-D-glucose. Chitosan as an immunostimulant is used in aquaculture for providing protection against bacterial diseases sin fish and shrimps and for controlled release of vaccines. Chitosan treatment enhances the phagocytic activity, nitrobluetetrazolium test (NBT) values, myeloperoxidase, lysozyme activity, neutrophil activity and Ig concentration.

Extracts from Marine Invertebrates

Extracts from some of the marine invertebrates are known to have immunomodulatory effects. Animal extract (Ete) from marine tunicate (*Ecteinascida turbinata*) and glycoprotein fraction from water extract(Hde) of marine gastropod called abalone (*Haliotis discus hannai*) show immunostimulatory effects leading to enhanced activity of phagocytes and natural killer (NK) cells, antitumor activity and also increased survival of eels against *A. hydrophila* infection. The heat extract from firefly squid (*Watasenia scintillans)* is immunostimulatory in nature and its effect on immune system is by stimulating the production of superoxide anion, potential killing activities by macrophages and the lymphoblastic transformation of lymphocytes.

Plant Extracts

Several plant extracts used successfully as immunostimulants in aquaculture have shown improved immune response and reduced disease problems. Some of the herbal extracts include from plants such as *Ocimumsanctum* (holy basil) *Emblica officinalis* (Indian gooseberry), *Cynodon dactylon* (Bermuda grass) *Adathoda vasica* (Malabar nut), *Allium sativum* (Garlic), *Zingiber officinale* (Ginger), *Nigella sativa* (balckcuminseed*),Echinacea purpurea* (purple cornflower), *Astragalus membranaceus* (Mongolian milkvetch) and *Lonicera japonica* (Japanese honeysuckle).

Use of garlic extract enhanced immune response and also improved overall health, survival rate, resistance against *Aeromonas hydrophila* and shelf life of cultured tilapia. Ginger extract showed significant increase in growth, feed conversion, protein conversion efficiency, protection against *Aeromonas hydrophila* infection and also proliferation of neutrophils, macrophages and lymphocytes and enhanced phagocytic, respiratory burst, lysozyme, bactericidal and antiprotease activities in rainbow trout. The extracts of Echinacea on Nile tilapia had improved growth rate, disease resistance, and increase in body weight, hematocrit values, lysozyme activity and total leukocyte counts. Methanol extracts of holy basil is reported to have significant improvement in phagocytic activity, serum bactericidal activity, albumin– globulin (A/G) ratio, and leukocrit incultured *Epinephelu stauvina* against *Vibrio harveyi* infection. Feeding tilapia with Mongolian and Japanese honeysuckle medicinal plant extracts in combination or alone significantly enhanced the phagocytic activity. The use of black cumin seeds as immunostimulants in tilapia improved health condition, growth and immune response against bacterial infection.

Glycyrrhizin is a sweet and aromatic flavouring compound obtained from the root extracts of a dicot flowering plant, *Glycyrrhizaglabra*, commonly called licorice. It is a glycosylated saponin containing one molecule of glycyrretinic acid. Glycyrrhizin has immune enhancement effects by enhancing the respiratory burst activity of macrophages and proliferative responses of lymphocytes in rainbow trout. It is also known to have anti-inflammatory and anti-tumour activities. Oral feeding of this compound to yellowtail showed increased protection against *E. seriola* infection. *In vitro* treatment of rainbow trout with glycyrrhizin enhanced respiratory burst activity of macrophages and proliferative responses of lymphocytes.

Nutritional Factors

Several nutritional factors such as vitamin C and E, carotenoids and trace elements involve in physiological functions including functioning of immune system. Vitamin C (ascorbic acid) plays an important role in several physiological functions including growth, resistant to infections, wound healing and response to stressors in most fishes. As an immunostimulant it involves in the activation of phagocytic cells and myeloperoxidase and inactivation of free radicals produced by normal cellular activity and various stressors. Using Vitamin C enhanced antibody production and complement activity in channel catfish and rainbow trout. The Indian major carps showed enhanced inflammatory response and disease resistance when supplemented with Vitamin C. High levels of vitamin C demonstrated increased resistance of channel catfish against *Edwardsiella tarda* and *E. ictaluri*, rainbow trout against *V. anguillarum*, infectious hematopoietic necrosis (IHN) virus and *Ichthyophthrius multifilis* infection and also showed enhanced complement activity in catfish and Atlantic salmon.

Vitamin E (Tocopherol): This nutritional factor plays important role in cell membrane structure and stability and found useful in enhancing humoral and cellular defences in mammals. It can enhance specific and cell-mediated immunity against infection in Japanese Flounder and macrophage phagocytosis in channel catfish, turbo and rainbow trout. Deficiency in Vitamin E can result in reduced protection against *Yersinia ruckeri* infection in fish.

Carotenoids

Carotenoids are naturally occurring coloured pigments and play an important role as antioxidants. Their role in immune response of animals is well established. In fish supplementing carotenoids in combination with Vitamin A, C, and E enhanced the complement, lysozyme, and phagocytic activities. Increased resistance to stress, salinity shock and improved antioxidant response was observed in shrimps fed with carotenoids.

Trace Elements

Trace elements are essential for vital body functions including metabolic functions, tissue repairs, growth and development. Supplementation of trace elements such as zinc, selenium and copper through feed showed improved immune response, growth and protection against infectious diseases in fish and shrimp.

Hormones

Hormones are involved in the regulation of neuroendocrine functions and thereby directly affect immunocompetent cells such as macrophages, lymphocytes and natural killer (NK) cells. Growth hormone (somatotropin) is a peptide hormone that regulates several physiological processes such as metabolism and growth and helps maintain tissues and organs throughout life in all organisms. Growth hormone also function as an immunostimulant in fishes. GH supplementation has resulted in increased protection against *V. anguillarum* infectionin rainbow trout. Use of recombinant bovine growth hormone was found to be effective in enhancing growth and immunity in shrimp larvae,

Lactoferrin (LF)

Lactoferrin in is a glycoprotein that exhibits antimicrobial activities thereby affects the proliferation of various pathogens such as bacteria, viruses, fungi and protozoa. Lactoferrin also regulates free radicals production by macrophages, granulocytes and neutrophil leucocytes. LF enhanced phagocytic activity and increased production of superoxide anion by macrophages demonstrated high resistance to *V. anguillarum* infection and elevated secretion of body mucus and resistance to Cryptocaryon infections in fish. Dietary supplementation of bovine lactoferrin to crustaceans enhanced their immune activity

Nucleic acids

Nucleotides that form the building blocks of RNA and DNA play vital roles in various physiological and biochemical functions including encoding genetic information, mediating energy metabolism and signal transudation. Besides enzymes and cofactors such as ATP, NAD and FAD also contain nucleotides as important constituent. Dietary nucleotides also show nitrogen sparing effect when absorbed nucleotides spare amino acids for use as precursors in nucleic acid synthesis. Besides their varied functions, nucleotides also have attained importance in recent years as immunostimulnts. In mammals, during early infancy stages nucleotides are considered semi-essential as some of the epithelial cells of the intestine or blood cells fail to synthesise *de novo* nucleotides and must derive them from other organs or from other dietary sources to meet their requirements. Inclusion of nucleotides in the diet of mammals is known to enhance cell mediated immunity, lymphocyte proliferation, interlukin-2 production and improve host resistance to bacterial infections.

Several studies on the use of nucleotides for shrimp and fish have also indicated their beneficial effectson immune functions and disease resistance. Immunostimulatory effects of dietary nucleotides showing increased resistance against stress and increased resistance towards pathogens has been reported in Atlantic salmon, coho salmon, rainbow trout, common carp, European sea bass and hybrid striped bass. However, similar beneficial effects of nucleotides have not been demonstrated in shrimps.

CpG oligodeoxynucleotides are short, single-stranded synthetic DNA molecules and known to exhibit immunostimulatory effect. These contain cytosine triphosphate deoxynucleotide 'C' followed by a guanine triphosphate deoxynucleotide 'G' with phosphodiester 'p' linking the two nucleotides. The CpG motifs are more prevalent in bacterial DNA and considered as pathogen-associated molecular patterns (PAMPs). When unmethylated, the CpG motifs are reported to function as immunostimulants.

Administration of Immunostimulants

The frequent outbreaks of diseases in culture and hatchery systems has necessitated the increased use of immunostimulants for rendering enhanced protection under rearing conditions. Immunostimulants administered as dietary supplement alone or together with vaccines has emerged as one of the promising approaches to prevent or control diseases in aquaculture. Administering immunostimulants and vaccines at the same time prior to the exposure of fish to pathogen is more advantageous as it helps the animal to prepare itself resist pathogen effectively. However, for the immunostimulant to be effective, factors such as timing, dosage, method of administration and health status of the animal needs to be considered.

Timing of immunostimulant administration is considered very important. Application of immunostimulants before the outbreak of diseases helps reduce disease related losses.

The route or method of administration determines the effectiveness of imunostimulant. Irrespective of the mode of application, the intended effect is achieved only when the right quantity of immunustmulant enters the animals to bring about immune response. Although, oral administration is found to be most practical method for delivery of immunostimulants, application through injection and immersion mode have also been found effective. Although administration of immunostimulants orally through feed is more practical as it is non-stressful and applicable to fish of all sizes, it requires to be fed at high dose to compensate for loss due to leaching and unconsumed feed. Administration through injection is cost effective method for larger and high value fish/ shrimp. For smaller fish / shrimp especially during hatchery rearing conditions immersion mode becomes very practical and effective. However, both injection and immersion methods cause stress to animals, time consuming and labour intensive.

Activation of immune response is mainly determined by whether the right dose of immunostimulant has been received by the animal or not. To be effective, required dose or quantity should enter the animal body. Both higher and lower dose may not prove effective as overdose may cause immunosuppression while, low doses fail to elicit immune response. Further, different culture systems and their feeding regime are known to complicate the effective dosage and exposure time.

Vaccines

Vaccines are considered as an effective, easy and preventive approach for controlling disease problems in aquaculture. Vaccine is any biological preparation containing antigens derived from pathogenic organisms which are rendered non-pathogenic through various means to establish or to improve immunity against particular disease. Vaccines have been developed and used for many years against various diseases in humans and animals.The functioning of vaccine is based on the principle of exposing the immune system of an animal to an antigen that may contain a part or whole pathogen and thereby rendering immune system to develop response and a memory component to enhance immune response during later infections by the targeted disease causing organism.

It is well established that in vertebrates, survivors of a pathogen infection exhibit resistance to subsequent infection by the same pathogen. This resistance called adaptive immunity is specific to the challenging pathogen and persists for a relatively longer period of time due to the involvement of memory component of immune system. It is because of adaptive changes in lymphocyte population of the host animal resulting from exposure to foreign molecules constituting the pathogen (antigen). Thus, specificity and memory are the two key elements of the adaptive immune system exploited in vaccination since adaptive immune response is stronger on secondary encounter with antigen. Vaccines as preparations of antigens derived from pathogenic organism are normally administered to healthy animals prior to a disease outbreak.

The Concept of Vaccination in Aquaculture

The concept of immunizing the fish using vaccines is possible only when it provokes a protective immune response in the host against the pathogen involved. The specific or adaptive immune response is mediated through humoral immunity (antibody production) and cell mediated immunity. Fishes beyond certain size/ age threshold are capable of mounting both antibody mediated and cellular immune responses. Thus, protecting the fishes from pathogen infection requires eliciting these immune responses. In this regard immunizing the fishes using antigens or a vaccine results in stimulation of lymphocytes responsible for antibody production or cell mediated immunity. The protective immune response elicited by use of vaccines generally persists for a long time by virtue of involvement of memory cells.

Though precise mechanisms involved in protective immunity in fish due to vaccine application are not clearly understood, several observations have indicated the possibility of protection against bacterial diseases by immunization. Vaccines targeting several bacterial, viral pathogens and parasites have been either commercialized or developed. Commercially available vaccines include those for infections caused by *Yersinia ruckeri* and *Vibrio anguillarum*. Vaccines against *Aeromonas salmonicida*, *Aeromonas hydrophila*, *Pseudomonas anguilliseptica* and *Renibacterium salmoninarum* have been claimed to be effective by laboratory challenge studies. Similarly several vaccines against viral pathogens of fish such as IPNV and IHNV have been developed. However, for various reasons, results with viral vaccines are not that encouraging as in the case of bacterial vaccines.

Immunizing Fish

Vaccination is an easy, effective and preventive method of protecting fish from disease. Vaccination is possible, if it is possible to provoke a protective immune response in the host against the pathogen involved. Fish, over a certain size/age threshold, are capable of mounting both antibody mediated and cellular immune responses. The specific or adaptive immune response has two arms; humoral immunity (antibody production) and cell mediated immunity (CMI). Exposure to an antigen or a vaccine results in stimulation of lymphocytes responsible for antibody production or CMI. Vaccine work by inducing protective immune response which by virtue of memory cells persist for a long time. It is establishment of memory state to which the term immunity refers. The role of these responses in the protection of fish against pathogenic microorganisms is in most cases not yet clear. For example, there appears to be no direct relationship between the antibody response of rainbow trout (*Salmo gairdneri*) to antigens derived from the bacterium *Vibrio anguillarum*, as measured by bacterial agglutination, passive haemagglutination or complement fixation, and the capacity of the fish to resist challenge. Likewise, the role of cellular immunity is not understood clearly.

In spite of the lack of information on the precise mechanisms of protective immunity in fish, it has been shown on numerous occasions that fish can be protected against bacterial diseases by immunization. Commercial vaccines to combat infections by *Yersinia ruckeri* and *Vibrio anguillarum* are available and experimental vaccines against *Aeromonas salmonicida, Aeromonas hydrophila, Pseudomonas anguilliseptica and Renibacterium salmoninarum* have all been claimed to be effective against laboratory challenge. Similarly several vaccines against viral pathogens of fish such as IPNV and IHNV have been developed. However, for various reasons, results with viral vaccines are not that encouraging as in the case of bacterial vaccines. In recent years attempts are also underway to develop vaccines against some of the important fish parasites.

Properties of an Ideal Vaccine

An ideal vaccine is expected to have following properties;

- Safe to fish, persons vaccinating the fish and the consumers without causing adverse reactions.
- Ensure protection for long time preferably for entire production cycle.
- Able to induce cell mediated, humoral and mucosal immunity.
- Effective against all strains and serotypes of pathogens.
- Effective in variety of fish species.
- Ease of administration to host.
- Commercially producible and economical.
- Possess reasonably long shelf life.
- Pose no problem for licensing and registration.

Vaccine Types

Vaccines used in aquaculture are of different types. They can be grouped as follows;

Killed Vaccine

Killed vaccines or bacterins are whole cell preparations of pathogenic microrganinsms (bacteria and virus) which are made avirulent or non-pathogenic by heat or chemical treatment. Such killed bacteria lose ability to cause infection but capable of eliciting immune response against antigenic determinants present in the vaccine preparation. These are easy to prepare and economical to use. However, killed vaccines have certain limitations. These include, possible destruction of productive antigens during inactivation occurrence of unwanted side-effects and antigenic competition because of presence of complex mixture of microbial products, and failure to induce correct type of immune response (humoral or local antibody, or cell-mediated). Thus, the vaccines produced by this approach are often far from perfect, and there is considerable interest in finding new alternatives. However, using this simple approach successful fish vaccines have been developed for commercial use against vibriosis and enteric red mouth disease.

Bacterial Biofilm Cells as Novel Oral Vaccine

Oral vaccines are the most preferred in commercial aquaculture. However, the currently used bacterial free cell oral vaccines provide poor and inconsistent immune response and protection to fish mainly due to their destruction in stomach/foregut before reaching the immune responsive areas of hind gut and lymphoid organs. This is also true with oral vaccine used in higher animals. In this context, biofilm cells of bacterial pathogens offer better alternative to free cell. Bacterial biofilm is unique characterized by high density of cells per unit area protected by a glycocalyx matrix/coat. Biofilm due to the glycocalyx coat is highly resistant to antibiotics/chemicals and this property has been explored to develop a novel gut destruction resistant oral vaccine for fish. As a model, biofilm of *Aeromonas hydrophila* - a pathogen of carp was developed and employed for oral vaccination of Indian major carps has given promising result with significant increase antibody titre and protection upon challenge for an extended duration compared to that with free cell vaccine. The better performance of the vaccine was reflected by longer retention of biofilm cells antigen in gut, spleen and kidney demonstrated by monoclonal antibody based immunofluorescence. In aquaculture herbivores, omnivores and carnivore fishes are cultivated and their gut enzyme and pH profiles are different. Against this background, biofilm vaccine of *A hydrophila* evaluated in herbivores(carps) omnivore (*Clarius batrachus*) and carnivore (*Channa marulius*, sea bass *Lates calcarifer*) provided higher antibody titre and protection upon challenge.

Besides whole cell vaccine, bacterial biofilm can also serve for oral delivery of recombinant DNA (sub unit) vaccines. Oral delivery of sub unit vaccine is often not effective due to their easy destruction in the gut and therefore delivery through biofilm could be promising. Biofilm cells of *E coli* transformed with sub unit of *A. hydrophila* expressed the sub unit indicating the potential for effective antigen delivery.

Furthermore, biofilm antigens as vaccine candidates have edge over that of free cells as they are considered mirror images of a pathogen *in vivo.* Antigen expression in biofilm is different compared to free cells of *A. hydrophila* and *Vibrio*. In a study on the antigen expression, biofilm of *A. hydrophila* showed 3 additional proteins, with repression of nine proteins, while biofilm of *V.alginolyticus* and *V.anguillarum*, each had three additional proteins and repression of four and ten proteins, respectively. And, in *A. hydrophila* the S layer protein was lost, while lipopolysaccharide possessed additional high molecular weight protein, which is known to improve protection. Earlier studies have shown a difference of 30–40% in proteins between biofilm and free planktonic cells. Normally a pathogen thrives in a host in protective biofilm mode. Therefore, antigen expression in biofilm mode is the mirror image of antigen expression *in vivo* and, hence, biofilm cells are is more appropriate as a vaccine antigen.

Live Attenuated Vaccines

These are attenuated or weakened form of disease causing microorganisms. Because of their avirulent nature live vaccines do not cause any infection but create situation similar to natural infection and stimulate a strong and long lasting immune response. Live vaccines are capable of stimulating both cell mediated and humoral immune responses.Attenuated vaccines are prepared by repeated *in vitro* culture or by the use of mutagens. Vaccines of this type are considered highly successful as it involves delivery of important protective antigen through *in vivo* microbial replication to the appropriate site in the form needed to stimulate the correct immune response. Although considered safe and superior, the major drawback with live attenuated vaccines is the possibility of reversion to a virulent form. Live vaccines are suitable for oral administration along with feed as it is economical, easy and practical approach to vaccinate fishes of all sizes. The live attenuated vaccines have been licensed for administration against bacterial diseases such as enteric septicemia of catfish disease, bacterial kidney disease and columnaris disease. The vaccine against koi herpesvirus (KHV) has also been commercialized.

Toxoids

Toxoids are altered or weakened form of exotoxins secreted by pathogenic bacteria. These heat or chemically treated toxins lose their disease causing ability but retain their antigenic potential. Thus, use of toxoid as vaccines is helpful in stimulating a protective immune response against pathogen.

Subunit Vaccines

Subunit vaccines comprise of a small antigenic portion of a pathogenic microorganism rather than the entire microorganism usually a particular protein or a peptide that can ideally stimulate an immune response in host against the entire pathogen. The subunit vaccines are advantageous as they are unable to replicate in host and pose no risk of pathogenicity to the host or non-target species.

The subunit vaccine can be produced by several approaches. Production of subunit vaccine involves isolation and purification of immunogenic component or specific proteins using various recombinant expression vectors and harvesting the potential antigen at the end of growth cycle. Subunit vaccines have been produced containing immunogenic proteins for bacterial and viral pathogens using prokaryotic as well as eukaryotic cellular systems. An *Escherichia coli* expression system is commonly been used for the propagation of plasmids carrying genes that encode a specific protective antigen. The expressed antigenic proteins are harvested and purified to use as vaccines. In spite of their ability to elicit specific immune response, often the subunit vaccines exhibit weaker immune responses than killed or live cell preparations. This is attributed to presence of limited number of antigenic components capable of stimulating an immune system when compared to multiple antigens represented in whole cell vaccine. Since the subunit vaccines as such lack immunogenicity they need to be administered with effective adjuvants and also to ensure long term protective immunity multiple booster immunizations are required. Subunit vaccines against several pathogens such as nodavirus rhabdovirus, birnavirus, infectious hematopoietic necrosis virus (IHNV) infectious pancreatic necrosis virus (IPNV), and viral hemorrhagic septicemia virus (VHS)have been successfully demonstrated. A subunit vaccine (peptide; VP2) against infectious pancreatic necrosis virus has been commercialized in Norway. Subunit vaccine development is being considered as a viable option to control and prevent several bacterial and viral pathogens in aquaculture.

DNA Vaccines

DNA vaccines or recombinant DNA vaccines are composed of circular portion of genetic material (gene) responsible for expression of a immune stimulating protein of a pathogen in the host continuously thereby providing protection against pathogen. DNA vaccine production involves inserting bacterial plasmids carrying the gene coding for the immunogenic protein under the control of eukaryotic promoter, an origin of replication site suitable for producing high yields of plasmid in *E. coli*, an antibiotic-resistant gene to permit growth of *E. coli* in presence of antibiotic, a strong enhancer/promoter and an mRNA transcript termination/polyadenylation sequence for directing expression in mammalian cells. The plasmids thus constructed are grown in *E. coli*, purified and suspended in saline and used for vaccinating the fish.

Administration of DNA vaccines is capable of inducing specific as well as non-specific immune response in the recipient host. Under clinical trials a high level of protection was observed due to the generation of specific antibodies and priming of T-cell responses. Following the injection of construct encoding the IHNV G protein a significant protection was observed against IHNV challenge in rainbow trout. Using this approach, the pathogen can be effectively targeted and destroyed by inserting a gene coding for an antibody instead of introducing part of the genome of pathogen coding for immunogenic protein.

DNA vaccines have been found to be more effective compared to other vaccine types. These are capable of inducing long lasting immune response and are economical and safe to use. Since injection is the most effective mode for vaccine delivery, young fishes can't be protected from pathogens during early stages because of difficulty in handling and injecting. However, brood stocks and fairly large fish can be easily immunized using DNA vaccines.

Routes of Vaccine Administration

In aquaculture, the immune response of aquatic animals and the extent of protection offered by the vaccines are greatly influenced by the method of vaccine delivery. Generallythe vaccine can be administered to culture animals (fish and shrimps) using three approaches such as oral, immersion and injection routes. A method of choice for vaccine administration is decided based on the pathogen and its natural route of infection, the life stage of fish to be vaccinated, vaccine production technique, and other practical considerations. However, to ensure adequate protection, a specific route of vaccine delivery or even multiple applications using different methods may be required.

Oral Administration

Oral vaccination involves delivery of antigen directly to the digestive system through feed. The vaccine is incorporated to the feed while preparing the feed or coated on prepared feed usually just before feeding. Oral route of vaccine delivery is the preferred method for eliciting immune response to entire pathogens. It has advantages over other modes of vaccine administration in that it is simple to perform, non-stressful, can be used to vaccinate fish of all sizes, and requires no extra labour or time. However, it has few limitations such as need for administering large dose of antigen to obtain effective immunity, higher antigen requirement than that needed in other modes of vaccine administration, elicit poor immune response lasting only for short time thus requiring multiple vaccination, and loss due to leaching from feed. The poor response is believed to be due to poor antigen delivery to immune responsive sites. Destruction of vaccine in foregut due to low pH and high enzyme activity causes poor antigen delivery to hind gut and other lymphoid organs of fish.

To reduce loss of vaccine due to leaching into the water and/or to provide some protection against breakdown of the vaccine by the fish's digestive processes a coating agent is often used. In this regard, possibility of oral delivery of bioencapsulated as well as nanoparticle coated vaccine have also been demonstrated and were found very effective. Bioencapsulation is the preferred method for oral delivery of vaccine to small fish of less than 5 gram and it involves adding live foods such as rotifers and brine shrimp to concentrated solution of vaccine, allowing them to take up vaccine and feeding this live food to small fishes (fry and fingerlings). Nanoparticles may be natural or synthetic polymers characterised by their uniform size and are found to be more efficient in antigen delivery. Some of the compounds used in oral vaccination include particles like alginate, chitosan, and polylactic co-glycolic acid (PGLA). Studies have indicated successful vaccination using chitosan nanoparticles. Chitosan being natural polysaccharide derived from crustaceans is non-toxic biodegradable and with their mucoadhesive property adhere well to skin and mucosal surface thereby enable effective antigen uptake.

Oral Vaccination with Bacterial Biofilm Cells

Oral vaccination is most ideal for aquaculture to immunise economically a large number of fishes of different age and size with least handling stress. Biofilm of *A hydrophila* - developed and evaluated for oral vaccination of Indian major carps has provided significant increase antibody titre and protection upon challenge for an extended duration compared to that with free cell vaccine. Biofilm cells have to be inactivated/killed by heat or formalin before incorporation in feed for safe oral vaccination. However, due to glycocalyx, inactivation of biofilm cells require higher temperature or formalin concentration compared to that for inactivation of free cells.

Besides appropriate antigen delivery for eliciting serum antibody titre, impact of biofilm oral vaccine on gut associated lymphoid tissue (GALT) and mucous antibody has been demonstrated which shows its potential for boosting the gut immunity for preventing enteric diseases in aquaculture. Developing an effective oral vaccine that resists gastric digestion and facilitate enhanced uptake for presentation of antigen to gut associated lymphoid tissue at the mucosal layer is a challenge. Enteropathogenic agents invade mucosal surface of gut and, therefore, mucosal immunity is very important. Leucocytes are extensively present in fish intestine where lymphocytes, plasma cells, macrophages and granulocytes are present under the epithelium. Common carp fed with *A. hydrophila* biofilm-incorporated feed at 10^{10} CFUg^{-1} fish day^{-1} for 10 days had higher antibody response in both gut mucosa and blood and the immune related genes were upregulated compared to control fish.

Bacterial biofilm also has potential for oral delivery of immunostimulant in shrimp culture. Oral route is reported to be the most beneficial for the administration of immunostimulants to shrimp. However, gut destruction of immunostimulants administered orally is a common setback in shrimp leading to short lasting poor response. Biofilm of *Vibrio harvei* as oral immunostimulant in *P monodon* has been demonstrated successfully with protection against the bacterium and against White spot virus. Immune response in *Penaeus monodon* fed with biofilm (BF) and free cells (FC) of *V. alginolyticus* 10^9CFU g^{-1} shrimp day^{-1} for two weeks had higher haemocyte count, phenoloxidase activity and antibacterial activity compared to FC-fed or control shrimp. BF-fed shrimp were more resistant to injection challenge with *V. alginolyticus* and white spot syndrome virus (WSSV) with significantly higher RPS compared to FC-fed and control shrimp. Better resistance was also reflected by rapid clearance of *V. alginolyticus* and WSSV from the haemolymph of biofilm-fed shrimp.

Immersion Vaccination

Immersion vaccination procedure involves immersing animal to be vaccinated in solutions containing vaccine for certain period of time. This simple vaccine delivery approach has potential of commercial application. This route of vaccine delivery permits exposure of immune cells located in the skin and gills of fish directly to antigens thereby mounting immune response to protect the fish from future infection. Further, the antigens are carried internally through other immune cell types present in skin and gills permitting development of more systemic responses. Vaccination by immersion can be performed by reducing the water level in the culture pond or by transferring the fish to a holding tank. Immersion vaccination is advantageous as it causes less or no stress because of minimum handling of fish, possibility of vaccinating large number of fishes in short time, suitability for fish of all sizes, and less labour intensive. However, it has certain limitations such as requirement of large quantity of vaccine and the protection offered does not last long thus warranting for second vaccination.

Immersion vaccination can be administered by direct immersion, hyperosmotic infiltration, bath treatment, spray vaccination and ultrasound. In direct immersion or dip treatment fishes are held in a holding tank containing the vaccine for a certain period of time and then transferred to culture area. This method is more practical and ideally suited for vaccinating small fish of less than 5 gram prior to stocking in culture ponds. In hyperosmotic infiltration method, prior to immersion in antigen solution fish are dipped for a short time in a hyperosmotic salt solution such as urea or sodium chloride to enhance antigen uptake. The procedure can also be performed by adding vaccine to the hyperosmotic solution and then exposing the fish. Although this method causes stress to fish, it is suitable for vaccinating large number of fishes as well as fish fry and fingerlings above critical size of immune responsiveness. This approach of vaccination is known to offer better protection to fish when compared to direct immersion method.

Bath treatment as a method of vaccine delivery involves exposing fish to lower concentration of vaccine for longer duration - an hour or more.This procedure involves addition of vaccine to holding tank/pond and then exposing the fish. This method is less stressful and can be used to vaccinate fishes of all sizes. Exposing fish to higher dilution of vaccine for longer duration is known to provide better protection. Spray vaccination is performed by spraying the vaccine under pressure on to the fish. The fishes are passed through shallow channel or conveyor belt for vaccination. The quantity of vaccine required for this method is less. Though it causes stress, it is suitable for larger fishes. Application of ultrasound to enhance vaccine uptake has been found to be successful. In this method fish are immunised in water containing vaccine and subjected to high frequency sound waves of about 20 kHz to enhance the permeability of the cells to antigen uptake. This method is found to be ideal for administering DNA vaccines. The fishes can be vaccinated using low concentration of antigen and protection offered is reported to be comparable to injection method.

Vaccination by Injection

Administering vaccine through injection is an effective method of provoking immune response in fish. It involves delivery of a small volume of antigen directly in to the body of fish for direct stimulation of systemic immune response. Vaccine can be administered into the muscle (intramuscular injection) or to the peritoneal cavity /body cavity (intraperitoneal injection) of the fish. However, intraperitoneal injection is known to offer better protection in terms of specific humoral antibody production.Vaccines are normally mixed with suitable oil or water based adjuvants before injection which helps in further stimulating the immune system.This vaccination approach produces specific memory response and booster doses help to produce higher antibody levels and protection. While primary vaccination induces T-helper cells that are short lived, the subsequent booster doses stimulate long lived memory cells. Because of these limitations, it is likely that injection of vaccine will be used only in fish of sufficiently high unit value to absorb the cost of the procedure.

Vaccination by injection is effective in offering protection against many pathogens that cause systemic disease, offers long term protection than other methods, gives assurance to farmer as every fish gets vaccinated, and permits delivery of multiple antigens at the same time. However, certain drawbacks associated with injection mode of vaccine delivery limit its potential use in the field. The limitations are; it is labour intensive, time consuming, requires skilled/ trained manpower, expensive to administer when large number of fish are to be vaccinated, requires handling of each fish which may cause stress, need for use of suitable drugs to anesthetize fish to reduce stress and struggling during handling, damage to fish when needle of right size is not used and not suitable for small sized fishes of below 10 gram. These limitations restrict wider application of injection mode of vaccine delivery except for high value fish. A large scale vaccine delivery can be achieved most effectively by employing hand held automatic syringes which deliver known volume of vaccine for each stroke of the piston or by using automated vaccination unit that injects right volume of vaccine to fish usually held in conveyor belt. Although, automated vaccination is less stressful to fish, it is suitable for relatively large fish and is also cost intensive.

Effectiveness of a Vaccine

The efficacy and success of vaccination is influenced by aquaculture husbandry practices as well as vaccine related factors. The effectiveness of the vaccines is high when fish are in prime health condition. But stress due to handling, poor water quality and high stocking density cause stress on immune system leading to reduced performance of vaccine. Use of a well-balanced diet with all essential nutrients helps in the synthesis of protective factors and activation of cell. Incorporation of Vitamin C and E to feed improves resistance of fish to infection and enhances antibody response and phagocytic activity. Pollutants including antibiotics even at sub-lethal levels affect defense mechanism leading to stress and poor antigen uptake thereby affecting success of vaccination. The season and water temperature also influence vaccination as many physiological processes vary on a daily, monthly or annual basis. Generally, immune response is high during warmer summer temperature than during cold seasons.

The success of the vaccination also depends on several vaccine related factors such as nature and dose of vaccine, route of administration and use of adjuvants. Delivery of appropriate vaccine of right dose is necessary to elicit immune response while both low and high dose are ineffective. As for the nature of vaccine, a particulate vaccine is found to be more effective than soluble vaccine. A vaccine administration route that ensures proper antigen uptake along with use of adjuvants and immunostimulants also determine the success or failure of vaccine. Further, if the pathogen has several variants, a polyvalent vaccine comprising all the isolates is more effective than a monovalent vaccine.

Vaccines used in Aquaculture

Several vaccines have been developed and commercialised against a number of bacterial and viral pathogens of fish. Vaccines have been developed for the prevention of bacterial diseases such as for streptoccocosis caused by *Streptococcus iniae* and related bacteria, pasteurellosis caused by *Photobacterium damsela* subsp. Piscicida, warmwater vibriosis caused by *V. alginolyticus*, *V. harveyi*, *V. vulnificus*, and *V. parahaemolyticus*, lactococcosis caused by *Lactococcus garvieae*, and yersiniosis caused by *Yersinia ruckeri*. A cost effective oral biofilm vaccine effective against *Aeromonas hydrophila* infection in Indian major carps is being commercialized in India. Vaccines for the prevention of viral diseases include those for koi herpesvirus (CyHV-3), infectious pancreatic necrosis (IPN), infectious hematopoietic necrosis (IHNV), viral hemorrhagic septicemia (VHS), salmon pancreatic disease (SPD), grass carp aquareovirus (GCR), iridovirus in *Seriola* spp. koi herpes virus (KHV)and spring viraemia of carp (SVC).

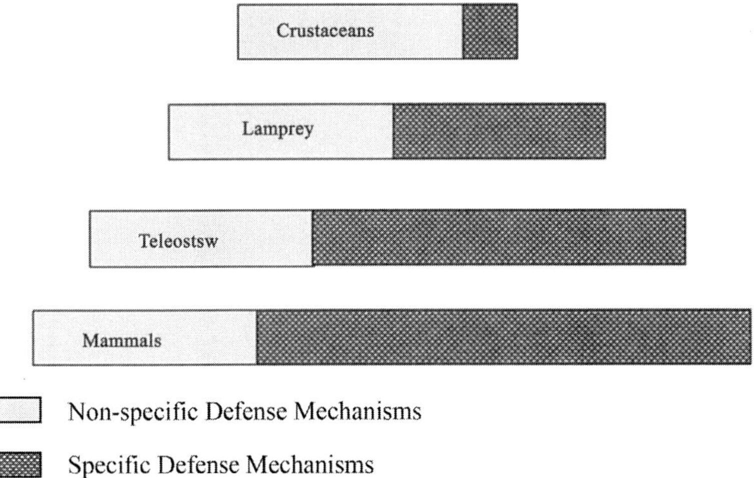

Fig. 1: Relative importance of specific and non-specific defense mechanisms in animals.

The sustainability and success of aquaculture industry is hampered by the several diseases. Maintaining good water quality and health of cultured animals following good husbandry practices only can ensure successful harvest. As culture animals are continuously exposed to several pathogens, it becomes necessary to boost the immunity to fight against pathogens. Adoption of prophylactic measures such as use of probiotics, immunostimulants as well as vaccines as reliable approaches along with good pond management practices are helpful in maintaining animals in good health thereby preventing disease outbreak and ensuring better production.

References

Azad I.S., Shankar K.M., Mohan C.V., Kalita B. 1999. Biofilm vaccine of Aeromonas hydrophila–standardization of dose and duration for oral vaccination of carps. Fish Shellfish Immunol. 9(7):519-28.

Compendium on Prophylaxis in Aquaculture. 2017. Editors: K.K. Vijayan et al. Publisher: ICAR-CIBA, Chennai. PP:220.

Cruz, P. M.,.Ibanez, A. L., Hermosillo, O.A. M. and H. C. R. Saad. 2012. Use of Probiotics in Aquaculture.ISRN Microbiology, Volume 2012.1-13.

Hai, N.V., 2015. The Use of Probiotics in Aquaculture. J. Appl. Microbiol. 119, 917-935.

Nayak D.K., Asha A., Shankar K.M., Mohan C.V. 2004. Evaluation of biofilm of Aeromonas hydrophila for oral vaccination of Claria sbatrachus—a carnivore model. Fish Shellfish Immunol. 16(5):613-9. 17.

Raa J., 1996. The use of immunostimulatory substances in fish and shellfish farming. Rev Fish Sci 4 (3):229-288.

Reddy, J. S (2015), Probiotics in aquaculture: importance, influence and future prospectives. Int. J. Bioassays, 4 (03), 3710-3718.

Sakai, M., 1999. Current research status of fish immunostimulants. Aquaculture, 172(1), 63-92.

Verschuere, L., Romaut, G., Sorgeloos, P., Verstraete, W., 2000. Probiotic bacteria as biological control agents in aquaculture. Microbiol.Mol. Biol. Rev. 64, 655– 671.

Wang, Y., Li, J., Lin, J., 2008. Probiotics in aquaculture: Challenges and outlook. Aquaculture 281, 1-4.

19

Biotechnology in Aquaculture Health Management

K.M. Shankar

Any technique that uses living organisms or substances from these organisms to make or modify a product or to improve plant or animal or to develop microorganism for specific uses, can be termed biotechnology. Biotechnology involves microbes or cells derived from plants and animals for efficient production of useful products for mankind. The use of microbes for production of alcohol and penicillin, to name a few, is well known. However, in the last 25 years, several novel organisms, yielding useful products have been created by genetic engineering. Today, there is a wide application of biotechnology in human medicine, agriculture and animal husbandry in the area of diagnosis prophylaxis and therapy.

Aquaculture industry has undergone a sea change world over in the last three decades. The important needs of the industry are development of fast growing disease resistant fish and shellfish varieties, development of cheap and effective vaccines, disease diagnostics, cell lines sand probiotics. Efforts on these lines have already begun since 3-4 decades and results are highly promising. Some of the biotechnologies employed in aquaculture health management (Plate 15) are fish cell lines, monoclonal antibodies and nucleic acids for diagnostics sub unit and nucleic acid vaccines for prophylaxis, transgenes for disease resistance, gene editing in fishes and their pathogens and probiotics for bioremedial measures.

A. Cell Lines for Health Management

Cell lines are transformed cells that are adapted well to grow and multiply outside the host in *in vitro* culture systems. Cells lines can be developed from fast growing tissues of young fish. There are two methods of developing cell lines (1) Explant method, (2) Enzymatic method. Cell lines are grown in sterile plastic or glass containers containing appropriate medium having nutrients salts, antibiotics, buffers, serum etc. Minimum essential medium (MEM) is the common medium employed for fish cell line. Cell lines are developed, characterised and deposited in national and international repositories, which can be procured for use in research and industry.

Several fish cell lines have been developed from a broad range of tissues, e.g., ovary, fin, swim bladder, heart, spleen, liver, eye muscle, vertebrae, brain, skin. More than 283 cell lines have been established from finfish from fresh, marine and brackish waters around the world, of which more than 60% are from species of Asia, which contributes more than 80% of total fish production. These cell lines have been updated scientifically, authenticated, checked for cross-contamination. Fish cell lines of fibroblastic and epitheliod types are commonly used world over (Table 1)

Table 1: Examples of widely used fish cell lines

Cell line designation	Cell morphology	Fish
RTG-2	Fibroblast	*Salmo gairdneri* (Rainbow trout)
CHSE-214	Epithelioid	*Oncorhynchus* (Chinook salmon)
BF-2	Fibroblast	*Lepomis macrochirus* (Bulegill)
CAR	Fibroblast	*Carassius auratus* (Gold fish)
EPC	Epithelioid	*Cyprinus carpio* (Carp)
FHM	Epithelioid	*Pimephales promelas* (Fathead minnow)
BB	Epithelioid	*Ictalurus punctatus* (Brown bullhead)

State of Art of Fish Cell Lines in India

Over the last decade there is tremendous growth in culture and breeding of carps, murrels, seabass, and a host of sport and ornamental fishes in India. Besides, an increased activity in transport and exchange of seed and brood stock from one place to another and even across international border has been witnessed. Under such circumstances, incidence of disease of viral etiology is common. Since cell cultures derived from the same species or a species closely related to that in which the disease occurs would be the most sensitive for virus isolation, cell lines derived from local species should be given high priority. Though there are cyprinid derived cell lines such as EPC and CAR, development of cell lines from the widely cultivated Indian major carps should be given top priority. In the last 20-30 years several institutions in India have developed cell lines from the various cultivable, sport and ornamental fishes of the country. In many cases, cell lines from several organs/tissues of each fish also developed. More than 75 cell lines have been developed in India and deposited in the National Repository for Fish cell lines (NRFC) at the National Bureau of Fish Genetics Research(NBFGR), Lucknow (Table 2) for use by researchers and industry.

Table 2: Cell lines developed from fishes of India available at the NRFC of NBFGR, Lucknow (*Courtesy* NBFGR, Lucknow)

Sl.No	NRFC Accession No.	Cell Line Code	Species	Tissue
1.	NRFC001	PCF	Puntius chelynoides	Fin
2.	NRFC002	SRF	Schizothorax richardsonii	Fin
3.	NRFC003	TTCF	Tor tor	Fin
4.	NRFC004	CCF	Cyprinus carpio	Fin
5.	NRFC005	WAF	Wallago attu	Fin
6.	NRFC06	RF	Labeo rohita	Fin
7.	NRFC007	CCKF	Cyprinus carpio koi	Fin
8.	NRFC008	HBF	Horabagrus brachysoma	Fin
9.	NRFC009	PDF	Sahyadria denisonii	Fin
10.	NRFC010	CFFN2	Amphiprion sebae	Dorsal Fin
11.	NRFC011	CFBR	Amphiprion sebae	Brain
12.	NRFC012	CFSP	Amphiprion sebae	Spleen
13.	NRFC013	CFCP1	Amphiprion sebae	Caudal Peduncle
14.	NRFC014	SISK	Lates calcarifer	Kidney
15.	NRFC015	SISS	Lates calcarifer	Spleen
16.	NRFC016	SIGE	Epinephelus coioides	Eye, Muscle
17.	NRFC017	IGK	Epinephelus coioides	Kidney
18.	NRFC018	SICE	Catla catla	Eye, Muscle
19.	NRFC019	SICH	Catla catla	Heart, Muscle
20.	NRFC020	CB	Catla catla	Brain
21.	NRFC021	ICG	Catla catla	Gills
22.	NRFC022	ICF	Clarias batrachus (magur)	Fin
23.	NRFC023	LRG	Labeo rohita	Gill
24.	NRFC024	DT1CPEx	Dascyllus trimaculatus	Caudal Peduncle
25.	NRFC025	DT1F4Ex	Dascyllus trimaculatus	Fin
26.	NRFC026	DT1CPTr	Dascyllus trimaculatus	Caudal Peduncle
27.	NRFC027	RC4H1Tr	Rachycentron canadum	Heart
28.	NRFC028	CTM	Catla catla	Thymus (Macrophage)
29.	NRFC029	CTE	Catla catla	Thymus (Epithelial)
30.	NRFC030	EM2HTr	Epinephelus malabaricus	Heart
31.	NRFC031	EM2GEx a	Epinephelus malabaricus	Gills
32.	NRFC032	EM3GEx	Epinephelus malabaricus	Gills
33.	NRFC033	EM4SpEx	Epinephelus malabaricus	Spleen
34.	NRFC034	CCM	Catla catla	Blood (Lymphocyte)
35.	NRFC035	PC1CpTr	Pomacentrus caeruleus	Caudal Peduncle
36.	NRFC036	PC1F1Ex	Pomacentrus caeruleus	Fin

Sl.No	NRFC Accession No.	Cell Line Code	Species	Tissue
37.	NRFC037	PC1L1Tr	Pomacentrus caeruleus	Liver
38.	NRFC038	HC2SPEx	Epinephelus merra	Spleen
39.	NRFC039	CFF	Pristolepis fasciata	Fin
40.	NRFC040	IEE	Etroplus suratensis	Eye
41.	NRFC041	IEK	Etroplus suratensis	Kidney
42.	NRFC042	IEG	Etroplus suratensis	Gill
43.	NRFC043	IEB	Etroplus suratensis	Brain
44.	NRFC044	RE	Labeo rohita	Eye
45.	NRFC045	CSK	Channa striatus	Kidney
46.	NRFC046	CSG	Channa striatus	Gill
47.	NRFC047	WAM	Wallago attu	Muscle
48.	NRFC048	WAG	Wallago attu	Gill
49.	NRFC049	CPG	Channa punctatus	Gill
50.	NRFC050	DRM	Danio rerio	Muscle
51.	NRFC051	AFF	Pterophyllum scalare	Fin
52.	NRFC052	OnlL	Oreochromis niloticus	Liver
53.	NRFC053	DRG	Danio rerio	Gill
54.	NRFC054	DrRPE	Danio rerio	Retinal
55.	NRFC055	OST	Channa striatus	Thymus macrophage
56.	NRFC056	CMP	Cirrhinus mrigala	Peritonal
57.	NRFC057	PHF	Pangasiodon hypothalmus	Fin
58.	NRFC058	FtGF	Carassius auratus	Caudal fin
59.	NRFC059	CIDu	Clarias dussumieri	Fin
60.	NRFC060	SREM-1	Schizothorax richardsonii	Eye
61.	NRFC061	BBdF-1	Barilius bendelisis	Fin
62.	NRFC062	CMgT-1	Clariasmagur	Testes
63.	NRFC063	LCF	Labeocalbasu	Fin
64.	NRFC064	CyCKG	Cyprinus carpio	Gill
65.	NRFC065	PSF	Etroplus suratensis	Fin
66.	NRFC066	CMgM-1	Claria smagur	Muscle
67.	NRFC067	CMgB-1	Clarias magur	Barbel
68.	NRFC069	ZFiM-1	Danio rerio	Muscle
69.	NRFC070	LRoF-1	Labeo rohita	Fin
70.	NRFC068	DDaF-1	Dario dario	Fin
71.	NRFC071	OnH	Oreochromis niloticus	Heart
72.	NRFC072	PFB	Paraneetroplus synspilus X Amphilophus citrinellus	Brain
73.	NRFC073	PFH	Paraneetroplus synspilus X Amphilophus citrinellus	Heart
74.	NRFC074	PFS	Paraneetroplus synspilus Amphilophus citrinellus	Spleen

Sl.No	NRFC Accession No.	Cell Line Code	Species	Tissue
75.	NRFC075	RBT-H	Oncorhynchus mykiss	Heart
76.	NRFC076	OCF	Pseudetroplus maculatus	Fin
77.	NRFC077	OCrF-1	Oreichthysore nuchoides	Fin
78.	NRFC078	PHT	Pangasiodon hypophthalmus	Thymus
79.	NRFC079	SRM-1	Schizothorax richardsonii	Muscle

Application of Fish Cell Lines

i) Diagnosis of fish viral diseases

Development of fish cell lines has several applications in aquaculture such as in study and diagnosis of fish viruses and production of viral vaccines. Fish cell lines have an important role in the expanding aquaculture industry of the world. Undoubtedly, the most widely employed application of fish cell cultures is for the study of fish viruses. Indeed, the primary requirement, for the development of many of the continuous fish cell lines was for isolating and identifying viruses that are the causative agents of epizootics of commercially important species. Cytopathic effect(CPE) where cell monolayer get destroyed due to virus infection is the basis for presence of virus and diagnosis. Each virus exhibits CPE in its own characteristic way, often provide a clue for type of the virus.

Besides diagnosis, cell lines are important in the much required national and international quarantine and certification programme for procuring virus free fish stock. Fish cell lines are essential in diagnosis and identification of virus. Further, cell lines are essential for production of viral vaccines.

ii. As In Vitro Models in Physiology and Toxicology

In recent years fish cell lines are being used as *in vitro* models to evaluate the toxicity of pollutants. This is a better alternative to use of live fish in toxicity studies. Cell line *in vitro* models provide specific information in toxicology at cellular level.

Besides, fish cell cultures also provide *in vitro* system for studying various cellular and physiological processes. For example, organ cultures of pituitary glands were derived from carp, tilapia, eel and rainbow trout for use in studying the production of growth hormones, prolactin and gonadotropins. These cells should provide a useful tool for investigating several aspects of the hypothalamo-hypophyseal system of fishes.

Fish cell lines besides major use in virology serve as valuable, fast, and economic *in vitro* tools for screening toxicity of chemicals and environmental samples, studies relating to immunology, genetic regulation, developmental biology and conservation of fish germ plasm. Cell lines are also used in vaccine production against pathogens, testing drug metabolism and antibody production, the study of gene function, and synthesis of biological compounds, e.g., therapeutic proteins.

iii) Production of Viral Vaccine

Viruses are grown in cell lines, purified and titre determined for number of infectious particles. The virus particles are inactivated by heat or formalin for development of vaccine. Dose and duration of vaccination in terms of antibody production and protection upon challenge in vaccinated fish are estimated. After safety studies, appropriate vaccine protocol is developed for vaccination.

B. Antibodies in Health Management

Antibodies are glycoproteins produced by animals against foreign material(antigen) such as microorganisms, toxins etc. Antibodies are highly specific to antigen at epitope level (length of 5-6 amino acids). Antibodies bind to antigen specifically and can be by protective neutralising the antigen. This specificity is made use in specific diagnosis, serotyping in epidemilogy besides in treatment

Monoclonal Antibodies (MAbs) in Health Management

Conventionally, antiserum (polyclonal antibodies) is prepared in rabbits for use in diagnostics, serotyping and vaccine development. Rabbits are immunized with a purified antigen with an adjuvant in 2 to 3 doses and the animal is killed and blood collected at the end of 30-40 days. The serum collected after blood clotting contains specific antibodies to the antigen injected. Rabbit antiserum contains different types of antibodies against different epitopes of the antigen derived from several plasma cell clones (polyclonal). Even in a hyperimmune animal, seldom are more than one-tenth of the circulating antibodies are specific for one antigen. The use of these mixed populations of antibodies creates a variety of different problems in immunochemical techniques such as background reaction and false positives. Besides, antiserum is not sensitive for detection of minute difference between antigens. In addition, availability of rabbit antiserum in limited quantities is also a serious drawback. Therefore, the preparation of homogenous monoclonal antibodies (MAbs) with a defined specificity was a long-standing goal in immunological applications. Monoclonal antibodies are being employed in disease diagnosis, pathogen classification, epidemiological analysis and development of vaccines/drugs.

The goal of producing monoclonal antibodies was achieved with the development of the of 1. Hybridoma technology, 2. Recombinant phage antibody system (RPAS)

Production of Monoclonal Antibodies (MAbs)

1. Hybridoma Technology

In animals, antibodies are synthesised primarily by plasma cells, a type of terminally differentiated B lymphocytes. Because plasma cells cannot be grown in tissue culture, they cannot be used as an *in vitro* source of antibodies. Kohler and Milstein (1975) developed a technique that allows the growth of clonal populations of cells secreting antibodies with a defined specificity. In this technique, an antibody-secreting cell, isolated from an immunised animal, is fused with a myeloma cell, a type of B-cell tumor. These hybrid cells or hybridomas can be cloned and maintained *in vitro* and will continue to secrete antibodies with a defined specificity. Antibodies that are produced by monoclonal hybridomas are known as monoclonal antibodies (MAbs).

The usefulness of monoclonal antibodies stems from three characteristics – their specificity of binding, their homogeneity, and their ability to be produced in unlimited quantities (Table 3). The production of monoclonal antibodies allows the isolation of reagents with a unique, chosen specificity. Because, all the antibodies produced by descendants of one hybridoma cell are identical, monoclonal antibodies are powerful reagents for testing for the presence of a desired epitope. Hybridoma cell lines also provide an unlimited supply of antibodies. Therefore, production of immortal hybridoma will be much cheaper in the long run compared to production of rabbit antisera. In addition, one unique advantage of hybridoma production is that impure antigens can be used in immunisation to produce specific antibodies. Because hybridomas are single-cell clone prior to use, mono-specific antibodies can be produced after immunisations with complex mixture of antigens.

Hybridoma secreting monoclonal antibodies specific for a wide range of epitopes of a pathogen have been prepared. Any substance that can elicit a humoral response can be used to prepare monoclonal antibodies. Their specificities range from proteins to carbohydrates to nucleic acids. Production of monoclonal antibodies although often more time-consuming and costly than rabbit antiserum(polyclonal antibodies) in the short term, the cloned hybridomas are immortal, providing unlimited specific antibody and hence cheaper in the long run. In theory, either as single antibody preparation or as pools, monoclonal antibodies can be used for all the tasks that require or benefit from the use of polyclonal antibodies. In practice, however, producing exactly the right set of monoclonal antibodies is often a difficult and laborious job.

Table 3: Comparison between antiserum (polyclonal) and monoclonal antibodies

Criteria	**Antiserum**	**Monoclonal antibody**
Antigenic determinant/ epitope recognition	Several	Single
Specificity	Variable with animal and bleed, Partial cross-reactions with common determinants, seldom too specific	Standard, highly specific Unexpected cross reactions may occur May be too specific for requirement
Affinity	Variable with bleed	May be selected during cloning
Yield of useful antibody	Up to 1 mg/ml	Up to 100µg/ml in tissue culture Up to 20mg/ml in ascitic fluid
Contaminating immunoglobulins	Up to 100%	None in cell culture, 10% in ascitic fluid
Purity of antigen required	Pure antigen is required	Some degree of antigen purification desirable but not essential
Approximate minimum cost (Rs.)	Usually below 5000 for limited quantity of antiserum	Capital cost – 10 lakhs, Running costs – 2 lakhs for unlimited quantity of highly specific antibody.

2. Recombinant Phage Antibody System (RPAS)

This is a technology of production of monoclonal antibodies without using animals.

Antibody producing cells are immortalised by immortalising their gene followed by cloning of antibody producing gene. It is a technique of mix of immunology and molecular biology. In reality, antibody producing spleen cells have potential against millions of epitopes in nature and hence a library of these antibody producing genes can be created for selective cloning and expansion for monoclone antibodies. In simple mRNA of antibody against several epitopes are extracted from spleen of an immunised mice or hybridoma and converted to cDNA library. From the antibody gene vL and vH are cut, separated and recombined with Phagemid. The recombined phagemid are transformed to *E coli*. The recombinant phages express and display antibody (only vH and vL) on the surface of transformed *E coli*. Antibody producing phage are enriched using Ag of choice.The transformed *E. coli* are further plated and cloned in presence of antibiotics and clones selected are transferred to 96 well plates and antibody produced by phage are selected by ELISA, followed by confirming with Western blot.

Advantages of RPAS are it does not require handling of animals, Ab producing genes are easily cloned and secreted, clones have stable genetic nature. RPAS is easy for production of human monoclonal antibodies. Disadvantages being the antibody obtained by RPAS are soluble products having only vL and vH without FC part. And the MAbs so produced by RPAS can not be patented for commercial purposes.

Application of MAbs in Aquaculture Health Management

By virtue of high specificity to antigen, MAbs are used in diagnosis, serotyping and treatment. Monoclonal antibodies to several viral and bacterial pathogens of fish and shellfish are available in the market today. In India, in the authors laboratory MAbs to *A. hydrophila*, *Vibrio*, EUS fungus *Aphanomyces invadans*, white spot virus of shrimp, protein of *Penaeus monodon*, IgM of catla and two antibiotics Oxytetracycline and Sulpha dimethoxine have been produced. These MAbs have been used for diagnostics, serotyping and evaluation of a novel bacterial biofilm oral vaccine. NBFGR, Lucknow has produced MAbs to IgM of *Channa marulius and* mrigal for diagnosis and serotyping.

i) **MAbs for development of Diagnostics:** Mabs have been used in ELISA, Immunodot, Western blot for specific and sensitive diagnosis at laboratory level. Besides MAbs are very ideal for development of field level diagnostics.In the authors laboratory several rapid, simple, cheap, specific and sensitive MAb based field level(farmer level) Flow through diagnostic kits have been developed for detection of microbial pathogens and antibiotics of aquaculture importance. Principle and methods of these antibody based diagnostics have been dealt in detail in Chapter 14.These MAb based farmer level Flow through diagnostic kits have sensitivity higher than I step PCR for detection of fish pathogens. Furthermore, minute serological difference among bacterial and viral pathogens of fish and shellfish could be detected by epitope analysis employing MAb based Immunodot which has helped significantly in serological and epidemiological studies. Monoclonal antibodies have also been extensively used elsewhere in detection of epitopes involved in pathogenesis for development of subunit vaccines.

ii) **Serotyping:** MAbs reacting at epitope level could be ideal tools for epitope analysis, serotyping, surveillance and for tracing movement of pathogens. These MAbs are employed in ELISA, Immunodot for epitope analysis. Information on epitopes of pathogen involved in pathogenesis can be derived which is ideal for designing and development of sub unit vaccines

iii) **Immunotherapy:** MAbs can be used for prophylaxis. The author laboratory produced a panel of neutralising and binding MAbs against white spot virus of shrimp. Employing the neutralising monoclonal antibodies through feed it was possible to neutralise shrimp white spot virus which could be demonstrated by low concentration of the virus in hepatoancreas and hemolymph along with improved protection against the virus upon challenge. Similar findings have been demonstrated to protect fish against virus. In future neutralising antibodies hold promise for prophylaxis in aquaculture.

C. Nucleic Acids in Health Management

i) **Gene probes for diagnostics:** Single strand of a gene can bind to complementary strand specifically and hence can serve as detection probe. Disease diagnosis by gene probes is the most sensitive and specific method.DNA probes also can be ideally used for genotyping of pathogens and used in epidemiological analysis. DNA based techniques such as DNA hybridization and PCR have revolutionised diagnostics and disease management, the details of which is given in the chapter 15. For large scale application in field there is a need of specific nucleic acid probes against several pathogens infecting fish and shellfish. Genetic engineering methods by DNA cloning in *E coli* is being used widely for large scale preparation of gene probes for diagnosis.

ii) **Recombinant Genes for Prophylactic Measures:** Immunising fish in aquaculture plays an important role in prevention of diseases in these cold blooded vertebrates. Prophylaxis is particularly gaining importance against use of antibiotics and drugs and emergence of antimicrobial resistance(AMR). Salmon culture in Norway has successfully demonstrated use of vaccines as an alternative to antibiotics in health management. Fishes can be immunised using whole cell killed vaccine, sub unit vaccine or Nucleic acid vaccines. Biotechnology has a crucial role in production of sub unit and nucleic acid vaccines

iia) **Development of Subunit Vaccine:** Conventionally whole cell vaccines are prepared against microbial pathogens. The major problems with the commonly employed whole cell vaccines are 1. protective antigens are often destroyed during inactivation, 2. unwanted side effects and antigenic competition may occur as the result of complex mixture of microbial products present, 3. correct type of immune response (humoral, local or cell mediated immunity(CMI) may not be always induced. Thus the vaccine produced by this approach are often far from perfect and hence need for alternatives such as subunit vaccines. The subunits are also useful where the pathogen is difficult to grow, where microbial products are in very small quantity, or in development of vaccines for multicellular parasites. Of late, reverse vaccinology is being adopted to design unique multi-epitope subunit vaccine against pathogens/ parasites. Once developed, subunits will become the cheapest, safest vaccine available in large quantity. Development of subunit vaccines against viral and bacterial diseases is the need of the hour for use in intensive aquaculture. Subunit vaccines are effective, safe, cheap and can be produced in large quantities. Reverse vaccinology was also used to design multi-epitope sub unit vaccines against pathogens.

Important Characteristics of Sub Unit Vaccines (SUs)

Sub Unit vaccines using only antigenic components for vaccination, cannot replicate in the host with no risk of pathogenicity to the host or non-target species. Subunit vaccines target immune responses toward specific microbial determinants. However, subunit vaccines stimulate a weak immune response than killed or live whole cell vaccine preparations due to the limited number of components represented and the lack of replication. Since the simplified, synthetic and highly purified antigenic components lack immunogenicity by themselves some subunit vaccines require adjuvants to elicit the appropriate immunity. Unnatural components can be easily incorporated in these vaccines, easily freeze-dried, transported and stored at room temperature. Often multiple booster immunizations required with sub units to ensure long- term protective immunity

Production of Sub Units (SUs)

Immunogenic antigens are isolated and purified directly from the target pathogen, or they are produced using various recombinant expression vectors in prokaryotic and eukaryotic systems. *Escherichia coli* expression system is commonly used for the propagation of plasmids carrying genes that encode a specific antigen, which is harvested at the end of the cycle. *E. coli*-based sub unit of infectious pancreatic necrosis virus (IPNV) used in Norway, a yeast-based sub unit against the ISA virus used in Chile are some of the successful subunit vaccines for aquaculture. Baculovirus and yeast for VHSV or infectious hematopoietic necrosis virus (IHNV) proteins have also been used in expression. Other expression system include IPN proteins produced via cabbage worms (*Trichoplusia ni*) and salmonid alphavirus 3 recombinant proteins produced in a fish cell line.

Overall as SU require adjuvants and often expensive to produce due to requirement of additional protein processing steps which limit their development and commercialisation for fish species. However despite these limitations, SU has the potential applications against pathogens that are difficult to cultivate in the laboratory.

ii) Virus-like Particles (VLPs)

VLPs are developed by the self-assembly of viral capsid proteins into particles which can mimic the natural structure of the virus. VLPs are considered components of an advanced subunit vaccine. As VLPs lack genetic material, reversion mutations or pathogenic infection is not possible and hence are very safe. VLPs can potentiate both innate and adaptive immune responses through the recognition of repetitive subunits. VLPs of both non-enveloped and enveloped viruses have been produced in *E coli*, yeast, transgenic plants, insects, mammals and cell-free platforms. Additionally, chimeric VLPs can be produced by genetic fusions or chemical conjugates to viral structural proteins. VLP vaccines have been developed for nervous necrosis virus (NNV) and IPN virus of fishes which could elicit an antibody response by either injection or oral vaccination. Recombinant VLPs construction with IPNV and IHNV elicited strong immunogenicity, can be safe alternative to inactivated or attenuated vaccines.

iii) Nucleic Acid Vaccines (NAV) for Genetic Immunisation of Fish

NAVs have the combined positive attributes of both live attenuated and subunit vaccines.NAV consists of DNA or RNA encoding the antigen(s) of interest and are simple to generate and safe. Furthermore, NAV production is easy to scale up, cost competitive, including enhanced immunogenicity and could be a revolution in medicine.

DNA Vaccine (DNAV)

DNA vaccines are considered as third generation vaccines. Direct injection of DNA in to skeletal muscle of animal leading to expression of epitope of a pathogen in the host with proper conformational folding of the antigen protein lead to presentation of peptides on cell surface serving as a vaccine. Gene construct are injected by gene gun delivering by bombarding gene coated gold particles into skin or muscle. Advantages of DNA vaccine include identical and inexpensive production processes, possibility of co-administration of multiple vaccines, and simplicity of storage due to high chemical stability of plasmid DNA. There is always possibility rapid modification of DNA sequences in the event of occurrence of new mutants. DNAVs could stimulate stimulating both humoral and cell mediated immunity with no risk of disease transmission. There is no need of adjuvants or boosting with DNAVs. DNA vaccines have been developed against a variety of aquatic pathogens causing disease in commercial fish species. DNA vaccines found to be more effective in protecting against fish rhabdoviruses and to intracellular bacterium, such as *Mycobacterium marinum*. However, only few DNA vaccines fish vaccines have been commercialized against-1. IHNV in Canada, and 2. pancreatic disease virus in the European Union.

RNA Vaccines (RNAV)

Two major RNA-based vaccines: conventional, non-amplifying mRNA and self-amplifying mRNA can be developed based on the translational capacity of the RNA. In both human and animals RNAV have been developed and demonstrated with encouraging results. RNAV are safe as RNA is non-infectious and easily degraded in animals by normal cellular processes.

Currently used self-replicating RNA vaccines use an alphavirus genome. During development of alphavirus vector vaccine, the single RNA gene responsible for RNA replication is retained as such but replacing structural gene with the antigen of interest. This antigen-encoding RNA replicon platform enables a large number of antigens from an extremely small dose of vaccine to be produced *in vitro* transcribed (IVT) from a DNA template. The viral replication is independent of host replication system as it takes place in the cytoplasm of the host cell. The replicase-based nucleic acid vaccines are very efficient and attractive delivery vehicle. The alphavirus replicase functions in a broad range of host cells including fish. Therefore, construction of RNA vaccine could be easier using alpha virus vector by replacing the structural genes with that of fish pathogen of interest, against a number of important fish diseases in aquaculture.

iv) Transgene Technology for Health Management Development of Transgenic Fish

Transgenics is transfer of foreign gene into the nucleus of a target cell where integration into the host genome takes place for expression. The gene constructed is transferred to fish eggs through microinjection and checked for its incorporation and successful expression in adults. More than 35 species had been genetically engineered worldwide. Fish transgenes developed are used in aquaculture and as model fish used in basic research. Transgenic carp (*Cyprinus* sp.), tilapia (*Oreochromis* sp.), salmon (*Salmo* sp., *Oncorhynchus* sp.) and channel catfish (*Ictalurus punctatus*) are used in aquaculture while zebrafish (*Danio rerio*), medaka (Oryzias latipes) and goldfish (*Carassius aur*atus) are used in basic research. Production of fast growing and disease resistant varieties of fish is one of the important requirements in aquaculture. Researchers have successfully produced fast growing transgenic salmon, trout and common carp with incorporation of growth hormone gene. Attempts have been made to produce fast growing transgenic Rohu and *Clarias sp* in India. On similar lines cold and disease resistant trouts and salmon have been produced by transgenesis. Several transgenic fishes such as common carp in China, hybrid tilapia in Cuba, Nile tilapia in the United Kingdom, Atlantic salmon in Canada and mud loach in Korea have been considered for commercial production due to their growth potential. Transgenic common carp in China, salmon in USA and tilapia in Cuba are already marketed.

Methods in the Production of Transgenic Fish

Microinjection method: Microinjection is the most established successful and commonly used method in the production of transgenic fish due to its simplicity, reliability and higher success rates. DNA injected into the cytoplasm of fertilized fish eggs could integrate into the fish genome and be inherited in the germ line up to as high as 20%. However, the method is time-consuming and labour intensive particularly for production of large number fish.

Electroporation method: Electroporation uses short electric pulses to permeate the cell membrane with temporary pores on the surface of the target cells through which the transgene enter into the cytoplasm and the nucleus of the cellular machinery. This method is preferred as it is ideal, quick and simple for most effective means of gene transfer in fish. However, the presence of a tough chorion layer around the fish eggs and its removal reduces efficiency and is additional stress on newly fertilized eggs.

Retroviral infection method: The successful use of retroviruses for gene delivery in fish has been reported. However, the preparation of retroviral particles including the transgene of interest is a very laborious process, increases costs and requires technology.

Sperm mediated gene transfer method: Spermatozoa are capable of binding DNA and carrying it into an egg. Fish spermatozoa can be stored in seminal plasma with little loss of viability for long periods. Therefore, this technique appears very promising for gene transfer in fish.

Use of Transgenic Fish in Aquaculture Health Management

a). Development of Disease Resistant Fish: Increased disease resistance could reduce use of antibiotics and production losses. Lysozyme which is found in blood, mucus, kidney of fin fish is a nonspecific antibacterial enzyme for fish defense. There is good correlation between level of lysozyme and disease resistance in salmon. Scientists have produced transgenic disease resistant salmon employing cDNA clone encoding trout lysozyme. Furthermore, there are promising results with disease resistant transgenic channel catfish. A transgene grass carp (*Ctenopharyngodon idella*) coding for human lactoferrin, with significant improve in survival rate relative to control fish after exposure to *Aeromonas* and grass carp hemorrhage virus has been developed. Cercopin has been used in channel catfish to enhance their protection against several pathogenic bacteria by 2–4 times. Besides, development of fast growing cold resistant, and metabolism modified transgenic fish indirectly promote overall health of fish details of which is provided below.

b) Other Application of Transgene Technology in Aquaculture Fast Growth Rate: There are several strategies for improved performances indirectly through transgenesis- growth, improved health, metabolism, cold resistance. Transgenic fish were produced mainly for their fast growth rate which is important in aquaculture. Two- to threefold increases in growth compared to nontransgenic fish have been reported for tilapia, Atlantic salmon and common carp.

Cold Resistance

In cold water aquaculture fish growth is poor during long winter. Hence "antifreeze" genes from fish that live in polar regions have been introduced in Atlantic salmon to resist cold for enhanced growth. This is still in an experimental phase. Transgenic goldfish with ocean pout type III antifreeze protein has shown a higher cold tolerance compared with controls.

Transgenic Fish with Metabolic Modification and Sterility

As aquaculture fish species require high protein diets, usually met by fish meal from wild-caught pelagic fish, alternatives are always looked for. By producing transgenic fish with improved carbohydrate metabolism, transgenic fish with a better use of diets based on ingredients of land origin, such as soybean meal and vegetable oils could be produced. The production of sterile fish would be an ideal means of containment of transgenic fish in production systems with chance of fish escape to the wild are high. Producing Transgenic fish that are deficient in a critical path leading to successful reproduction will be highly benefecial. Although this is a hypothetical possibility, research in this area has been initiated and some progress has already been made with rainbow trout.

Use of Transgenic Fish in Research

Broadly transgene fish are used in following five area of research.a) Enhancing the traits of commercial fish, b) as bioreactors for the development of bio-medically important proteins, c) as indicators of aquatic pollutants, d) developing new non-mammalian animal models, e) Functional genomics studies

Concerns About Transgenic Fish

Several concerns on transgenic fish have been expressed by consumers, environmental groups, industry and even scientific experts.

a. **Ethical Issues in Aquaculture:** Ttransgenic fish is unnatural as the DNA manipulation is by human intervention. Hence genetic modification are uncontrollable, unanswerable and may be risks for human health, animal welfare and environment.

b. **Escape of Transgenic Fish to Wild:** The impacts of escapes are of substantial concern as containment of aquatic animals can be much more problematic than that of their terrestrial counterparts. There could be long-term consequences through the introgression of transgenes into wild populations. The primary ecological concerns regarding utilization of transgenic fish are the loss of genetic diversity and loss of biodiversity and reduction in species richness. Escape or introduction of transgenic fish into natural communities is a major ecological concern. Concerns range from interbreeding with native fish populations to ecosystem and effects resulting from heightened competition for food and prey species. Some evidence suggests that transgenic fish are generally less fit than wild fish, so the effects of escapes may not be as great as anticipated. They generally show reduced swimming ability and lower reproductive performance than non-transgenic fish. The transgenic fish are also more active and aggressive when feeding, and more willing to risk exposure to predation. Changes in cognitive abilities or brain function and structure have occurred in transgenic fish.

c. **Safety for Human Consumption:** Transgenic fish could produce new or modified proteins that could be toxic to humans and risky for human consumption. Continuos studies in this direction required to ensure there is no human intolerance to the new proteins.

v. Gene Editing of Aquatic Organisms in Health Management

Earlier employing transgene technology, fast growing and disease resistant fishes were produced and even commercialised. However, considering ethics, safety to human consumption and possible escape of transgenics to wild, gene editing employing CRISPR technology for desirable qualities in fish and pathogens is becoming popular. Intense research has begun gene editing of both cultivable aquatic organisms and their pathogens.

Gene editing is a type of genetic engineering in which DNA is inserted, deleted, modified or replaced in the genome of a living organism. Unlike early gene engineering that randomly inserts genetic material into a host genome, gene editing targets the insertions to site specific locations. And mutations in germ cell lines are transmitted to off springs. Use of CRISPR (Clustered Regularly Interspaced Short Palindromic Repeats) technology is common and popularly applied in fishes. Gene editing by CRISPR is being employed for a range of traits in a variety of farmed aquatic species- salmonids, carps and crustaceans. There are good gene editing possibilities for several advantages in aquaculture -growth, disease resistance, sterility, controlled reproduction and colour enhancement. Both fish and its pathogens can be gene edited for management of disease. In the years to come with the wider application of gene editing development of disease resistant fish is soon reality. However, gene edited technology has a long way to go to be financially and regulatorily viable for commercial aquaculture.

So far, gene editing technology has been successfully applied to economically valuable fish, such as tilapia, carp (*Cyprinus carpio*), grass carp (*Ctenopharyngodon idellus*), Atlantic salmon, and tongue sole (*Cynoglossus semilaevis*). Fast growing gene edited fishes have been created and even marketed eg Aquaadvantage salmon, Japanese tiger puffer fish and Red sea bream.

D. Probiotics for Bioremediation in Aquaculture

Probiotics for bioaugmentation are being increasingly used in aquaculture. In bio-augmentation useful microbes are inoculated and enriched to improve pond bottom for decreasing accumulated waste. Use of biomats for improving water quality for reducing NH3 level is in practice in fish and shellfish aquaculture. Useful microbes are immobilised on inert or biodegradable material for use in fish and shrimp ponds. Probiotics to augment gut flora is another area having potential in aquaculture. Probiotics encourage beneficial bacterial species in the host or in water while suppressing the other, helping in disease prevention and growth promotion. Details on probiotics and bioremediation are given in the chapter 18.

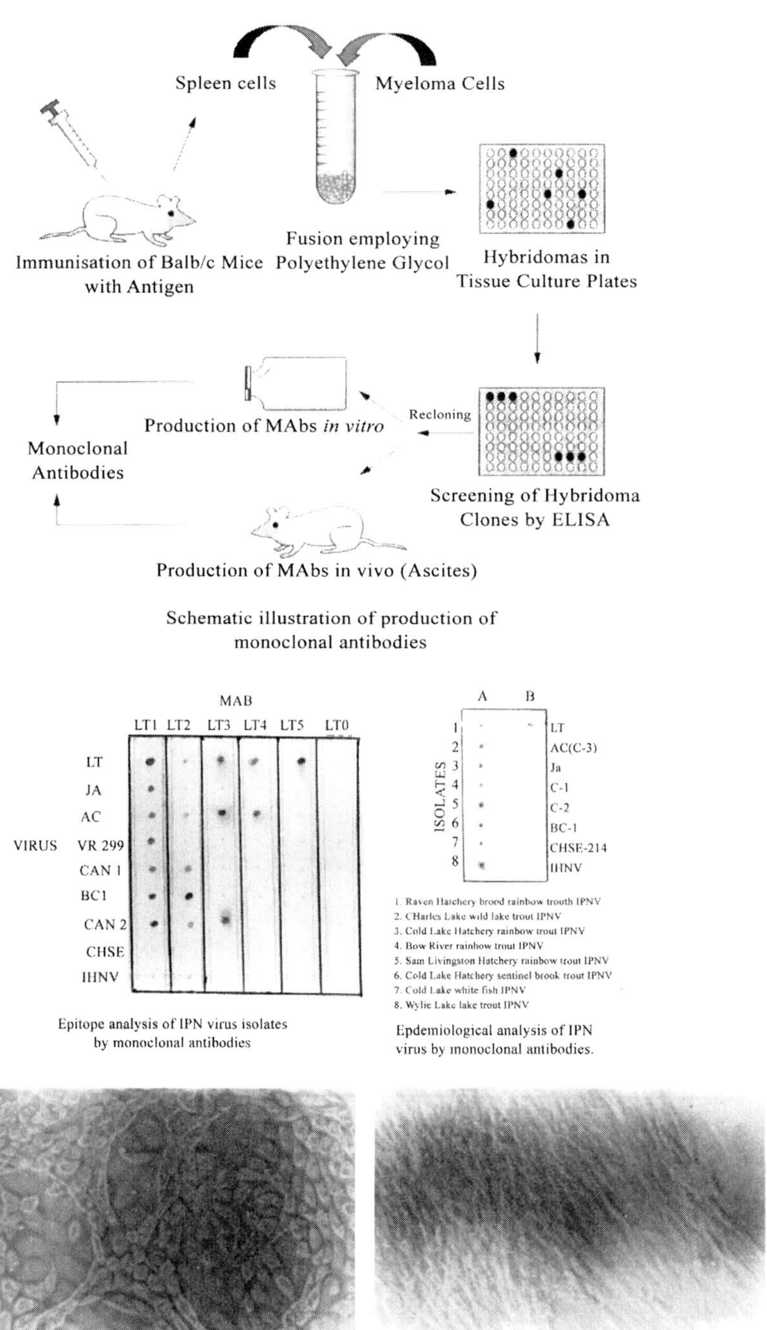

Schematic illustration of production of monoclonal antibodies

Epitope analysis of IPN virus isolates by monoclonal antibodies

Epdemiological analysis of IPN virus by monoclonal antibodies.

Cell culture (Epithelioid) from catla.

Cell culture (Fibroblastic) from catla.

Plate 15: Biotechnology in Aquaculture Health Management (See colour version on page 379)

References

A. Gupta, R Gupta 2021, Gene Editing Technology for Fish Health Management, In Biotechnological Advances in Aquaculture, PP 101-122, Springer

Catherine Colins, Niel lorenzen, Betrand Collect 2019, DNA vaccination for fin fish aquaculture, Fish and Shell Fish Immunology, 85, 106-125

Jie Ma, Timothy J. Bruce, Evan M. Jones, and Kenneth D. Cain,2019. A Review of Fish Vaccine Development Strategies: Conventional Methods and Modern Biotechnological approaches, Microorganisms, 7 (11), 569

Lucy Towers, 2016 Importance of Transgenic Fish to Global Aquaculture: A Review Environment, Sustainability Breeding and Genetics, The Fish site

W.S. Lakra, T Raja Swaminathan, K P Joy, 2011, Development, characterization, conservation and storage of fish cell lines: A review, Fish Physiology and Biochemistry, 37, 1-20

20

Welfare of Fish and Shellfish in Aquaculture

K.M. Shankar

Fish are Sentient Animals

There is strong increasing evidence that finfish, like other vertebrates are sentient animals. This means that fish are self-aware, they can feel pain and distress, they have long term and short-term memory and to some extent, they can experience emotions.

Fish are intelligent, sensitive creatures and like many other animals, they explore, travel, socialise, hunt and play. Some species care for their young and use tools as humans do. Fish are sentient animals capable of suffering and feeling pain. Most fish have highly developed senses with excellent taste, smell, hearing and colour vision. Until fairly recently, many people didn't realise that fish were sentient or feel pain, and the mental abilities of fish were given limited attention by the scientific community. Now, recent discoveries open up a new world of understanding. Far more complex than we ever realised, fish live rich social lives: communicating; hunting cooperatively; and, in some cases, developing cultural traits. Consensus arrived in the mid-2000s around science demonstrating high-level pain perception in fish, and examinations of fish behaviour are beginning to reveal a picture of fish as complex, social and emotional beings. Appreciation of these facts are not widespread in civil society, but identified in time for the principle of fish welfare to be recognised by legislation by the European Union.

Evidence of fishes' sentience, and emergent knowledge of their behaviours and emotions, has given credibility to divers' experiences and anecdotal evidence of the world of fish. These are increasingly coming through in media coverage and documentaries exploring the ocean environment and our understanding of fish. This progress is mirrored in the animal welfare movement where organisations focused specifically on fish and aquatic creatures are increasingly visible, and multi-issue organisations increasingly include fish within their scope of work.

Fish have the same stress response and powers of noiception as mammals. Their behavioural responses to a variety of situations suggest a considerable ability for higher level neural processing – a level of consciousness equivalent perhaps to that attributed to mammals.

Stress in fish in Aquaculture and Ornamental Industry

Fish and shellfishes are cold blooded animals which are subjected to environmental changes to a great extent than the warm blooded vertebrates on land. In the wild, fish enjoys living at low density with plenty of place to move around, adequate oxygen and food of choice. Fish in the natural waters live with least stress surrounded with few pathogens. On the contrary, fish in aquaculture live at high density and although fed well, suffer from inadequate space for movement, low oxygen, accumulated waste and loads of pathogen and hence likely to live a stressful life. Stress reduce their immunity, encourages the pathogen to cause diseases. The stress in aquaculture affects sustainable production in two ways 1) by suppressing growth and production, 2) increasing susceptibility to disease resulting reduction and loss of production. In ornamental fish industry many of the same stressors occur as those in the aquaculture, including poor water quality, handling, transportation, confinement, poor social and physical environment and disease. Transportation and handling of fishes imposes a range of stressors that can result in stress and mortality. Given the numbers of ornamental fishes traded, any of the estimated mortality rates potentially incur significant financial losses and serious welfare issues.

The stress response in fish, as in other animals, is a protective mechanism which allows the fish to cope with changes imposed on it. Stress itself does not necessarily have a negative effect on fish production, it is usually only when there is continuous or repeated stress that this becomes damaging to the fish and increases the risk of clinical disease, poor growth and poor survival which is then reflected in increased mortalities and poorer productivity in terms of increased disease incidence, reduced growth performance and poorer food conversion. By ensuring the best welfare at all stages of the production cycle, stress and damage are minimised resulting in greater chances of successful fish management and productivity

Need for Fish Welfare

There is an increasing importance of aquaculture as a source of protein for human consumption in the world. Why fish welfare?:An estimated 70-180 billion fish are farmed in the world each year. That makes them by far the most farmed vertebrate in the world. And little is done to safeguard their welfare.

Fish have needs that are not always obvious or emotionally compelling to us. However, just like humans, fish have a serious interest in avoiding pain and living a good life.

"The welfare of an animal is its state as regards its attempts to cope with its environment." In fish farming, for example, some fish may tolerate poor water quality conditions for a short period but when the conditions become too challenging or prolonged they cannot maintain homeostasis. Today, huge numbers of fish are reared in underwater factory farms. Just like on land, these farms are crowded, grim places where the animals suffer immensely. Fish are often killed inhumanely and many endure slow, painful deaths by asphyxiation, crushing or even being gutted alive.

Fortunately, there is increasing awareness in consumers about status of welfare of fish in aquaculture who would like to know, how well their animals are reared. Till recently, fish was not listed for consideration of ethical treatment, only large mammals were under the list. It is a good trend where in the world has recognised the stress fish undergo during intensive culture and there are attempts to grow them more humane. Fish grown under such system are certified and there are markets recognising this value. More and more certification, traceability efforts are being designed for the future. Furthermore, welfare is important for successful fish production management have propsed legislation for general welfare of fish in aquaculture particularly fortransport and slaughter. Welfare is fundamental to all aspects of fish management and has a profound effect on the health, survival, productivity and final product quality. Every management procedure we impose on our fish stocks from the egg to the final harvest will have an impact on the welfare of the fish by causing stress or even damage.

An appropriate interpretation of welfare in aquaculture system is required considering good understanding of both physiological and behavioural measures. There is enough knowledge on the physiological consequence of the many aquacultural practices, but information on the behavioural responses is lacking or limited. It is possible to interpret accumulated data on physiological consequence of diseases, handling, transport, food deprivation and slaughter technique on fish welfare. Physiological effect of stocking density is complex to understand as it is dependent on a number of interacting and case specific factors. However, investigation of these factors through behavioural studies helps to improve welfare measures. A significant effect on stress responses due to stocking density, diet, feeding technique and other management procedures and subsequent stress tolerance, health and expression of aggressive behaviour have been well studied. Therefore strategies to minimise disease susceptibility, stress responses, and avoiding aggression is very important. However, it is important to understand and differentiate normal behaviour from abnormal. Philosophy of concept of welfare need to be analysed and discussed with more practical areas of fish welfare comprising all husbandry and management activities.

Underlying Basis for Fish Welfare in Aquaculture

According to the United Nations Food andAgriculture Organization (FAO), 43% of fish for human consumption in 2005 came from aquaculture. The increasing importance of fish welfare in aquaculture comes from ethical considerations as well as from the perspective of improving standards and quality of fish production technologies and aquaculture products. Production and sustenance of industry as whole is linked to welfare of fish. Under appropriate welfare, fish are less susceptible to diseases, need less chemicals, drugs and treatment , have better growth rate and feed conversion and provide better quality product. Finally, the economic benefits are obvious. In addition, consumers care about welfare issues potentially associated with intensive production practices and they expect from the fish farmers that the welfare of farmed fish is addressed. To develop regulations and recommendation for fish welfare is challenging as there is lack of both scientific knowledge and practical experience in this area. It is not well-understood to what extent welfare concepts and definitions as they have been developed for common warm-blooded farm animal species can be applied to fish.

As stated by the European Food Safety Authority (2009), the welfare concept is same for all farm animals including fish, under the treaty of Amsterdam. However, welfare of fish in aqauculture is not studied well , with respect to welfare needs of large number of species farmed. Aquaculture has large number of species, each with several life stages with its requirement and there is a need to understand welfare relevant biology which is currently very limited. And this is the main reason for not developing and validating Operational welfare indicators(OWI) in the growing aquaculture. Only by providing basic biological requirement (food, water and better health) fish can be produced ethically which is also economically sustainable. Fish under suboptimal welfare conditions will perform poorly; for instance, they may show slow growth and increased susceptibility to disease. This negatively affects fish production and compromises the image of the industry. Therefore, it is in the self-interest of the aquaculture industry to optimize culture conditions in order to support and promote fish welfare. Fish welfare in culture also extends to the practices associated with harvesting and killing (harvest, stunning, killing, exsanguination and evisceration), aiming at reducing stress to the animal and therefore also optimizing the quality of the product. It is required to develop certain guide lines to develop welfare measures under various aquaculture systems for further monitoring it.

Measurement Indices of Welfare in Farmed Fish

There are several indices of farmed fish welfare such as changes in colour, ventilation rate, changes in swimming patterns, reduced food intake, reduced growth rate, body injuries, disease status and impaired reproductive performance.

Darkening of the skin colour is known to be associated with stress and may be used as an indicator of poor welfare. The rate of opercular (gill cover) beat is increased during stress (e.g. low oxygen levels). Abnormal swimming can be used as a sign of poor welfare in fish. These may include excessive activity, lethargic swimming high in the water column, rubbing to dislodge ectoparasites. Feeding patterns of fish are disturbed during and after stressful procedures, therefore an unexpected loss of appetite can be used as a sign of poor welfare. Growth of fish can be measured and compared with an expected growth curve. Deviations from the expected growth can be indicative of welfare problems. Damage to the body of the fish may be used as a sign of adverse welfare e.g. damage due to predator attacks. Increased incidence of disease can occur due to a combination of environmental and management factors. Levels of both infectious and non-infectious diseases should be monitored and recorded. Good welfare of valuable brood stock is essential as stress is known to affect the reproductive capacity in fish.

Using Biomolecular Marker to Measure Welfare

There is a need to establish new welfare markers that will promote productivity and consumer's acceptance of farmed fish. The establishment of a new type of welfare biomarkers using cutting□edge technologies like proteomics and other omics technologies is proposed as a solution to this issue. Discovery of fish welfare biomarker: The proteomic approach was applied also for confirming the presence of specific proteins or peptides in the proteome content of fish organs. The peptide identities of purified natterin like protein were confirmed in the organs of Atlantic cod via LC-MS/MS proteomic approach and its expression was higher in the skin, head kidney, liver, and spleen

Welfare Concepts

It has been proven by hormonal changes that fish do experience stress and fight or flight responses. However, although there are substantial neuroanatomical, physiological, and ethological evidences to support that fish are sentient, still it is controversial, whether they experience feeling and emotion, and are conscious of pain and fear. Animal welfare is defined with 3 broad concept definition-nature based, functional based and feelings based , although each with different view points, but these are not mutually exclusive. As nature based definition considers animal welfare from natural behavioural point, but this is limited because we can not determine extentness of fish sentience and emotional state. Therefore, it is pragmatic to identify the factors and conditions that impact biological functions and health status. This concept easily allow us to identify " operation welfare indicators or OWI" to monitor welfare in various farm activities such as stocking, transport or slaughter. A most reliable welfare measure is ensuring physical health of fish which to some extent support mental health as we still lack understanding of fish emotion and sentience.

Processes and Factors in Aquaculture that can be Critical for Fish Welfare

Fish farming economically involves good understanding of biology, technology and personnel management. As many of the husbandry practices are directly or indirectly linked to fish welfare, safeguarding good fish welfare should be routine concern in farm management. As a first step for safeguarding welfare in aquaculture, the following critical processes and factors for fish welfare have to be considered.

a) **Education, Training, Responsibility of Staff:** Well-trained staff is a critical factor for providing good welfare. The staff should be aware of general biological needs as well as specific requirement of each species, and their life stages.

b) **Fish species, Life Stages and Domestication:** World over a large number of fish and shellfish species have been adapted for cultivation in farms. FAO (2018) data shows more than 350 fin fish species are farmed which are chosen from different habitats and environments. Each species and each stage of life has its own specific requirement from the rearing system.

c) **Husbandry Techniques:** A large number of species are grown in aquaculture and requirement of each species need to be studied to develop husbandry and other practices to provide good welfare measures.

d) **Water Quality (Physical and Chemical Parameters):** Water quality has to be maintained at optimum level under all circumstances. Water quality parameters include temperature, PH, dissolved oxygen, nitrogenous compounds such as nitrite, nitrate and ammonia which are most important and critical for fish welfare. Each species has its own water quality requirement. Poor water qualities or sudden change of water parameters can lead to both acute and chronic health and welfare problems. Maintaining good water quality is a challenge in intensive and super intensive farming systems.

Flow rate or exchange rate of water is another important and critical factor as it ensures fresh oxygen, dilute and disperse metabolic wastes. Flow speed and direction helps to avoid dead unexchanged zones with low oxygen. Flow also aids removal of accumulated NH3, unwanted sediments and feed waste. It should be mentioned that exposure to sunlight, noise, vibration also impact welfare and these factors need monitoring.

e) **Stocking Density:** Determining stocking density (kg fish/M3 or fish biomass per unit of water) is very crucial in aquaculture particularly under intensive system. Stocking density influences water quality, growth, stress, social interaction and hence impact welfare of fish. There is always a need to determine optimum stocking density based on species, age, life stages of each species.

f) **Rearing Environment:** Rearing tanks, ponds should be ideally constructed to avoid/preventing injury to skin and fin of fish. Priority for easy removal of feed waste and feces by appropriate construction of tank. Avoiding disturbances due to noise from pumps and other machines, and predators is important for welfare.

h) **Temperature:** Fish are cold blooded vertebrates very much influenced by temperature. They are very sensitive and get disturbed by sudden change, extremes of heat and cold beyond optimum range. Handling fish during high and low temperature should be avoided always to avoid injury.

g) **Slaughter:** Various steps involved in harvesting for slaughter in reality are very stressful and affect fish welfare. Steps such as crowding the stock, removal for from rearing system, transport to abattoir are very stressful to fish. Therefore, it is important to properly organise harvesting well and minimise time at each step. Further it is always advised to stun the fish before killing. Effective and faster stunning by manual or automated methods which induces unconsciousness until killing ensuring a humane slaughter.

h) **Disease prevention and prophylaxis:** Prevention of diseases is major step safeguarding welfare. First step in this direction is preventing entry of pathogens to a farm through infected fish, equipment and personnel. In case of any disease outbreak, it is always advisable to segregate the healthy stock from affected one. And instead of using/misusing of drugs and chemicals which also affects the fish, vaccination against pathogen is the best alternative to prevent disease s which is a desired strategy for better welfare.

Organisations Addressing Fish Welfare

Welfare and wellbeing of fish in aquaculture besides addressing ethics also ensure sustainable and profitable aquaculture. As fish suffer in aquaculture, their welfare should be our responsibility. Two major world organisations.

1. Food and Agriculture Organisation (FAO) and 2.World Animal Health Organisation (OIE) have played key role in developing welfare guide lines in aquaculture. Other world Organisations working on fish welfare are 1).World Wildlife Fund- mostly for fishing, 2) " Fish Welfare initiative" an organisation initiating research on fish welfare- Fish Welfare Initiative aims to improve the welfare outcomes of billions of fish through researching and executing targeted, highly-scalable welfare interventions.3). Global Aquaculture Alliance, 4). "Ornamental Aquatic Trade Association" (OATA), with established standards and codes of best practice for handling fishes.

Training Fish Doctors for a Better Fish Welfare

Fish also feel pain and crave for comfort from a doctor. Therefore training the fisheries professional in fish health and welfare is essential in aquaculture production. These well trained 'Fish Doctors' have important role in welfare of fish and shellfish in aquaculture

Aquaculture is one of the fastest growing food sectors in Asia which contributes 80 % of world aquaculture production. The aquaculture system of Asia is quite diverse and different from that of the West, in terms of number and type of species used, husbandry practiced, microbes, parasites, diseases, soil and water chemistry and climatic conditions. At the present, drug prescription and certification in fisheries/aquaculture in India and Asia in general are addressed by veterinarians as in the West. The veterinarians world over receive only most basic and bear minimum training in fisheries and as such they have least knowledge of the biology of the cold blooded vertebrates and invertebrates, aquatic environment, leave alone the vast aquaculture husbandry practices of today. Therefore, this practice of aquaculture and fisheries dependence on veterinarians should change in India and Asia as soon as possible. To achieve this, ideally the Bachelor of Fisheries Science (BFSc) programme in addition to fish biology, aquaculture and aquatic environment should include 'Aquatic Medicine' with pharmacology, toxicology and immunology and Fish welfare. Most of the Asian countries have adopted this type of training in the last 30-40 years. Eventually, these groomed fish doctors can replace the veterinarians for prescription of drugs, prophylactic measures and health certification for fisheries and aquaculture.

Aquatic health management is unique, differs from that of higher warm blooded vertebrates. Health management of aquatic animals with inclusion of 'Aquatic medicine, will be a beginning of a new era. As such aquatic animals have poor immune system compared to higher vertebrates. Furthermore, no land animal get influenced by the environment as much as fish and crustaceans. Hence these delicate animals husbandry differs from that of land vertebrates. Kinetics of drugs in aquatic animals is time temperature dependent. Added to these, there is insufficient information on chemotherapy and effect of these chemicals on water chemistry, microbes and antimicrobial resistance(AMR).

In Asia several institutions/universities are offering Bachelor of Fisheries Science (BFSc) programme. Usually these 3-4 years programme giving theoretical and practical training in fisheries such as fisheries biology, aquaculture, fishery Engineering, Environment and Ecology, has a good amount of syllabus on fish parasitology, microbial diseases, histopathology, and diagnostics in fish health management but poor in Aquatic medicine, Pharmacology, Toxicology and Immunology. Realising the vacuum and need, during 2010 the College of Fisheries, Mangalore introduced for the first time in the country Aquatic medicine with Pharmacology, Toxicology and Immunology in health management in the BFSc degree programme. Furthermore, during 2015 the College took the lead to introduce these courses in the BFSc programme at the national level through the ICAR, New Delhi. And with this comprehensive training in health management comparable with that of veterinary, a BFSc graduate could be an ideal "Fish Doctor" capable of prescribing drugs, certifying health while ensuring welfare of the fish. And this new approach is a deviation from the practice in the West where veterinarians still are entrusted with responsibility in fish health management.

Against this background, health managers in aquaculture should ideally be Bachelors of Fisheries with adequate knowledge, training and experience in fish biology, environment and aquaculture package of practices with aquatic medicine. Ultimately, an improved fish welfare and health, address ethics and support sustainable production adding to human health and prosperity. Hence the strategy and effort of these "Fish Doctors" is to develop practices for improving fish welfare and health which should reflect in the higher sustainable production leading to farmers wealth and consumer health.

Research on Welfare

Finfish welfare is considered based on the same principles as that for terrestrial vertebrate species. The OIE defines animal welfare by the way in which an animal copes with the conditions in which it lives ; an animal is in a good state of welfare if, as indicated by scientific evidence is healthy, comfortable, well nourished, safe, able to express innate behavior and not suffering from unpleasant states such as pain, fear and distress.

Despite the considerable progress made to date, there is still relatively limited knowledge and research regarding sentience and pain in aquatic animals and the welfare needs of aquatic animals. It is thus important for the OIE to continue its work in the field of aquatic animal welfare in order to establish an appropriate framework for protection of these animals, as has already been done for terrestrial animals.

Legislation for Welfare of Fishes

Several countries have proposed legislation for general welfare of fish in aquaculture particularly for transport and slaughter. Protections afforded to animals during transport and slaughter should similarly apply to fish. Using the World Organization for Animal Health's Aquatic Animal Health Code as a model, model legislation for fish transport have been proposed: with regulations to ensure that fish are humanely slaughtered. At this point, it is heartening that the welfare of poultry birds in India has caught the attention of animal lovers, public and the government. And the Government of India is bringing rules to improve the living conditions in poultry birds (Times of India , April 9 th 2019). A similar scenario in aquaculture in the near future in India and world at large can be expected. Already traceability including welfare of aquatic animals in aquaculture is being linked to market in developed countries.

Continuing Fish Health Education (CFE)

"Continuing fish health education"(CFE) is essential as practiced in human and animal health systems. CFEs help to update latest information on research developments, technologies to the working professionals for better management. In the last 30 years with expansion in aquaculture in India, there have been spurt in fish and shellfish diseases. A number of ICAR institutions, state agriculture universities and traditional universities in India are engaged in research in health management. Research area broadly covers causative agents, pathology, diagnosis, medicines and chemicals, remedial and preventive measures - vaccine, Immunostimulants and probiotics. In addition, research in supporting areas such as Fish biochemistry, Immunology and Genetics has increased. The number of publications on health management in India in national and international peer reviewed journals in the last 25 years is an indication of world class research activity. Although research on curative measures and vaccines have been initiated, there is a long way to go to accomplish the targets. However, in diagnostics conventional and modern with molecular biology touch have added a number of tools, several of which transferred to industry for commercialisation. There is urgent need for research on war footing on drugs/antibiotics for each disease of important species fish and shellfish.

Research activity in health in aquaculture of Asia is also strengthened by work in Europe, Australia and North America. Research and discussion are going on in Asia through several regional and national level seminars and symposia. A very important Asian symposium "Diseases in Asian Aquaculture (DAA)" held once in 3 years in different countries of Asia since 1989 has gained significance. The 8 th DAA was held in India for the first time in 2011 at the College of Fisheries, Mangalore.

Overall Health management to be successful, knowledge should reach the stake holders- farmers, farm workers and government personnel. Accumulated knowledge on health management is being disseminated to the stake holders through training and orientation programmes, field visits and demonstration by various ICAR, State agricultural universities and State government fisheries departments.

References

Ashley, P.J. 2007. Fish welfare: Current issues in aquaculture. Applied Animal Behaviour Science 104, 199-235.

Broom, D.M. 1986. Indicators of poor welfare. British Veterinary Journal, 142, p. 524.

Edward J Branson(editor), "Fish Welfare" ISBN 978-1-405-14629-6, Jan 2008, Wiley-Blackwell, 316 pages, Chapter II. Fish Health & Welfare (welfare index)_ 1 The farmed salmonid health handbook chapter II : Fish Health and Welfare

EFSA 2009. General approach to fish welfare and to the concept of sentience in fish1 Scientific. Opinion of the Panel on Animal Health and Welfare. The EFSA Journal 954, 1-27

Ellis, T., Yildiz, H.Y., López-Olmeda, J., Spedicato, M.T., Tort, L., Øverli, Ø, and Martins, C. I. M. 2012. Cortisol and finfish welfare. Fish Physiology and Biochemistry 38,163–188.

FAO. 2018. The State of World Fisheries and Aquaculture 2018 - Meeting the sustainable development goals. Rome. Licence: CC BY-NC-SA 3.0 IGO.

Farm Animal Welfare Advisory Committee 1979. "Press Statement". 12.05.1979. https://webarchive.nationalarchives.gov.uk/20121010012428/http://www.fawc.org.uk/pdf/fivefreedo ms1979.pdf1.

Huntingford, F. A., Adams, C., Braithwaite, V. A., Kadri, S., Pottinger, T. G., Sandøe, P. & Turnbull, J. F. 2006. Current issues in fish welfare. Journal of Fish Biology 68, 332- 372.
Segner, H. Reiser, S. Ruane, N., Rosch, R., Steinhagen D and Vahnen T 2019, Welfare of fishes in Aquaculture, FAO Fisheries and Aqauculture circular No 1189, Budapest, FAO

Colour Plates

Chapter 3: Parasitic Diseases of Fish and Shellfish

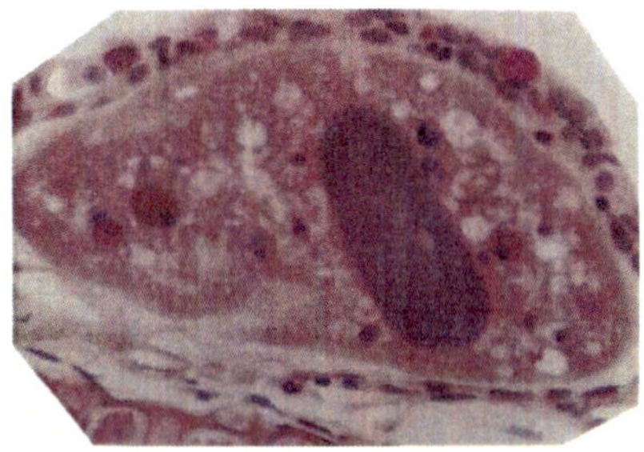

Adult *Ichthyophthirus multifiliis* with gill epithelium

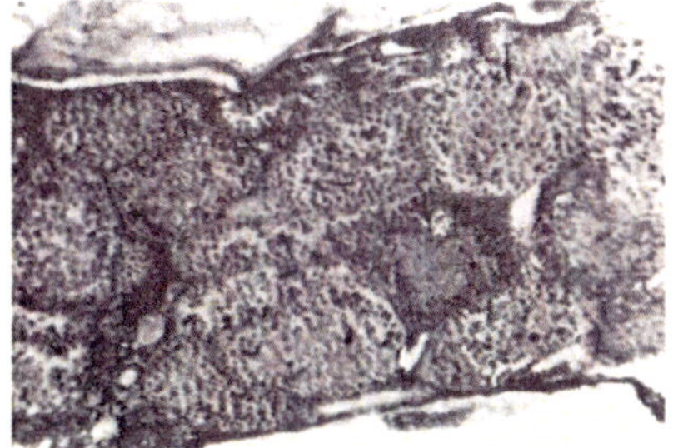

Cysts of *Myxobulus* sp. in the kidney of catla

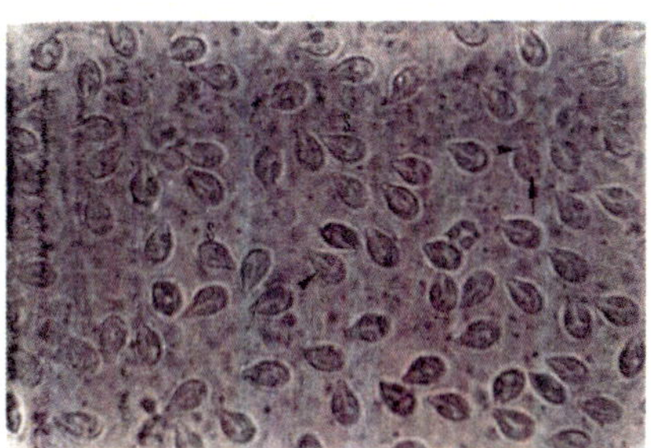

Spores of *Myxobulus* sp.

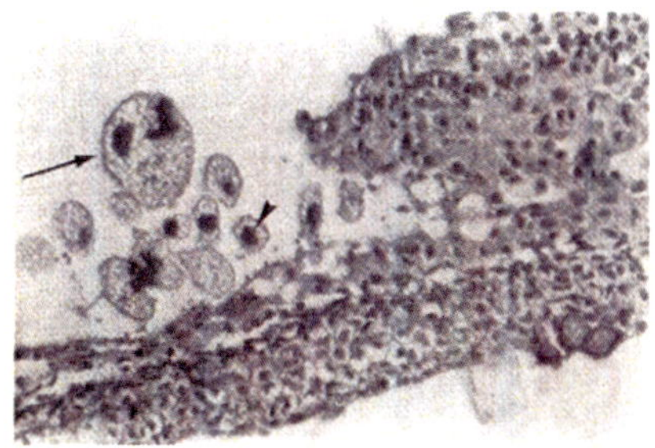

Piscinoodinium sp. in the epidermis of Tilapia

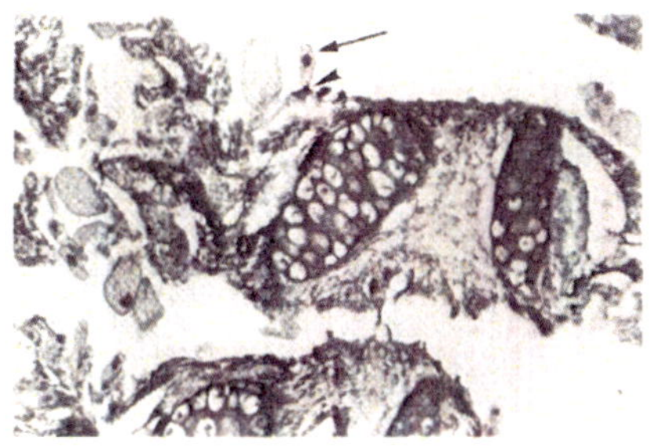

Flagellates in the gills of carp.

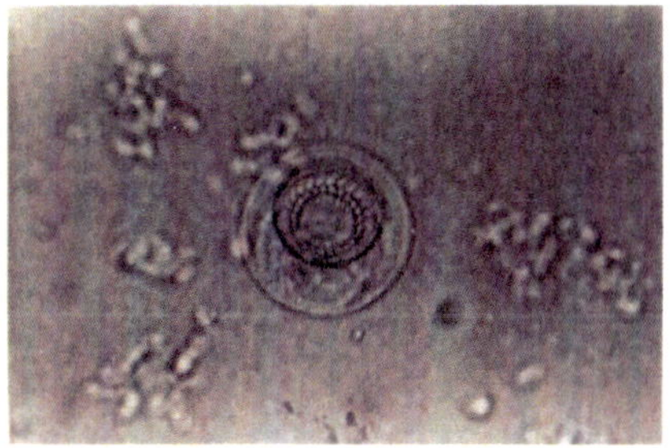

Trichodina in fresh samears of fish skin

Plate 2: Parasites of fish

Chapter 4: Bacterial Diseases of Fishes

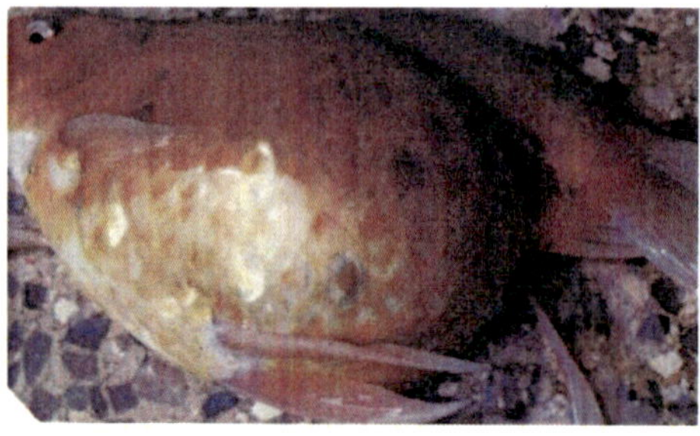

Accumlation of exudate in the body cavity of gold fish (dropsy)

Dropsy (arrow) in cultivated mrigal

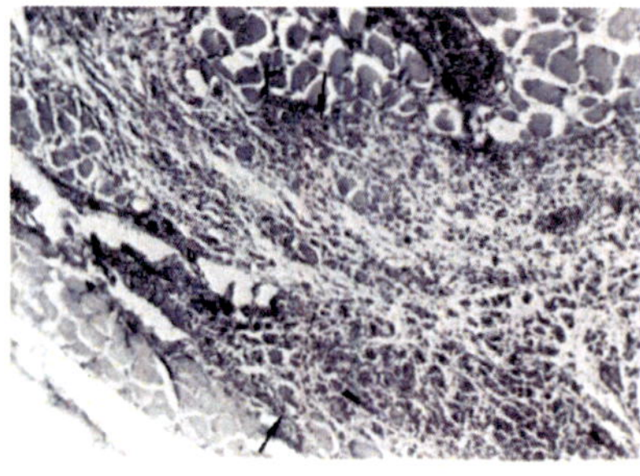

Liquifactive necrosis in the skeletal musculature associated infection with motile aeromonad infection

Surface ulcer associated with aeromonad

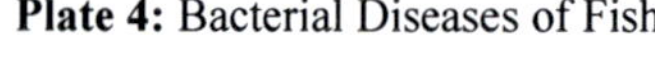

Plate 4: Bacterial Diseases of Fish

Suspected *A. hydrophil* infection of skin

Suspected *A. hydrophil* infection of gills

Plate 5: Bacterial diseases affecting Carps

Chapter 5: Bacterial and Other Diseases of Significance of Shellfishes

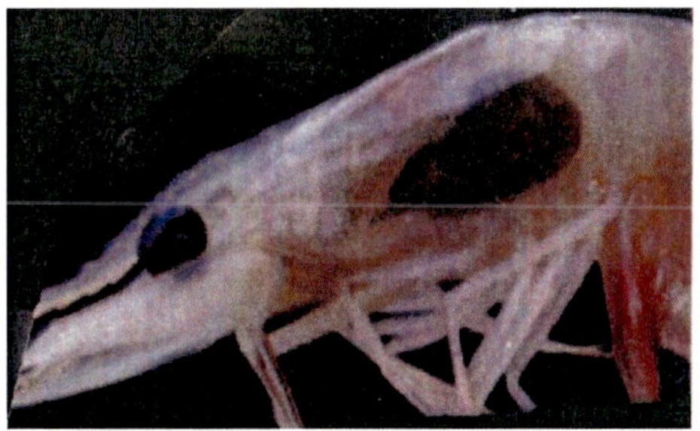

Gross melanised lesion on the gills due to bacterial infection (black gill)

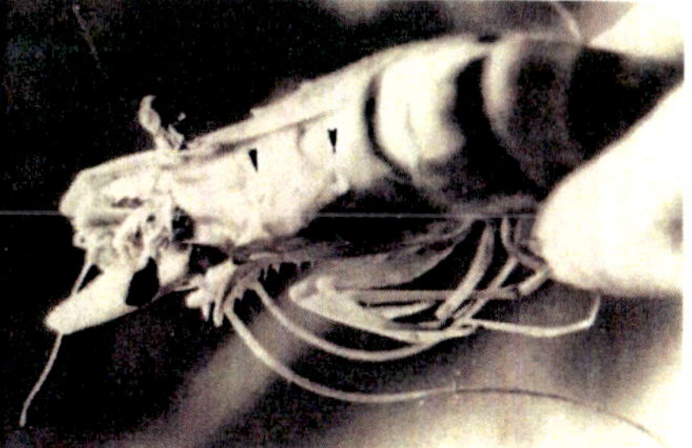

Damaged shell due to cuticular vibriosis

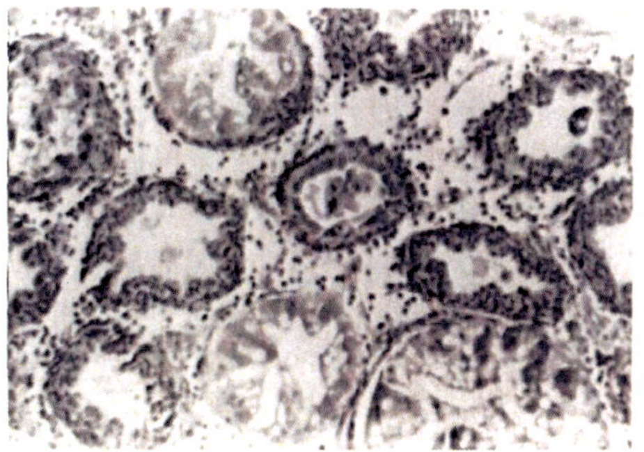

Hepatopancreatic pathology associated with enteric vibrios

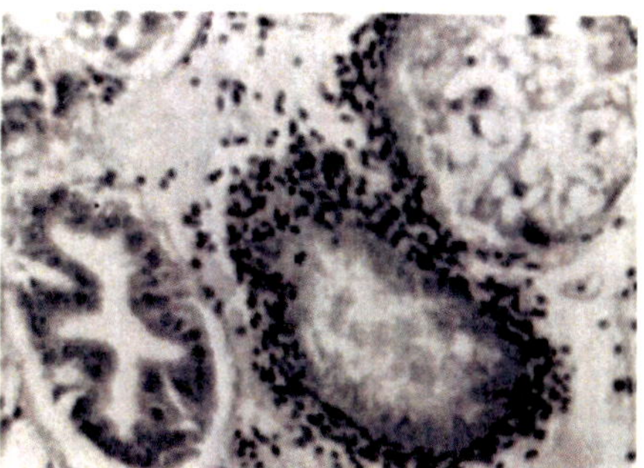

Hemocytic nodule in the hepatopancreas

Hemocytic aggregation and nodule formation *P. monodon*

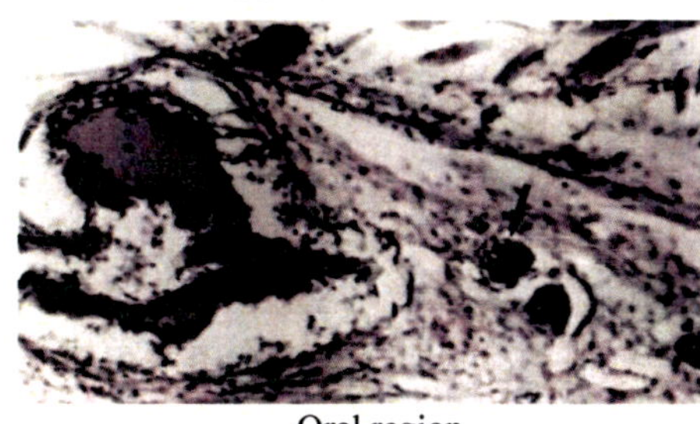

Oral region

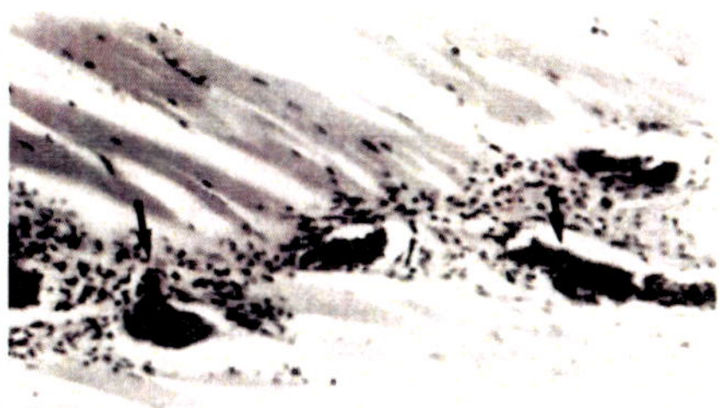

Cephalothoracic muscle.

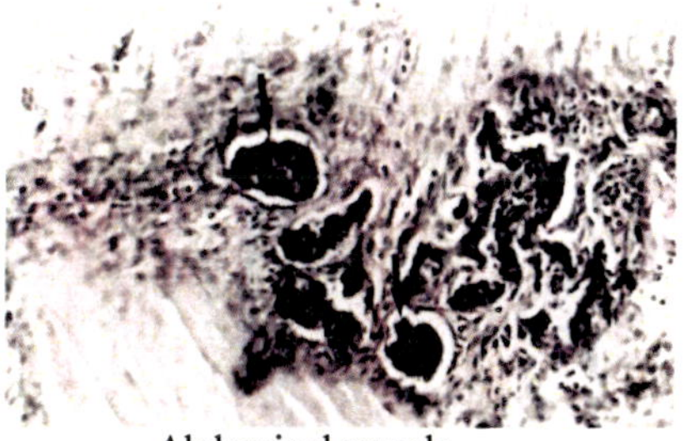

Abdominal muscle

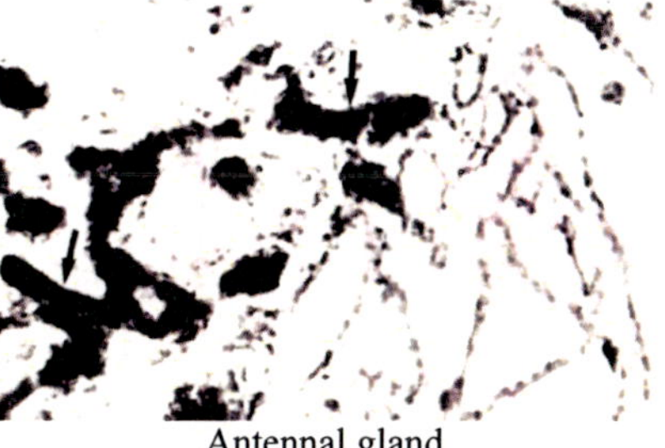

Antennal gland

Plate 6: Bacterial Diseases of Shrimp

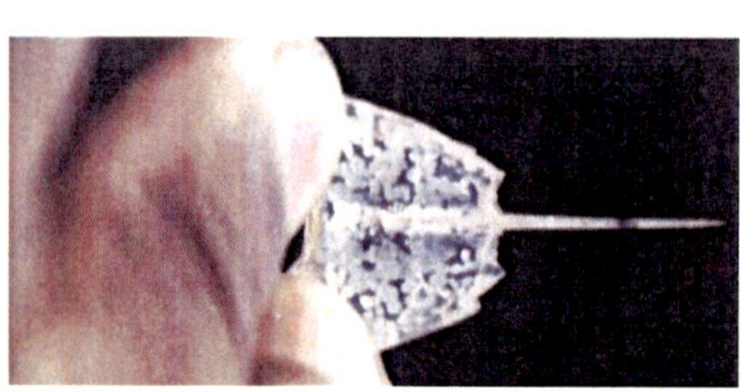

Clinical white spots associated with WSS Vinfection in *P. monodon.*

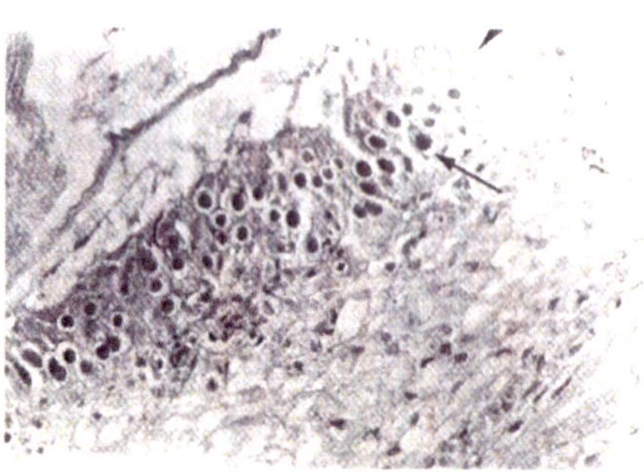

Basophilic intranuclear inclusions of WSSV in foregut cuticular epithelium

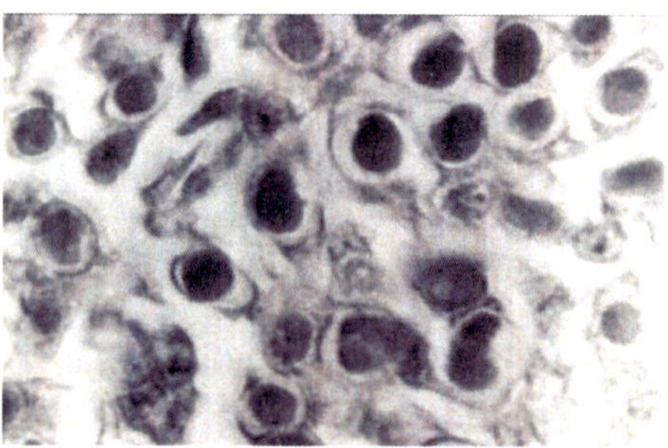

WSSV inclusions in connective tissue

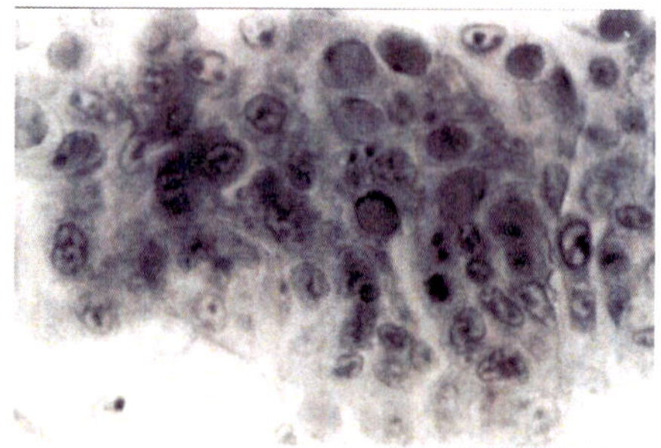

Dual infection in *P. monodon* with WSSV (intranuclear inclusion) and YHV (dense intracytoplasmic inclusion)

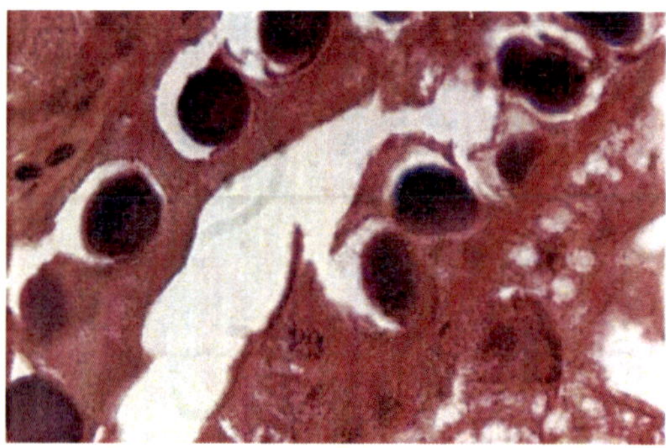

Prominent intranuclear basophilic inclusions of HPV in the hepatopancreas

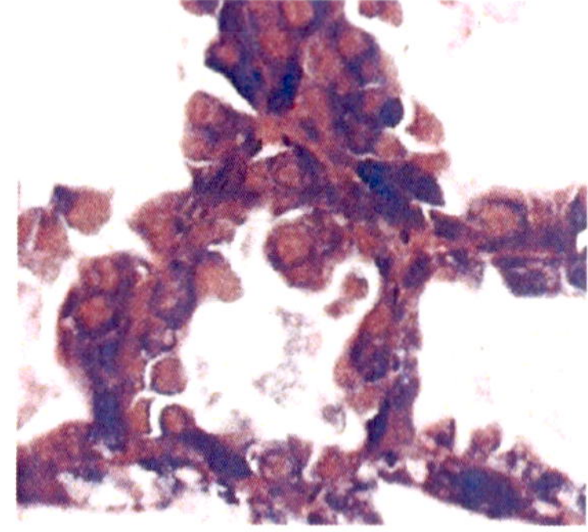

Occlusion bodies of MBV in the hepatopancreas

Plate 8: Shrimp Viral Diseases

Chapter 8: Oomycete and Fungal Diseases in Finfishes

Open dermal ulcers in *Puntius* sp.

Dermal uclear in *Channa* sp.

Healing ulcer in *Channa* sp.

Plate 9: Gross Surface Lesions Associated with Epizootic Ulcerative Syndrome (EUS)

Plate 10: Gross surface lesions associated with Epizootic Ulcerative Syndrome (EUS)

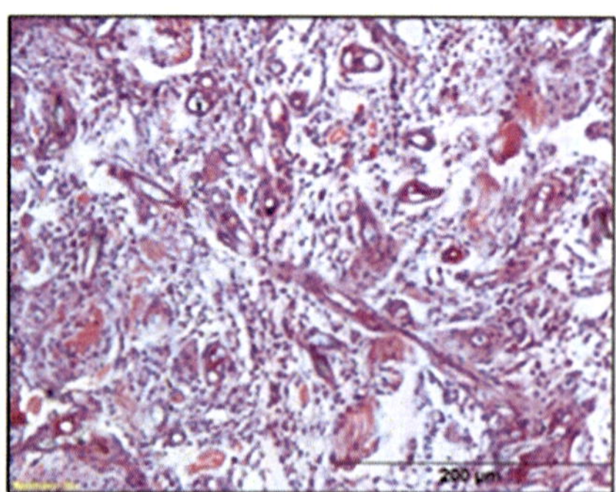

Histological section showing pathology in a susceptible fishinfected with *A. invadans*

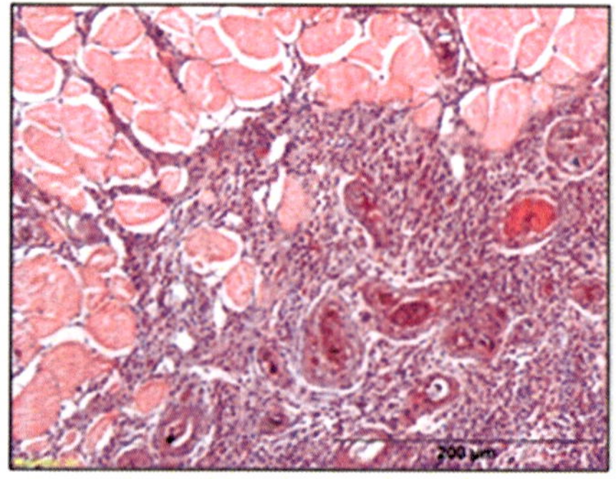

Histological section showing pathology in a fish which is able to resist the infection with *A. invadans*

Plate 11: Histopathology of EUS

Chapter 14: Antibody Based Disease Diagnosis

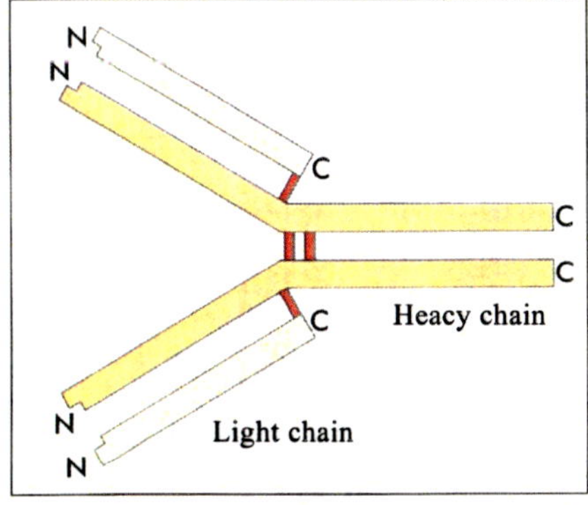

Basic structure of Immunoglobulin.

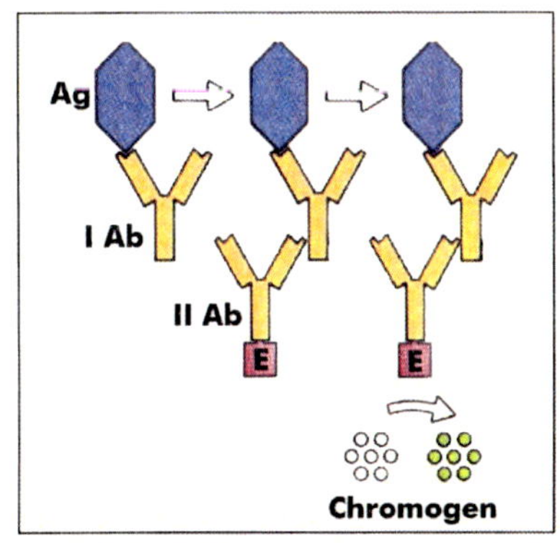

Principle of solid phase immunoassay.

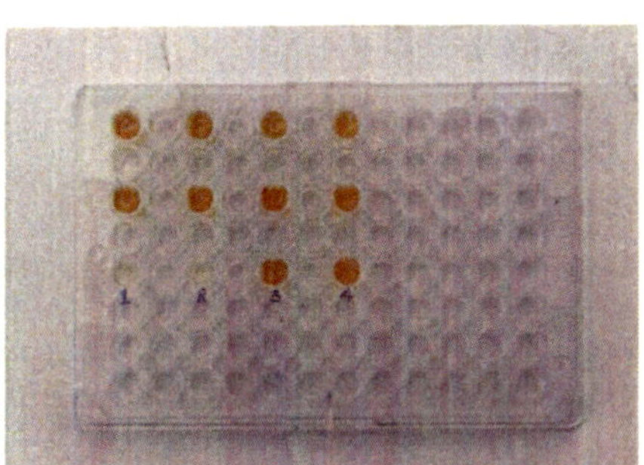

ELISA of WSSV.

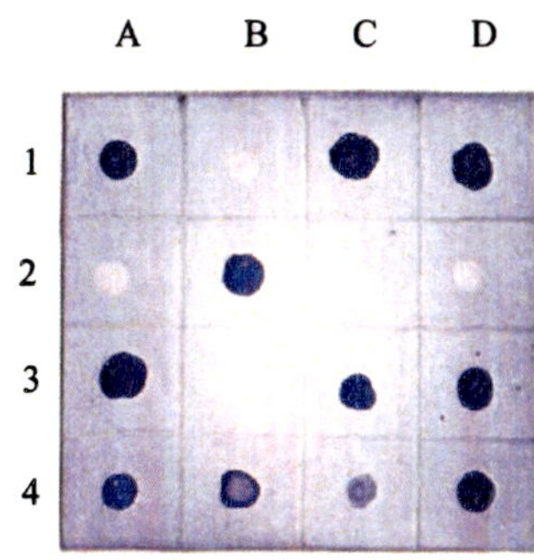

Immunodot of WSSV

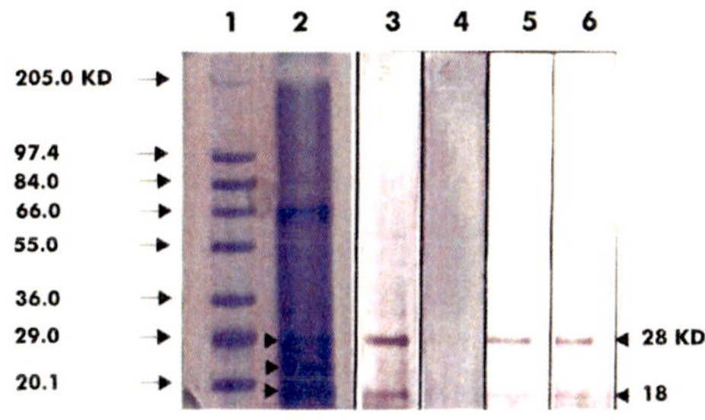

Western blot analysis of WSSV by Monoclonal antibodies

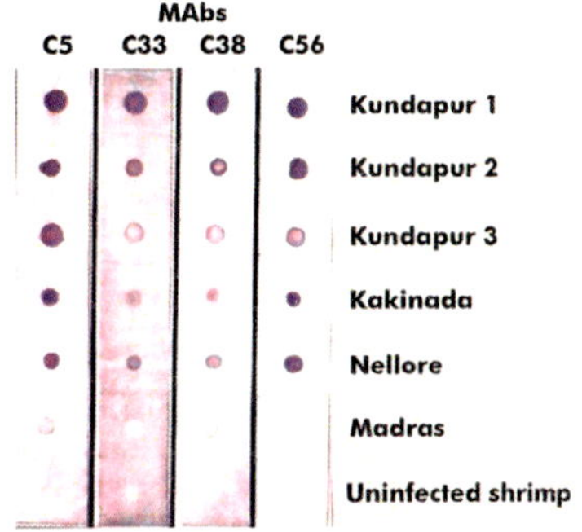

Serotyping of WSSV isolates by MAb based immunodot

Plate 12: Antibody based diagnostics (Solid Phase)

Monoclonal antibody (MAb) based immunofluorescence of *A. Hydrophila* in carp gut

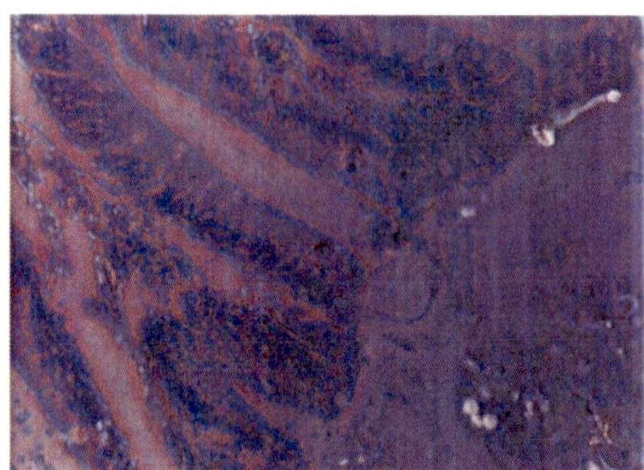

MAb. based immunoperxidase of *A. hydrophila* in carp gut

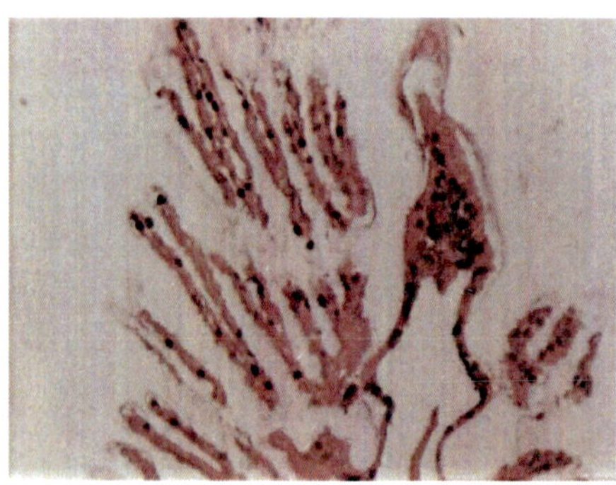

MAb based immunoperoxidase of WSSV in gills

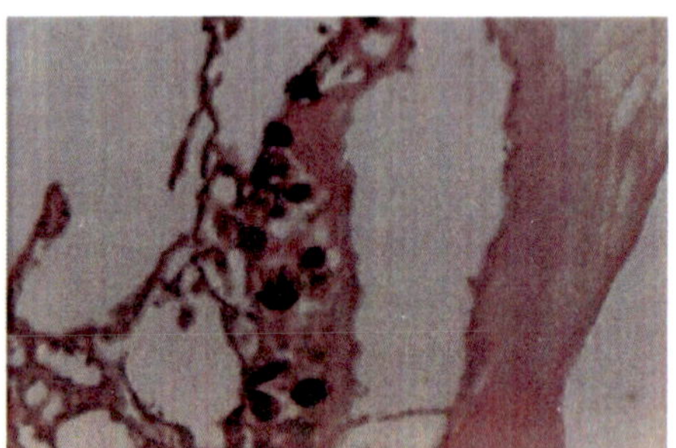

MAb based immunoperoxidase of WSSV in cuticular epithelium

Plate 13: Antibody based diagnostics (*in situ*)

Chapter 19: Biotechnology in Aquaculture Health Management

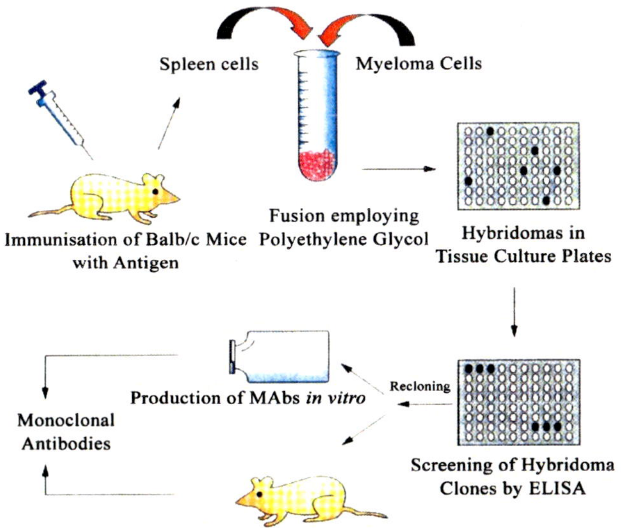

Schematic illustration of production of monoclonal antibodies

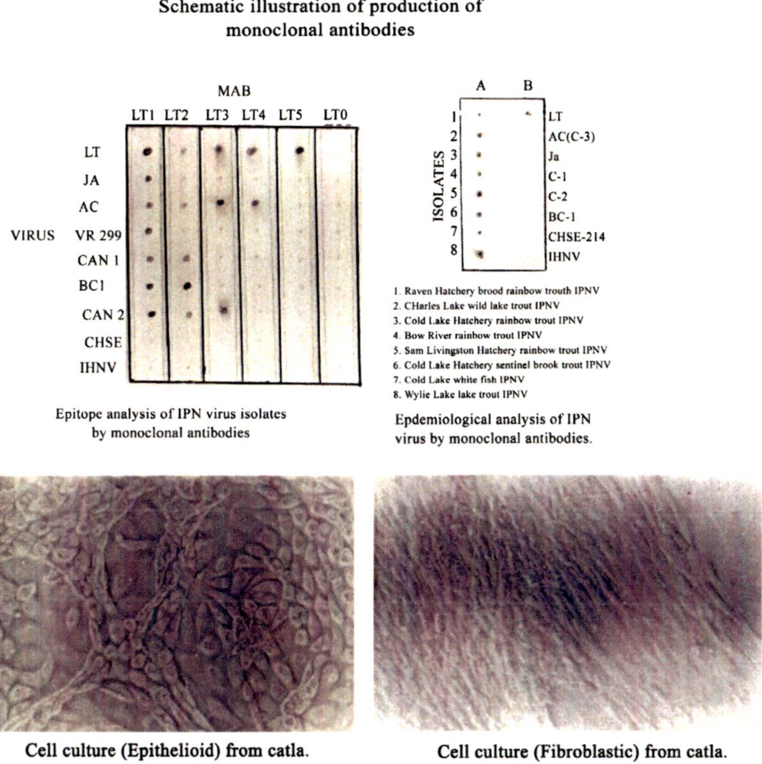

Epitope analysis of IPN virus isolates by monoclonal antibodies

Epdemiological analysis of IPN virus by monoclonal antibodies.

Cell culture (Epithelioid) from catla.

Cell culture (Fibroblastic) from catla.

Plate 15: Biotechnology in Aquaculture Health Management

Index

About the Editor

Dr K. M. Shankar with Bachelor and Master in Fisheries Sciences-Aquaculture (1972-1978) from the UAS, Bangalore and Doctorate in Microbiology-Virology from the University of Alberta, Canada, served in KVAFS University, Mangalore from 1979 to 2016. He received two national awards, 1.Bharatha Rathna Dr C Subramanyam Award for Outstanding Teachers 2008, ICAR, New Delhi, 2. Biotechnology Award ,2012, from the Ministry of Science and Technology, New Delhi and 10 other awards from professional societies and university. Served as an Expert in the ICAR Research Advisory Boards of CIFE,CIFA, CIBA, MPEDA, as Chairman, ICAR 5 th Deans' Committee for review of fisheries program syllabus and as ICAR Peer Review Team Member for evaluating agricultural universities. Presented 25 invited talks on Fish and shellfish health and Biotechnology. Held assignments in International institutes such as NACA, FAO and SAARC. Fellow of NAAS, New Delhi and KSTA, Bangalore

Actively engaged in research in fish and shellfish health and biotechnology with projects funding from national agencies ICAR, DBT, MPEDA, NBFGR, and international agencies from Sweden, Norway, UK, Australia and European Union. Published 45 research papers in peer reviewed International journals, 55 in Asia regional /national journals besides two text books,10 popular articles, 6 chapters in books and four manuals. Serving as Referee for six international and regional peer review research journals.

Developed a novel "bacterial biofilm oral vaccine" for fish for the first time in the world and five farmer level monoclonal antibody based diagnostic kits for detection of aquaculture pathogens, some of which transferred to industry. Widely travelled in aquafarms of India. Visited 15 countries for higher studies, research collaboration, workshops and meetings.

Post retirement (2016) served as Emeritus Scientist of ICAR (2017-2020). Serving as Expert Member in Biotechnology and Genetics area of National Agricultural Science Foundation of ICAR, N. Delhi. A consultant in fish health management, reviewer for four peer review international research journals, has delivered several invited lectures on research and education in agricultural universities. Since 2024 serving as Vice Chairman of Asian Fisheries Society (India Branch).